The Physical Chemistry
of
Biological Organization

An Introduction to

The Physical Chemistry of Biological Organization

A. R. Peacocke

CLARENDON PRESS · OXFORD

Oxford University Press, Walton Street, Oxford OX2 6DP

Oxford   New York   Toronto
Delhi   Bombay   Calcutta   Madras   Karachi
Petaling Jaya   Singapore   Hong Kong   Tokyo
Nairobi   Dar es Salaam   Cape Town
Melbourne   Auckland

and associated companies in
Berlin   Ibadan

© A. R. Peacocke 1983
First published 1983
First published in paperback (with corrections) 1989

All rights reserved. No part of this publication may be reproduced,
stored in a retrieval system, or transmitted, in any form or by any means,
electronic, mechanical, photocopying, recording, or otherwise, without
the prior permission of Oxford University Press

This is sold subject to the condition that it shall not, by way
of trade or otherwise, be lent, re-sold, hired out, or otherwise circulated
without the publisher's prior consent in any form of binding or cover
other than that in which it is published and without a similar condition
including this condition being imposed on the subsequent purchaser

British Library Cataloguing in Publication Data

Peacocke, A. R. (Arthur Robert, 1924–)
An introduction to the physical chemistry of biological oganisation

I. Title
574.

ISBN 0-19-855359-5
ISBN 0-19-855557-1 (pbk)

Printed in Northern Ireland at The Universities Press (Belfast) Ltd

Preface to the 1989 reprinting

The reprinting of this book has afforded a welcome opportunity not only of correcting a number of misprints but also of bringing up to date the references to work published since 1982 (and to a few important articles published before then but not mentioned in the 1983 edition). The variety and extent of these supplementary references testify to the increasing development and application in the last decade of new physico-chemical concepts for understanding the nature of biological organization and complexity.

Amongst the areas covered in this book the whole study of non-linear phenomena (of 'chaos theory', for example) has notoriously undergone an almost explosive growth, to which the supplementary references related to Chapter 2 can bear only a partial witness, principally selecting biological applications. These and other developments to which the list of supplementary references draw attention only reinforce the statement originally made on the last page of the main text. There I made a plea for the widening of biological education to enable the rising generation of biological scientists first to cope with and then to develop these increasingly enlightening and fruitful concepts and intellectual tools. This edition is offered as a contribution to that process which has still only fitfully begun.

A.R.P.

St Cross College, Oxford
May, 1989

Preface to the hardback edition

SCIENTISTS reflecting on the course of their research lives have not infrequently been known to remark on the decisive effect upon the main thrust of their work of the very first piece of research they ever undertook. Such has certainly been my own experience. For it was as a chemistry graduate in the late 1940s with a primary interest in physical chemistry and starting my first year of research—then, as now, a farsighted requirement for honours in the Oxford chemistry school—that I was initiated by Sir Cyril Hinshelwood in the Physical Chemistry Laboratory at Oxford into both an awareness of the absorbing intricacies of the complexity of biological organisms and of the challenge it poses to physico-chemical interpretation. I have referred elsewhere in this volume

(Chapter 4, Section 2) to some of the broad features of the chemical kinetic approach that he and his research pupils and colleagues developed for interpretation of the dynamics of biological organisms. It matters less now, some decades since this approach was initiated, whether it was, or was not, sometimes over-ambitious in its forays into the preserves of cellular biologists and bacterial geneticists than that it began to break down the conceptual barriers that then tended to separate the world of physical chemistry from that of biology. Hinshelwood was not the only pioneer in this respect and, in the section of the book I mentioned, it has been appropriate to draw attention also to the work of others (A. J. Lotka, V. Volterra, A. M. Turing, H. Kacser, K. Denbigh) on the application, or rather extension, of physico-chemical concepts to biological systems.

Subsequent to this early initiation of mine, and in step with the rapidly expanding field of molecular biology, I became more interested in the application of the methods of physical chemistry to the study of biological macromolecules in solution. But one was always aware that this necessary, but methodologically reductionist, approach could never do adequate justice to the many-levelled complexity of biological organisms. Indeed, the application of physical and chemical techniques to the study of living organisms served increasingly to make the intricacy of their organized complexity more manifest. At the same time, this elaboration of the multiple complexity to be found in biological organisms rendered them more intractable to conceptual elucidation in terms of the physics and chemistry developed for the study of the non-living world.

This book expounds recent, and some not so recent, thermodynamic and kinetic interpretations of living systems that have involved major conceptual developments in the application of physico-chemical ideas. These include the thermodynamics of dissipative structures, network thermodynamics, the kinetics of biological self-organization, both in space and time, and the selection and evolution of biological macromolecules. A concluding chapter on the interpretation of biological organization serves to relate these physico-chemical developments to other widely canvassed ideas relevant to living systems (*inter alia,* analysis of the modes of biological complexity, hierarchies of organization, thermodynamic aspects of evolution, integrality, information theory, and the reduction of biological concepts)—not to mention general systems theory referred to in the first chapter. It is hoped that this (mathematically simplified) introduction to the extension of new and amplified physico-chemical concepts to biological organization may serve not only to disseminate them more widely, but also to provide a starting point for the emergence of yet more comprehensive concepts applicable to the complexities of living organisms. (An earlier version of part of this work was presented at the 1976 Oxford International Symposium and was subsequently published in *New approaches to genetics*, edited by Dr P. W. Kent (Oriel Press, Stocksfield & London, 1978) as Chapter 16 'The nature and evolution of biological hierarchies', pp. 245–304).

Some of the exposition in this volume, especially that of Chapter 3, developed as a result of the stimulus provided by members of the Network Thermodynamics Study Group which met intermittently for nearly four years and which originated in a meeting of the Physical Biochemistry Group of the British Biophysical Society that I was responsible for organizing, 20 December 1976, on 'Some aspects of modern thermodynamics in relation to biochemistry and biology'. Among the members of that group I am particularly indebted to Dr John Blanshard, of the University of Nottingham School of Agriculture, who spent a sabbatical leave in Cambridge during which we discovered together many of the themes of Chapter 2. I have been much encouraged and helped in detail by the comments made by a number of notable investigators in this field who were generous enough to read drafts of particular chapters, namely, Dr H. Atlan at Jerusalem, Dr A. Goldbeter and Professor G. Nicolis at Brussels, Dr N. Weiss at Cambridge, and Professors M. Eigen and P. Schuster at Göttingen. As always, the editorial and technical staff of the Oxford University Press have been skilled and patient guides to one who has imposed heavy burdens on them. I am particularly grateful to Mrs J. Pike for transforming into type an often untidy and wayward manuscript. I am glad too to be able to acknowledge here both the warm hospitality and excellent facilities afforded me in the Department of Biochemistry of the University of Cambridge by Professor Sir Hans Kornberg, F.R.S., with the friendly co-operation of Dr. Paley Johnson, and the opportunity for interdisciplinary study provided me by an enlightened college.

A.R.P.

Clare College, Cambridge
January, 1983

Acknowledgments

Figures 1.1 and 1.4 are reproduced by permission of Springer-Verlag, New York;
Figures 1.2 and 1.3 are reproduced by permission of McGraw-Hill Book Co.;
Figures 2.1, 2.3, 2.5, and 2.7 are reproduced by permission of John Wiley & Sons, Inc., New York;
Figures 2.9, 2.10, 3.16, 3.17, 3.18, 4.6, 4.7, and 4.10 are reproduced by permission of Academic Press Inc. (London) Ltd.;
Figures 3.2, 3.5, and 4.11 are reproduced by permission of MacMillan Journals Ltd from *Nature*;
Figures 3.3, 3.11, 3.12, 3.13, 3.14, and 3.15 are reproduced by permission of Pergamon Press Ltd from *J. Franklin Inst.*;
Figures 3.4, 3.6, 3.7, 3.8, 4.16, 4.17, 4.18, 5.2, 5.5, 5.6, 5.7, 5.8, and 5.9 are reproduced by permission of Cambridge University Press from *Quart. Rev. Biophysics*;
Figure 3.9 is reproduced by permission of Prof. A. S. Perelson and The Rockefeller University Press;
Figures 4.9 and 4.15 are reproduced by permission of Drs A. Goldbeter and R. Lefever and The Rockefeller University Press;
Figure 3.19 is reproduced by permission of Dr J. M. V. Blanshard;
Figures 4.1, 4.2, and 4.3 are reproduced by permission of The Society for Experimental Biology and Cambridge University Press;
Figure 4.4 is reproduced by permission of Dover Publications;
Figure 4.8 is reproduced by permission of Prof. G. Nicolis;
Figure 4.12 is reproduced by permission of The American Chemical Society from *Ind. Eng. Chem.*;
Figure 4.13 is reproduced by permission of Dr P. Decker and The New York Academy of Sciences:
Figure 4.14 is reproduced by permission of John Wiley & Sons Ltd;
Figures 4.19 and 4.20 are reproduced by permission of Dr A. Goldbeter, Prof. G. Nicolis and Academic Press, Inc.;
Figure 4.21 is reproduced by permission of Dr A. Goldbeter;
Figures 4.22 and 6.3 are reproduced by permission of W. H. Freeman & Co., from *Scientific American*;
Figures 4.23, 4.25, and 4.34 are reproduced by permission of The Company of Biologists Ltd from the *J. exp. Biol.*;
Figure 4.24 is reproduced by permission of Prof. B. Hess and Academic Press, Inc.;
Figure 4.26 is reproduced by permission of Prof. B. Hess and *Berichte der Bunsen-Gesellschaft*;
Figure 4.27 is reproduced by permission of Springer Verlag (New York) Inc.;
Figures 4.28, 4.31, and 4.33 are reproduced by permission of Dr P. C. Newell and Pergamon Press Ltd;
Figures 4.29 and 4.30 are reproduced by permission of Dr P. C. Newell and Marcel Dekker, Inc.;
Figure 4.32 is reproduced by permission of Dr P. C. Newell and Chapman and Hall;
Figures 5.1 and 5.3 are reproduced by permission of Pergamon Press Ltd;
Figures 5.4 and 5.10 are reproduced by permission of Prof. M. Eigen, Prof. P. Schuster and Springer-Verlag, Heidelberg;
Figures 6.1 and 6.2 are reproduced by permission of D. Reidel Publishing Co., Dordrecht, The Netherlands, and Boston, Mass., U.S.A.

Contents

1 Introduction   **1**
    1.1 Science and complexity   1
    1.2 Biological complexity and general systems theory   3
    1.3 Biological hierarchies   8
    1.4 The 'central problem of biology': the contribution of physico-chemical ideas   11

PART 1 · THERMODYNAMIC INTERPRETATIONS OF LIVING SYSTEMS

2 The thermodynamics of dissipative structures   **17**
    2.1 Thermodynamics and living systems   17
    2.2 Open systems   20
    2.3 'Local equilibrium' and entropy production   23
    2.4 Stability of the equilibrium state   28
    2.5 Linear thermodynamics of irreversible processes   32
    Flows and forces—Non-equilibrium stationary (steady) states in the linear range
    2.6 Non-linear thermodynamics of irreversible processes   40
    Dissipative structures and 'order through fluctuations'—A general evolution criterion—Criteria of stability of non-equilibrium states—Fluctuations and instability—Chemical instability—A critique of the Glansdorff–Prigogine stability criterion
    2.7 Dissipative structures and evolution   56
    Dissipation and 'evolutionary feedback'—Stability and evolution
    2.8 Dissipative systems in biology   62
    Enzyme reactions with feedback—Cellular systems
    2.9 The role of irreversible thermodynamics in biological interpretation   70

3 Network thermodynamics   **73**
    3.1 Thermodynamics and electrical circuits   73
    3.2 Energy storage and dissipation: bond graphs   75
    Ideal elements—'Ports'—Basic procedure: state variables—Energy bonds—Ideal junctions
    3.3 Diffusion through a membrane   81
    3.4 Causality   82
    3.5 Chemical reactions: the 'transducer'   84

CONTENTS

3.6 Network properties 90
Tellegen's theorem—Stability—Evolving networks

3.7 Applications and assessments 94
Generation of heat in biological systems—Coupling of ionic transport, ATP reactions and protein synthesis—Skeletal muscle glucose metabolism—Starch gelatinization—Other topological graph representations

3.8 Conclusion 107

PART II · KINETIC INTERPRETATIONS OF LIVING SYSTEMS

4 The kinetics of biological self-organization — 111

4.1 Chemical kinetics and living systems 111

4.2 Pioneers of the kinetic approach 114
Living matter—Oscillations and symmetry-breaking

4.3 Dissipative structures and bifurcations in reacting systems—theoretical considerations 137
Reaction in the absence of diffusion—Reaction with diffusion—The role of fluctuations

4.4 Conditions for self-organization and oscillations 148
Self-organization—Oscillations

4.5 Models of kinetic self-organization 151

    4.5.1 Temporal self-organization 152
    Autocatalysis—End-product inhibition—Substrate inhibition—Activation by product—Cooperativity in allosteric enzyme action—Covalent enzyme modification

    4.5.2 Spatial self-organization 171
    The 'Brusselator'—Substrate inhibition—Allosteric enzymes—Coupled oscillations—Pattern formation

4.6 Kinetic self-organization in biochemical and biological systems 187

    4.6.1 Temporal self-organization 189
    Glycolytic oscillations

    4.6.2 Spatial self-organization 195
    Dynamic compartmentation in glycolysis—A cellular oscillator: aggregation and communication in a cellular slime mould

4.7 Concluding remarks 212

5 Selection and evolution of biological macromolecules — 214

5.1 The origin of life 214

5.2 Phenomenological deterministic treatment 218
The model system—Selection under constraint—'Selection equilibrium'

5.3 Stochastic approach 224
Limitations of the deterministic treatment—Stochastic models of fluctuation and stability

5.4 Self-organizing systems of macromolecules 228
Hierarchies of cyclic reaction systems—Homogeneous self-organizing cycles—Self-organization by encoded catalytic function: the hypercycle

5.5 Reality and theory 238

PART III

6 The interpretation of biological complexity — **245**

    6.1 Modes of biological complexity — 245

    6.2 Complexity and hierarchy — 249
Constraints of time and space—Structure–function relationships—Control—Specifying descriptions—Operative constraints—Emergence

    6.3 Can complexity and organization be quantified? — 255
Entropy and order—Integrality—Information

    6.4 The evolution of biological complexity — 263

    6.5 Biological complexity and the reduction of biological concepts — 268

    6.6 The thermodynamic and kinetic interpretations — 272

    6.7 New bottles for new wine? — 274

References — **280**

Appendix: Supplementary references — 295

Name index — 303

Subject index — 307

To

J. R. P. O'Brien

without whose encouragement this book
would not have been possible.

1 Introduction

1.1 Science and complexity

IT is over thirty years since Warren Weaver, who was one of the founders of information theory, drew attention in an influential paper to certain features in the way the sciences, as he discerned them in 1947, dealt with complexity in nature (Weaver 1948). Weaver had had some experience of reflecting on the state of the sciences as one of the principal administrators of the Rockefeller Foundation, which was extensively funding the revival in fundamental biological science in the western world after the Second World War. He pointed out that, roughly speaking, the physical sciences in the seventeenth, eighteenth, and nineteenth centuries had been largely concerned with two-variable *problems of simplicity* and had achieved considerable intellectual sophistication and practical application. On the other hand, the life sciences up to 1900 had been largely concerned with 'necessary preliminary stages in the application of the scientific method' (Weaver 1948, p. 536)—stages involving collection, description, classification, and observation of correlated effects (no doubt what the physicist Rutherford would have called the 'stamp-collecting' phase). This was not surprising since biological and medical problems usually involve the consideration of most complexly organized wholes, that are influenced by multiple variables, and so the life sciences had not yet been able to become highly quantitative or analytical in character—in spite of the great stimulus given to the them by the Darwinian perspective.

Weaver went on to recall that around 1900, consequent upon the pioneering work of J. W. Gibbs and L. Boltzmann, the physical sciences found a way, through probability theory and statistical mechanics and thermodynamics, of handling problems of *disorganized complexity*. These are systems characterized by enormous numbers of variables (at least three per unit, and numbers of units in powers of tens of millions) but whose average properties were of interest and were empirically accessible (like the pressure of millions of gas molecules on the walls of a containing vessel). Successful and powerful though these methods and concepts have proved when applied to problems of simplicity and disorganized complexity, they nevertheless, Weaver urged, left untouched a 'great middle region', namely the *problems of organized complexity*, in which the number of variables to be considered is very much greater than two, but still very much less than a million. All these problems are concerned with organization and Weaver instanced a number of such problems. These included, amongst others: 'What makes an evening primrose open when it does?'; 'What is the description of ageing in biochemical terms?'; 'What meaning is to be assigned to the question: "Is a virus a living organism?"?'. All these problems, let it be noted, are concerned with the organized complexity of living organisms.

These questions were being asked before the double-helical structure of DNA had been discovered and before the rise of molecular biology. However to read all of his questions again over thirty years later shows not only how far we have gone, but how far we still have to go in tackling

> problems which involve dealing simultaneously with a *sizable number of factors which are interrelated into an organic whole*. They are all ... problems of *organized complexity*. ... These problems ... are just too complicated to yield to the old nineteenth-century techniques which were so dramatically successful on two-, three-, or four-variable problems of simplicity ... [and] cannot be handled with the statistical techniques so effective in describing average behavior in problems of disorganized complexity. These new problems ... require science to make a third great advance ... [it] must, over the next 50 years learn to deal with these problems of organized complexity. [Weaver 1948, pp. 539, 540.]

More than thirty of the fifty years of which Weaver then wrote have now passed and it now seems suitable for an assessment to be made of how far science, or rather the sciences, have been able to cope with the organized complexity of the living world. The time is ripe for such an appraisal because of two factors. On the one hand, the veritable deluge, called 'molecular biology' (which was unleashed with the discovery of the structure of DNA) has now widened into a broad torrent of continuing publications which serve increasingly to acquaint us with the formidable intricacies of the ramified complexity of living organisms, with respect to both their internal and external relationships. On the other hand, new conceptual tools and ways of analysing and modelling biological complexity have begun to be devised. Weaver himself, on the basis of wartime experience, put great store on research by mixed teams of experts into problems of organized complexity, aided by the application of computers. These two features of modern research have undoubtedly contributed to the increase in our knowledge of the extraordinary mechanisms that constitute biological processes, and his predictions have to this extent been fulfilled. However, the very increase in the richness of our knowledge of biological complexity has served only to emphasize even more starkly the paucity of our conceptual tools for handling it and making it accessible to the critical intelligence. These conceptual barriers that remain to be leaped may well prove to be more daunting than the

1.2 Biological complexity and general systems theory

From one regard, what characterizes any system in the natural world as 'living' can be confined to a few apparently simple properties, such as those in the short list of A. I. Oparin (1969), namely: metabolism, self-reproductivity, and mutability. However, as soon as one attempts to indicate the kind of organization of matter that has such properties, the complexity of the living has to find expression. This may be in the physico-chemical terminology of C. N. Hinshelwood (1951):

> One might in very general terms regard a mass of living matter as a macromolecular, polyfunctional free radical system, of low entropy in virtue of its order, with low activation energy for various reactions in virtue of its active centres, and possessing a degree of permanence in virtue of a relatively rigid structure.

Or, it may be in the more systems-orientated terminology of R. W. Gerard (1957, p. 433):

> The attributes that help define living orgs [= material systems or entities which are individuals at a given level but are composed of sub-ordinate units ... and which serve as units in super-ordinate individuals, p. 430] are:— (i) highly ordered and clearly bounded heterogeneity ...; (ii) dependable mechanisms for reproducing units and patterns ...; (iii) powerful homeostatic mechanisms for maintaining and regaining equilibrium ....

Gerard (1957, p. 432) regards the 'repetitive production of ordered heterogeneity' as especially characteristic of life and emphasizes that living organisms are living *systems.*

This concept seems to go back at least as far as Rudolph Virchow in the nineteenth century who published in 1862 an important essay on 'Atoms and Individuals'

> Life must be the collective product of the activity of all its individual parts, and all of these parts must in themselves have something general as well as something special. ... Scientific investigation reveals the individual to be composed of an array of systems; one attends to sensations, another to movements, others to the intake of nourishment and air, some support the parts and others bind them together, and so on. Every one of these systems comprises a certain number of special organs; every organ includes a number—usually limited—of tissues, and every tissue is in the end composed of cell regions and cells. [Virchow 1862 and 1959, p. 138, 9.]

The more recent emphasis of Gerard on *systems* is characteristic of a whole mode of approach to biological organization of which note must certainly be taken, even though its actual fruitfulness in providing *new* conceptual tools may remain somewhat in doubt. It may eventually provide a broader framework with which the more physico-chemical approaches, to be described later, can be fitted and it has certainly served to re-emphasize more 'holistic' approaches to biology than some of the more reductionist ones, a dichotomy and a controversy to which we shall refer again in the last chapter.

General systems theory is a general mathematical theory which is concerned, amongst other things, with the foundation of such disciplines as the theory of control systems, of automata, of information-processing systems, and of games theory. The origin of these developments is usually traced back to an article by L. Bertalanffy (1950) in which he urged that there exist

> *general system laws* which apply to any system of a certain type, irrespective of the particular properties of the system or the elements involved. We may say also that there is a structural correspondence or logical homology of systems in which the entities concerned are of wholly different nature. This is the reason why we find isomorphic laws in different fields. The need for a general super-structure of science, developing principles and models that are common to different fields, has often been emphasized in recent years, ... But a clear statement of the problem and a systematic elaboration has apparently never been made. ... Such considerations lead us to postulate a *new basic scientific discipline* which we call *General System Theory*. It is a logical-mathematical field, the subject matter of which is the formulation and deduction of those principles which are valid for 'systems' in general. There are principles which apply to systems in general, whatever the nature of their component elements or the relations or 'forces' between them. ... General System Theory is a logical-mathematical discipline, which is in itself purely formal, but that is applicable to all sciences concerned with systems

The kind of thinking which Bertalanffy is urging here is that which we indulge in when we notice how the exponential law, the law of compound interest, applies to a wide variety of natural and social phenomena (radioactive decay of radium, first-order reactions, killing of bacteria, changes of population, growth of bacterial cultures or human populations, the rate of increase of scientific publications). The entities concerned in these various systems are all very different and so are the causes or mechanisms involved, yet the mathematical law is the same: other examples are given by Bertalanffy. One naturally asks what is the origin of these isomorphisms? General Systems Theory is based on the belief that there are isomorphic laws characterized by their holding generally for certain classes of complexes or systems, irrespective of the special kinds of entities involved. As it has developed it has become a theory of formal (mathematical) models of actual or conceptual systems, based on the assumptions that any theory of real phenomea is always founded on a 'model' and that invariant aspects of that model can be represented as a

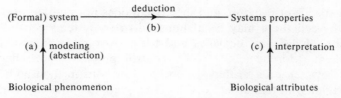

Fig. 1.1. Methodology of applying systems theory in biology. (a) Formalization (abstraction)—construction of a system S and providing of a constructive specification for S. (b) Deduction—study of the properties of S using deductive methods. (c) Interpretation—study of the meanings of the deductively derived properties in the context of the biological phenomenon under consideration. (From M. D. Mesarovic 1968, p. 62.)

mathematical relation. This relation is then termed a 'system'. The relationships between the formal system and biological phenomena may be depicted along the lines of Fig. 1.1. The process (a) of modelling an abstract formulation often has to be selective among the attributes of the biological phenomena, choosing those that are the focus of interest, and the system is formalized so that process (c) of interpretation gives the attributes of interest. There is a 'constitutive specification' involved in process (a), usually in the form of relations between some attributes (inputs, causes) and others (outputs, effects) and these can be numerical or non-numerical.

Mesarovic (1968) has noted certain particular problems that arise when one tries to apply Systems Theory to biology. Complexity, for example, necessitates the selection of the attributes considered in (a) and there are individual variations between members of a biological species. Moreover, a given system may have various 'constitutive specifications' giving different attributes at (c). These can include 'goal-seeking' descriptions and, sometimes, somewhat surprisingly, this character of descriptions is necessitated for mathematical and formal reasons only. Mesarovic (1968) gives an example of input–output relationships which necessitate a goal-seeking representation. Often simple feed-back concepts are inadequate, for in a complex goal-seeking system there exists a hierarchical interdependence of goals and sub-systems, especially in biology. For biological systems are, in general, multi-level, multi-goal systems and any attempt to represent them as single-level (even if multi-variable) systems, or even more as an input–output system, might lead to a model which is valid only over particularly narrow sets of conditions. One naturally thinks of the control of protein synthesis at the genetic level by means of repressors and the operon system; co-repressors or inducers act as feed-forward signals while there exist also negative feed-back signals from the end-products. These two actions (feed-forward and feed-back) occur, in general, with different speeds of response and the system does not operate in isolation. Within the cell there may be a thousand or so systems controlling the production of different enzymes and there exist many interactions between them. Moreover, the same control processes are affected by contacts between cells so that the control functions in thousands of cells are interrelated in tissues and organs. The overall behaviour of organs in turn can again be described as controlled systems on a different biological level. A biological system appears to be formed of an extremely large family of interrelated goal-seeking sub-systems whose components also consist of a family of interrelated goal-seeking systems which in turn have again goal-seeking components, etc., down the biological scale. So, in the application of Systems Theory to biological organisms, one has to recognize the importance of (a) interactions, (b) the multi-level character of biological systems, and (c) their goal-seeking character.

The problem of interactions has been tackled by several authors, interesting interacting control systems have been discussed (Goodwin 1963), and it is claimed that some general principles of organization for multi-level systems can be described from more detailed mathematical investigations (see Mesarovic 1968).

It may be instructive to give just one piece of early 'systems theorizing' which was developed by H. A. Simon in an important article (1962) in which he showed the evolutionary advantages of the modularization that can occur in the assembly of a hierarchy, for this has had an important influence on biological reflection. This is the much-quoted story of Hora and Tempus.

> There once were two watchmakers, named Hora and Tempus, who manufactured very fine watches. Both of them were highly regarded, and the phones in their shops rang frequently—new customers were constantly calling them. However, Hora prospered while Tempus became poorer and poorer and finally lost his shop. What was the reason? The watches the men made consisted of about one thousand parts each. Tempus had so constructed his that if he had one partly assembled and had to put it down—to answer the phone say—it immediately fell to pieces and had to be reassembled from the elements. The better the customers liked his watches, the more they phoned him the more difficult it became for him to find enough uninterrupted time to finish a watch. The watches that Hora made were no less complex than those of Tempus. But he had designed them so that he could put together sub-assemblies of about ten elements each. Ten of these sub-assemblies, again, could be put together into a larger sub-assembly; and a system of ten of the latter sub-assemblies constituted the whole watch. Hence, when Hora had to put down a partly assembled watch in order to answer the phone, he lost only a small part of his work, and he assembled his watches in only a fraction of the man-hours it took Tempus.

Simon went on to make a more quantitative analysis (see Section 6.4) of the relative difficulty of the tasks of Hora and Tempus. If there is one chance in a hundred that either watchmaker would be interrupted while adding one part to an assembly, then a straightforward calculation showed that it would take Tempus, on the average, about four thousand times as long to assemble a watch as Hora! Although, Simon points out, the numerical estimate cannot be taken too seriously (and indeed it can be shown to be an underestimate) the lesson for biological evolution is quite clear. The time required for the evolution of a complex form from simple elements depends critically on the number and distribution of potential, intermediate, stable forms, in particular if there exists a hierarchy of potentially stable 'sub-assemblies'. This story is an example of how careful thinking about assemblies and structures may in fact be illuminating about how apparently impossible structures could nevertheless have evolved by piecemeal and separate evolution of their finally integrated components.

The kind of integration actually to be found in living systems has been comprehensively analysed and classified in a book by J. G. Miller (1978), *Living systems*, which cannot but be regarded as the apogee of the systems-theoretical approach to biological and indeed practically all natural living systems, including the social. Whether or not one follows him in subsuming all the systems that may be categorized as 'living' under his one classificatory scheme, his far-ranging attempt to do so—and the extraordinary range of knowledge the one volume encompasses—does at least serve to emphasize, as perhaps no other attempt could have done, the actual complexity of all living systems. Figure 1.2 illustrates Miller's diagram for a 'generalized living system interacting and intercommunicating with two others in its environment'. This generalized system consists of nineteen critical sub-systems essential to life: two which process both matter–energy and information (reproducer, boundary); eight which process matter–energy (ingestor, distributor, converter (or decomposer), producer, matter–energy storage, extruder, motor, supporter); and nine which process information (input transducer, internal transducer, channel and net, decoder, associator, memory, decider, encoder, output transducer). This subdividing of a system into nineteen critical sub-systems is applied, with the same categorization, to systems at the seven different levels in the natural hierarchy of complexity, namely: cell, organ, organism, group, organization, society, supra-national system. Cross-comparisons are made (Fig. 1.3) between the same sub-system as it functions in the seven different levels in a process he calls 'shredding-out'—'as if each strand of a many-stranded rope has unravelled progressively into more and more pieces' (Miller 1978, p. 1033).

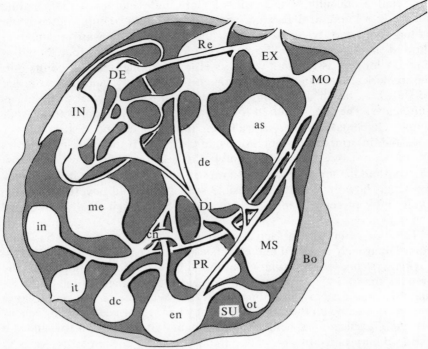

Fig. 1.2. A generalized living system interacting and intercommunicating with its environment.
Subsystems which process both matter–energy and information: Reproducer, *Re*; Boundary, *Bo*.
Matter–energy processing subsystems: Ingestor, *IN*; Distributor, *DI*; Decomposer, *DE*; Producer, *PR*; Matter–energy storage, *MS*; Extruder, *EX*; Motor, *MO*; Supporter, *SU*.
Information processing subsystems: Input transducer, *it*; Internal transducer, *in*; Channel and net, *cn*; Decoder, *dc*; Associator, *as*; Memory, *me*; Decider, *de*; Encoder, *en*; Output transducer, *ot*. (From J. G. Miller 1978, Fig. 1.1.)

1.3 Biological hierarchies

Miller has chosen these seven systems as reference levels since each of them is living, that is, it is a self-reproducing organized complex but a molecular biologist would include lower non-living entities in the biological hierarchy and would perhaps more naturally think of a sequence such as: atom–molecule–macromolecule–subcellular organelle–living cell–multicellular functioning organ–whole living organisms–a population of organisms–ecosystem (e.g. Gerard 1957). Each successive member of this series is a single 'whole' constituted by 'parts' which are the preceding members; the 'wholes' are 'organized systems of dynamically interrelated parts' (Nagel 1952). (It is common in speaking of such a series to refer to

BIOLOGICAL HIERARCHIES

Level	Approximate number of years since period of origin		Approximate median diameter size (in metres)
Cell	3×10^9		10^{-6}
Organ	$5 \times 10^8 +?$		10^{-3}
Organism	5×10^8		10^{-2}
Group	$5 \times 10^8 ?$		10^{-1}
Organization	1.1×10^4		10^2
Society	7×10^3		10^6
Supranational system	4.5×10^3		5×10^6

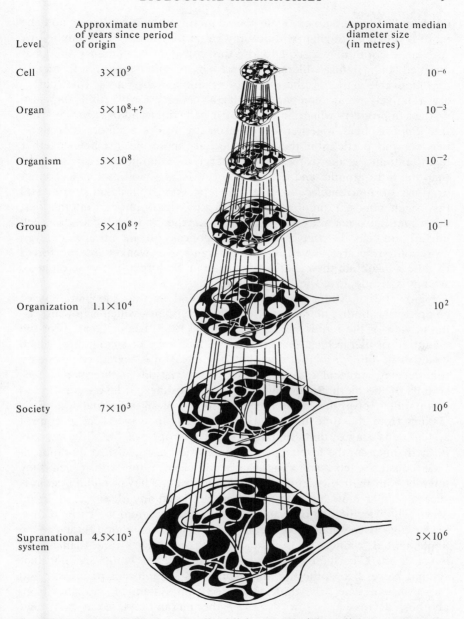

Fig. 1.3. Shred-out. Miller's generalized living system (Fig. 1.2) is here shown at each level. The diagram indicates that the nineteen subsystems at the level of the cell 'shred out' to form the next more 'advanced' level of system, the organ, which still has nineteen subsystems, each being more complex. A similar 'shredding-out' occurs to each form of the more 'advanced' levels. (After J. G. Miller 1978, Fig. 13.1.)

the first and less complex, as in the sequence above, as 'lower' and the more complex as 'higher' without any value judgement being implied by such a description; we shall follow this convention.) Simon (1962) has affirmed that natural hierarchies are, more often than not, 'nearly decomposable'—that is, the interactions among sub-systems (the 'parts') are relatively weak compared with the interactions within the sub-systems, a property which simplifies their behaviour and description. Thus the *atom* is held together by electrostatic forces weaker than those between the 'particles' in the atomic nucleus; molecules are held together by interatomic weaker forces (lower energies—coulombic, exchange, etc.) than the intra-atomic; and *aggregates of molecules*, and particular configurations of *macromolecules*, are held together by weaker forces still (hydrogen bonds, London dispersion forces, 'hydrophobic' interactions, etc.); and more complex larger units, organelles (e.g. mitochondria, cell nuclei) and cells themselves rely on the cohesion of membranes or even more solid structures, as well as on longer-range even weaker ordering forces in gels, to maintain their spatial organization. So forces of a wide range of strengths are deployed in any one system.

Figure 1.3 of Miller serves also to remind us that the hierarchy of complexity is both a hierarchy of evolutionary span—which is not surprising in view of the requirements that Simon (1962) has delineated for the evolution of hierarchical systems—and a hierarchy of spatial size, which would still be so if the diagram (Fig. 1.3) was extended to macromolecules, molecules, and atoms. This variation in the spatial constraints of the hierarchy of natural systems at different levels is reflected in a parallel constraint in the time of response of an individual system, as distinct from the time it took to be evolved. In a hierarchical control system consisting of units similar in intrinsic properties but differing only in their function, the levels of control are often distinguished by different time constants, corresponding to the different rates with which they interact with their own kind. For example, the medieval feudal system is such a 'similar units hierarchy' (Bastin 1969) and any observer, say from Mars, who was looking objectively at that system would have found that, because each command from each level above has many details to be implemented, changes at a higher level take a longer time than at any lower level. Clearly, in biological hierarchies, the units are not that similar. Nevertheless the time of response is a significant parameter and the 'transient time', the time of change of a system to one state from another, decreases as one descends the hierarchy of complexity and correlates with space (for example, in a cell, Fig. 1.4). The requirement of slow processes for more space to generate structural order is not surprising because, as Morowitz (1974) has pointed out, biological organizations can only maintain themselves in existence if there is a flow of energy and

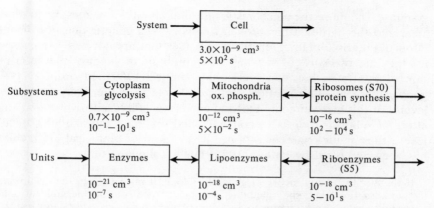

Fig. 1.4. Functional and structural order of cellular metabolic systems: mean operational volume in cm^3, mean transient times in s. (From B. Hess, in Mesarovic 1968, p. 89.)

this flow requires that the system be not in equilibrium and therefore spatially inhomogeneous. The precise physico-chemical basis for this relation between the spatial and temporal features of biological organization is a challenging problem which, as we shall see (Chapter 4) is only now beginning to be resolved when the interplay of chemical kinetic (temporal) and diffusional (spatial) mechanisms can occur.

1.4 The 'central problem of biology': the contribution of physico-chemical ideas

There would be general recognition of the success of the natural sciences in answering questions of the kind 'How does it work?', the reduction type of question about mechanisms. This question is usually separated from questions of the kind 'How did it arise?', the genetic question of origins. The first, mechanistic question, views the lower level from the upper level and takes the constraints imposed by the upper level for granted; it emphasizes structure. The second, genetic question, which may now be described as equivalent to the question 'How did the constraints arise?', is a question asked from the lower level and assumes lower-level laws; it emphasizes function, which is never uniquely determined by a particular structure but by the total context of the organism (for example different structures can have the same function in different organisms). How are the biological sciences faring in handling their own particular kinds of complexity? Can they look for any help from physico-chemical concepts, as distinct from physico-chemical techniques, that have undoubtedly enhanced our knowledge of the biochemical mechanisms in all living systems? The more the mechanistic question about 'How does it work?' is answered the harder it has proved to answer the genetic question about 'How did it arise?'. However, these two questions are not really separable for they are two ways of viewing the same basic problem of the description of interfaces between hierarchies (Pattee 1969). In ordinary language, there is subordination of lower levels in a hierarchy to higher levels but this is never complete, and each level still has its own rules. The effect of subordination is to constrain and integrate the activities of all levels, thus giving coherent action. The problem of biological hierarchies has been described by Pattee (1970) in the following terms:

> If there is to be any theory of general biology, it must explain the origin and operation (including the reliability and persistence) of the hierarchical constraints which harness matter to perform coherent functions. This is not just the problem of why certain amino acids are strung together to catalyse a specific reaction. The problem is universal and characteristic of all living matter. It occurs at every level of biological organization, from the molecule to the brain. It is the central problem of the origin of life, when aggregations of matter obeying only elementary physical laws first began to constrain individual molecules to a functional, collective behaviour. It is the central problem of development where collections of cells control the growth or genetic expression of individual cells. It is the central problem of biological evolution in which groups of cells form larger and larger organizations by generating hierarchical constraints on sub-groups. ... These are all problems of hierarchical organization. Theoretical biology must face this problem as fundamental, since hierarchical control is the essential and distinguishing characteristic of life.

But how is this 'central problem' of biology to be tackled? The temptation is immediately to espouse a 'holistic' approach, such as that which characterizes general systems theory, with the hope that overarching principles and laws will become apparent from this 'higher' and more generalized viewpoint. However, unfortunately what can only too easily result from such considerations is rather vapid generalizations and the whole exercise transpires to be over-optimistic, if not over-inflated (cf. Berlinski's (1976) attack—shrewd, if not always fair). Such 'holistic' approaches are often quite properly initiated by a desire not to be sucked into the vortex of the whirlpool of a reductionist ontology of living organisms and to reaffirm the genuine, emergent features of complete living systems that require for their description and articulation their own distinctive, non-epistemologically reducible, concepts and language (cf. Peacocke 1976).

This is perfectly proper and, indeed, essential for any considered epistemology of the biological sciences, in this author's view (Peacocke 1976, 1979, Chapter IV). However, it must also be recognized that it is by a reductionist methodology, which first breaks up and analyses even a biological system with its component units and processes, that the greatest advances have been made in the last few decades of molecular biology; and that, such a methodology having yielded up its information, it is only as the actual complexity of the system so articulated is clarified that new concepts and ways of thinking about the systems as a whole can hope to be reliably formulated. So the question underlying the exposition in this book is: To what extent can physico-chemical considerations contribute to an understanding of the specific differentia of biological complexity? It will not be expected that the concepts and suggestive ideas which will be described would have arisen without the stimulus of the baffling challenge of the nature of biological complexity nor is it to be expected that these concepts will necessarily be directly applicable in all their aspects to simpler physical and chemical systems. It will always be a matter for consideration of each system in question. So there is no attempt here to 'reduce', in the strict sense of that term (Nagel 1952; Ayala 1974) biology to physics and chemistry—rather, to see what extension of physico-chemical concepts is required to cope with biological complexity. Nor will such an aim in any way detract from that of more 'holistic' biologists who (e.g. Medawar 1974; Thorpe 1974) argue that many biological concepts and language are often *sui generis* and not reducible to physics and chemistry, certainly not in the form to which they apply to simpler and restricted atomic and molecular systems. There are certainly grounds for arguing that such concepts and language can refer to genuine features of reality (cf. Wimsatt 1976, 1981) as much as do physico-chemical conceptualization and descriptions. These more philosophical questions will be reverted to in the last chapter of the book (6.5), but for the moment there is good sense, from the general experience of the sciences in tackling complex systems, to attempt to tackle living systems by a kind of biological equivalent of the *aufbau* method in atomic physics. (In this method the electronic levels available to electrons around a charged nucleus are conceived of as gradually being filled up, under the restrictions of the Pauli exclusion principle, so that the sequence of atomic structures gradually unfolds and with it the arrangement of the periodic table of the elements emerges as defining a pattern of both electronic configuration and chemical properties.) Such a method of proceeding might hope to commend itself to biochemists, molecular biologists, biophysicists, and physiologists who are not only aware of the minutiae of the complexity of biological processes and mechanisms but also know the ability of many hierarchically organized biological complexes (e.g. viruses, muscle, bacteriophage: see K. Holmes 1974) of assembling themselves out of their component units by means of an observable *aufbau* process. It might also help to clarify the limitation of such interpretations.

One is encouraged to consider the potentialities of physico-chemical interpretations of biological complexity by a certain congruence between the foci of interest of physical chemistry and of biology—both physical chemistry and biology are concerned with *processes* occurring in modifiable *structures* by the operation of *mechanisms*, and the dimension of time is an inherent parameter for both physico-chemical and biological systems. Furthermore, physico-chemical processes, structures, and mechanisms are, as we have indicated, built hierarchically into the corresponding processes, structures and mechanisms of biological organisms—so theoretical concepts generated by the analysis of the former might at least provide clues and shaping hypotheses, and guidelines for the concepts needed to understand and to categorize the latter. Moreover, it is only by pressing and developing the physico-chemical concepts applicable to biological complexity to their limits that it will become clear precisely at which points entirely new concepts will be required, whether derived from other sources or from themselves.

The physico-chemical contributions to conceptualizing biological systems can be broadly divided into those derived from chemical thermodynamics and those stemming from reflection on the kinetics of chemical processes in networks of reaction and diffusion, that do, or might, occur in living sytems. Parts I and II will give accounts, respectively, of these two broad areas of development, in the last few decades, of physico-chemical ideas relevant to biological complexity and, in the last chapter, we shall reflect again on the nature of biological complexity.

Part I Thermodynamic interpretations of living systems

2 The thermodynamics of dissipative structures

2.1 Thermodynamics and living systems

ALTHOUGH politics may or may not be the 'art of the possible' in human affairs, there would be little dispute that thermodynamics is the 'science of the possible' in the natural world. It generates no mechanisms but, rather, prescribes relationships to which all natural phenomena are expected to conform—indeed, in some ways, thermodynamics is to physical chemistry what logic is to philosophy. There is a paradox here: for classical thermodynamics has been most successful in deriving the relationships that characterize systems at equilibrium, that is, just those systems in which all processes are reversible and which themselves are limits to which actual systems tend; whereas all *natural* processes are, in practice, irreversible.

But the achievements of classical thermodynamics cannot be gainsaid. Through an admittedly 'black box' approach—concerned with relations between external measurements, such as the exchanges of heat and other forms of energy and of matter between a system and its surroundings—it has nevertheless been able to deduce fruitful and powerful relationships between the 'internal' parameters of a system, e.g. the concentrations of substances in a reaction mixture at equilibrium. It is still a remarkable achievement that the equilibrium constant of a chemical process at all temperatures and pressures is, in principle, deducible by classical thermodynamics from purely *thermal* measurements (heat capacities, heats of transition, heats of reaction, etc.) and only one equilibrium constant measurement, at a given temperature and pressure (and even this latter is in certain cases itself deducible from statistical thermodynamic relationships combined with spectroscopic data for the reacting molecules). Indeed this 'black box' approach has been the source of its success. For with its second law—and its consequential definition of the entropy (S) as a property of a system that always increases in natural, irreversible processes in isolated systems, its relation of changes in S to heat uptake and its subsequent elaboration in statistical thermodynamics—thermodynamics may be regarded as one of the best established pillars of modern science, ranging in its applicability from systems as widely different as electromagnetic radiation in equilibrium with matter in a containing vessel (the famous black body, thermodynamic consideration of which led Planck to postulate the quantization of energy), through all known terrestrial systems, to the exchanges of energy that occur in the vicinity of black holes (consideration of which, starting from purely thermodynamic considerations, may lead to important new general connections between gravity, thermodynamics, and quantum theory (Hawking 1975; Davies 1976)).

The edifice of classical equilibrium thermodynamics has rightly been likened by G. N. Lewis and M. Randall to that of a medieval cathedral, through which, in their own classic exposition (1923), they conceived of themselves as conducting their readers:

> There are ancient cathedrals which, apart from their consecrated purpose, inspire solemnity and awe. Even the curious visitor speaks of serious things, with hushed voice, and as each whisper reverberates through the vaulted nave, the returning echo seems to bear a message of mystery. The labor of generations of architects and artisans has been forgotten, the scaffolding erected for their toil has long since been removed, their mistakes have been erased, or have become hidden by the dust of centuries. Seeing only the perfection of the completed whole, we are impressed as by some superhuman agency. But sometimes we enter such an edifice that is still partly under construction; then the sound of hammers, the reek of tabacco, the trivial jests bandied from workman to workman, enable us to realize that these great structures are but the result of giving to ordinary human effort a direction and a purpose.
>
> Science has its cathedrals, built by the efforts of a few architects and of many workers. In these loftier monuments of scientific thought a tradition has arisen whereby the friendly usages of colloquial speech give way to a certain severity and formality. While this may sometimes promote precise thinking, it more often results in the intimidation of the neophyte. Therefore we have attempted, while conducting the reader through the classic edifice of thermodynamics, into the workshops were construction is now in progress, to temper the customary severity of the science in so far as is compatible with clarity of thought.

One of the central pillars of this great edifice—the second law itself—was curiously at variance with another great discovery of the nineteenth century, that of biological evolution. For according to the second law all natural processes, including those in biological systems are, or must have been, irreversible; that is, they must have occurred with a net increase in entropy, if the systems were isolated, and must finally have attained a static condition of maximum entropy—equilibrium, a state which came, through Boltzmann, to be interpreted as maximal 'disorder' or 'randomness'. For Boltzmann pointed out that the entropy S of a system could be regarded as proportional to the logarithm of the 'number (W) of complexions', i.e. the number of possible dispositions of matter over the (equiprobable) available energy states, i.e. the number of microstates corresponding to a given macrostate identified through its macroscopic properties; so $S = k_B \ln W$, where k_B = Boltzmann constant (gas constant (R)/Avogadro's number (N_0)). Whether or not decrease in entropy as a measure of the kind of organization that characterizes biological systems will prove to be adequate for this purpose is a question we must defer but the kind of *dis*order (W) associated with *in*crease in entropy is clearly inimical both to 'orderliness' (whose ideal reference state is 'order'), as it is applied to purely physical systems via the Boltzmann relationship, and to the 'organization' that characterizes biological systems (see section 6.3). Hence this second law of thermodynamics seemed to be in conflict with the understanding of biological evolution, after Darwin, as characterized by an *increase* in organization and by the coming into existence of structures of greater and more intricate complexity, both functional, structural, and behavioural. Of course, in non-living physical systems, order *can* be generated by random processes in non-isolated systems near to equilibrium and at sufficiently low temperatures, when only the lowest energy states are occupied. But could this ever explain the formation of biological structures, the chance of formation of which by random processes is vanishingly small at ordinary temperatures, starting with a macroscopic number of the molecules ultimately to be assembled (see penultimate paragraph of this section)? The formation of living systems was not going to be explicable in terms of classical equilibrium thermodynamics, and statistical thermodynamics, although, as we shall see below (2.2), the actual functioning of a living biological system, once it exists, can be shown to be consistent with the second law.

In the twentieth century, classical thermodynamics, especially in the formulations of de Donder and Duhem, has been extended to irreversible processes at the hands of Onsager, Meixner, and Prigogine and these ideas have been further amplified by A. Katchalsky, de Groot, and members of the Brussels group such as Glansdorff and Nicolis. Although this extension reaped many rewards in other fields, it did not assist much in the interpretation of biological processes. The reason was that this new thermodynamics of irreversible processes was developed for situations in which the flows and rates of the processes were linear functions of the 'forces' (temperature, concentration, chemical potential, gradients) that drove them. Such linear, non-equilibrium, processes can, in fact, lead to the formation of configurations of net higher entropy and higher order (e.g. when a thermal gradient is applied to a mixture of two different gases and one component concentrates at the hot wall and the other at the cold)—so that non-equilibrium can be a source of order. But the 'order' so created is not really structural and far from the organized intricacies of biology (see section 6.3 for a fuller discussion of 'order' in this context). Moreover, and more pertinently, biological processes depend ultimately on biochemical ones and these, like all chemical reactions not at equilibrium, are usually non-linear (e.g. eqns (2.33) and (2.53)), and fall outside the scope of irreversible thermodynamics as at first developed.

It has been the study, chiefly by the Brussels school, of situations further away from equilibrium, in the non-linear range, that has provided the clues that are applicable to the *coming into existence*, as distinct from the maintenance, of biological order. These considerations apply, as we shall see, to open systems (see section 2.2) and they lead to the conclusion that new ordered forms of such a system can come into existence and be stable, if matter and energy are flowing through to maintain them. These new ordered forms are called *dissipative structures* and are radically different from the 'equilibrium structures' studied in classical thermodynamics, whose 'order' is attained at low enough temperatures in accordance with Boltzmann's ordering principle (as described above, and coupled with $e^{-E/k_B T}$ being proportional to the probability of occupation of a state of energy E at temperature T). In this non-linear range, non-equilibrium can indeed be the source of an order that would not be predictable by the application of Boltzmann's principle.

To show that the Boltzmann ordering principle is inadequate to explain the origin of biological structures it suffices to take an example of Eigen (1971a). Consider a protein chain of 100 amino acids, of which there are 20 kinds. The number of permutations of the possible order of amino acids in such a protein, and on which its biological activity and function depend, is $20^{100} \simeq 10^{130}$, assuming all sequences equally probable. This is the number of permutations necessary to obtain a given arrangement starting from an arbitrary initial distribution. If a change of structure occurred at the (impossibly) high rate of one every 10^{-8} seconds, then 10^{122} seconds would be needed—yet the age of the earth is 10^{17} seconds. So the chance of 'spontaneous' formation of the protein, by the Boltzmann ordering principle, through processes at equilibrium, is indeed negligible. Equilibrium cannot give rise to biological order.

'Order' of biologial significance arises, as we shall see, through the amplification of unstable fluctuations and so is a case of 'order through fluctuation'. In this chapter we examine these ideas, developed by I. Prigogine, P. Glansdorff, G. Nicolis, and others at Brussels, and their relevance to biological processes, as well as their relation to more mechanistic, kinetic interpretations.

2.2 Open systems

The second law of thermodynamics, to which we have already referred, can be expressed more formally in terms of changes in entropy (S) in a system in relation to its internal state and external interactions. It is essential to remember that the possibility of defining S as a property of the state of a system and its relation to heat exchanges (S change due to reversible heat intake is q/T, where q = heat absorbed) is only possible because of the validity of the second law, and that S is defined so as to increase in all irreversible (i.e. all natural) processes in an isolated system (a system which no matter or energy enters or leaves). The second law can be expressed more formally in the following way. During any small time interval, dt, the change in entropy (S) of any system *open* to exchange of energy and matter from its surroundings is

$$dS = d_iS + d_eS \qquad (2.1)$$

where d_eS is the change in S due to exchanges with the surroundings *external* to the system; and d_iS is the change in S resulting from irreversible processes occurring *inside* the system, such as chemical reaction, diffusion, heat conduction, etc. (Fig. 2.1). The second law states that always $d_iS \geq 0$, the equality applying only at equilibrium, i.e. when only reversible changes occur in the system. If the system is *isolated*, there is no intake of energy or matter so d_eS is zero, and for such systems

$$dS = d_iS \geq 0. \qquad (2.2)$$

It has long been clear that the existence of highly organized structures, such as living organisms, could be reconciled in the following way with the second law, provided they were open systems. Although d_iS is never negative, the sign of d_eS is not definite, depending as it does on the exchange of matter and energy of the system (the living organism in this case) with its surroundings. So an evolving open system may reach a state where the entropy is smaller at the end than at the beginning. Moreover, this relatively more improbable state may persist provided it can reach a *steady state* in which S remains unchanged, so that over a small interval of time, dt

$$dS = 0 = d_iS + d_eS \qquad (2.3)$$

and

$$d_eS = -d_iS \leq 0. \qquad (2.4)$$

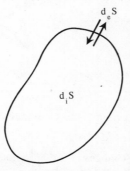

Fig. 2.1. Entropy flux and entropy production in an open system. (From Nicolis and Prigogine 1977, Fig. 1.1.)

Thus an open system can be maintained in an ordered state by giving up positive entropy to its surroundings, as heat and degraded chemical substances. This process is sometimes regarded as equivalent to the system taking in 'negative entropy' (or 'negentropy' as Schrödinger (1944) called it). This can only happen if the system (the living organism) is *not* in thermodynamic equilibrium, so that d_iS is not zero for its internal

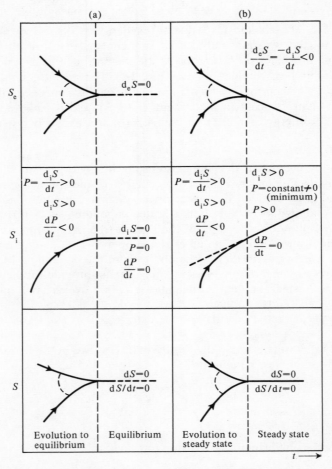

Fig. 2.2. Entropy changes in evolution to (a) equilibrium and (b) a steady state

$$dS = d_iS + d_eS$$

where S = total entropy; d_iS = changes in S due to internal, irreversible processes; d_eS = changes in S due to interaction with the external surroundings (through exchanges of matter and/or heat). Internal entropy production rate = $d_iS/dt = P = \int \sigma \, dV$.

processes, that is, internal irreversible processes must be occurring. Such a non-equilibrium constraint is, indeed, the pre-condition of the existence of the whole biosphere, in the form of the need for a gradient of solar energy from the sun penetrating into the earth and its surface layers. At a more detailed level, all living cells are fed with energy and chemical substances and so are all metabolic processes. Thus living organisms are members of the class of systems which are both open and disspative, that is, in a steady state in which their internal processes are occurring irreversibly with an increase in entropy (d_iS) while their total entropy (S) is maintained constant, at some low level characteristic of their relatively high state of order, through interaction with their surroundings. This situation, and its contrast with the state of thermodynamic equilibrium, is illustrated in Fig. 2.2, which also shows how the *internal entropy production*, P ($=d_iS/dt$), of the whole system varies during the approach to equilibrium or a steady state. The quantity P applies to the whole system and, in many applications, it is more convenient to define the internal entropy production *per unit volume* (σ) as

$$\frac{d_iS}{dt} \equiv P = \int \sigma \cdot dV. \tag{2.5}$$

So, if the system is homogeneous, the second law can be expressed as either

$$P \geqslant 0 \tag{2.6a}$$

or,

$$\sigma \geqslant 0. \tag{2.6b}$$

Equations (2.6) are equivalent to (2.2). With the 'local equilibrium' assumption, discussed below (section 2.3), eqns (2.5) and (2.6b) can refer to non-homogeneous systems. If the interaction with the environment involves only heat exchanges then the usual result of classical thermodynamics applies, namely that

$$d_eS = q/T, \tag{2.7}$$

where q is the heat absorbed by the system.

2.3 'Local equilibrium' and entropy production

The foregoing has denoted a quantity, the entropy S, whose definition has referred to reversible processes in isolated systems, i.e. to processes occurring at equilibrium. We have seen, moreover, that entropy has a key role in defining irreversibility. But what are the variables on which entropy depends both in open systems and in situations far from equilibrium, the ones pertinent to biological systems? Clearly entropy will depend on the variables defining the composition of any given system (i.e. the molar amount of component i (x_i) per unit volume), but dependence on the composition gradients with respect to space and time would also, in general, be anticipated. This entails the possibility that arbitrarily high values for such gradients (i.e. abrupt changes in x_i), in both space and time might vitiate the possibility of attributing any macroscopic meaning to entropy far from equilibrium. Prigogine in a key paper (1949) has been able to demonstrate the conditions under which a macroscopic evaluation of entropy production and entropy flow is possible. Basically, he showed that the equations of macroscopic thermodynamics can be applied to the system as a whole, even under conditions far from equilibrium, if a state of *local equilibrium* prevails in each small mass element of the system, so that the local entropy s (S per unit volume) is the same function of the local macroscopic variables as it would be in the equilibrium state. The conditions for this to be so are: (i) that the range of variations imposed by external constraints (i.e. gradients) must be larger than the mean free path, for spatial constraints, or the relaxation frequency of thermal-molecular motion, for temporal ones (i.e. dissipative processes arising mainly from elastic collisions constitute the dominant contribution to the molecular distribution function); and (ii) that collisions leading to chemical reaction are sufficiently rare that elastic collisions may restore the Maxwell–Boltzmann distribution to a good approximation—i.e. the energies of activation of reaction should be large compared with thermal energies. Under these conditions the entropy of a non-equilibrium state can be defined and includes a dominant contribution from its values at various locations in the system as a function of local variables, which will be the same variables as those for ordinary macroscopic thermodynamics, e.g. those occurring in the Gibbs formula

$$T \, dS = dU + p \, dV - \sum_i \mu_i \, dN_i, \qquad (2.8)$$

that is, initially U, V, N_i and then, derivatively, T, p, and chemical potential, μ_i (N_i is the number of moles of component i). The local states are then completely described by equations of state independent of gradients.

This domain of validity of local equilibrium thermodynamics, that is, of a local macroscopic description of matter, is the domain in which dissipative processes, mainly collisional ones, are sufficiently dominant to exclude large deviations from statistical equilibrium, which are the possible effects of imposed gradients, or of chemical affinities. Systems for which these conditions are fulfilled and yet, macroscopically, display far-from-equilibrium behaviour include: quite complex systems of chemical reactions with highly non-linear kinetics in which the rate of elastic collisions is larger than the rate of reactive collisions—and this is generally true for reactions in not-too-rarefied media, and so for all biological processes in liquids; and all convective and transport effects described by the ordinary equations (e.g. Navier–Stokes) of fluid dynamics. These conditions will, however, not be fulfilled by certain other processes—e.g. shock waves, highly rarefied gases and, indeed, by any dense systems in which there is a negligible change of pressure, density, energy density, etc., over the range of the intermolecular energy (see, for example J. D. van der Waals 1894, and for a recent discussion, Harrington and Rowlinson 1979, and Rowlinson and Widom 1982). Many, if not all, biological systems have gradients of density, energy density, etc. over distance of less than 10 Å (e.g. material near or in a membrane or a cell nucleus) and so cannot, strictly speaking, have a 'local thermodynamics'. This limitation of Prigogine's assumption has to be acknowledged while accepting, at this point in the argument, that it is a necessary one for making progress to at least qualitative conclusions.

These results may be summarized by saying that it *is* possible to define an entropy density s (entropy *per unit volume*) such that

$$\text{Total } S = \int s \cdot dV \tag{2.9a}$$

and that

$$s = s(u, V, x_1, x_2 \ldots x_i \ldots x_n), \tag{2.9b}$$

where u is the internal energy per unit volume (almost equivalently, the temperature), and the x_i are the composition variables (mole numbers per unit volume) of the n components and are themselves variable with spatial location and time (V is the volume, here unit volume). Entropy then becomes an implicit function of time through $x_1 \ldots x_i \ldots x_n$ varying with time, for differentiating (2.9b) with respect to time for a given location yields

$$\sigma = \left(\frac{\partial s}{\partial t}\right)_{u,V} = \sum_i \left(\frac{\partial s}{\partial x_i}\right)_{u,V,x_j} \left(\frac{\partial x_i}{\partial t}\right) = -\sum_i \mu_i T^{-1} \cdot \left(\frac{\partial x_i}{\partial t}\right) \quad \text{(constant } T\text{)}. \tag{2.10}$$

Here the subscripts u, V mean the differentiation refers to conditions of constant internal energy per unit volume, and so constant T, and constant volume (again, unit volume, in this case). The relation of macroscopic thermodynamics, that $(\partial s/\partial x_i)_{u,(T),V,x_j} = -\mu_i T^{-1}$, can be introduced because of the local equilibrium assumption. The $(\partial x_i/\partial t)$ with respect to any volume element are constituted by a balance of the net flows of components i in and out of that volume element and their rate of appearance or destruction in chemical processes, which may be expressed formally as

$$\frac{\partial x_i}{\partial t} = -\text{div}\,\mathbf{j}_i + \sum_\rho \nu_{i\rho} w_\rho \qquad (2.11)\dagger$$

where $\mathbf{j}_i$ is the diffusion flux of component i; ρ is the ρth reaction process in which $\nu_{i\rho}$ is the stoichiometric coefficient of i, and w_ρ is the reaction rate per unit volume—so that the overall reaction rate is

$$W_\rho = \int w_\rho \cdot \mathrm{d}V.$$

Substitution of (2.11) in (2.10) yields an expression of the form:

$$\frac{\partial s}{\partial t} = \begin{array}{l}\text{term due to flux of}\\\text{matter and energy into}\\\text{the whole volume } V,\\\text{i.e. into the system}\\\text{as a whole}\end{array} + \begin{array}{l}\text{entropy production}\\\text{due to irreversible}\\\text{processes in the}\\\text{system } (\sigma)\end{array}$$

Here

$$\sigma = \begin{array}{l}\text{entropy production}\\\text{per unit } volume \text{ due}\\\text{to irreversible}\\\text{processes in the system}\end{array} = \sum_\alpha (J'^{\text{diffusion}} X'^{\text{diffusion}} + J^{\text{reaction}} X^{\text{reaction}})$$

where σ is now $\mathrm{d}s/\mathrm{d}t$, as in equation (2.10) (and also satisfies equation (2.5) which now, with the local equilibrium assumption, need no longer be confined to homogeneous systems); the J_α are flows ($\mathbf{j}_i$ and w_ρ) associated, respectively, with 'diffusion' and chemical reaction (without heat flows); and the X_α are the corresponding generalized forces, respectively, $-\nabla \mu_i \cdot T^{-1}$ for 'diffusion' of each component i and $A_\rho T^{-1}$ for each chemical reaction ρ, and the summation is taken over all appropriate pairs (α) of flows and 'forces' (i.e. over all components i and all reaction

---

† The *divergence*, denoted by 'div', of a vector, say $\mathbf{a}$ (here, in equation (2.11), $\mathbf{a}$ is $\mathbf{j}_i$) is a scalar quantity given by

$$\text{div}\,\mathbf{a} = \frac{\partial a_x}{\partial x} + \frac{\partial a_y}{\partial y} + \frac{\partial a_z}{\partial z}.$$

The divergence of the vector $\mathbf{a}$ represents the net flux per unit volume of $\mathbf{a}$ out of a small volume V surrounding any particular point. The flux of $\mathbf{a}$ *out* of any element $\mathrm{d}\mathbf{S}$ of the surface $\mathbf{S}$ of this small volume is $\mathbf{a} \cdot \mathrm{d}\mathbf{S}$, and over the whole surface the net outward flux is $\int \mathbf{a} \cdot \mathrm{d}\mathbf{S}$. The divergence of $\mathbf{a}$ is then also given by

$$\text{div}\,\mathbf{a} = \lim_{V \to 0} \frac{1}{V} \int \mathbf{a} \cdot \mathrm{d}\mathbf{S},$$

and this can be shown (e.g. Riley 1974, p. 98 ff.) to be equivalent to the definitions above.

processes ρ). Here 'diffusion' is a shorthand for the sum of all flow processes in which components i contribute to the change in entropy, $(\partial s/\partial t)$, given by (2.10). The quantity A_ρ is the affinity (negative of the free energy change) of the chemical process ρ, given by

$$A_\rho = -\sum_i \mu_i \nu_{i\rho}, \qquad (2.12)$$

where μ_i is the chemical potential of component i, whose stoichiometric coefficient in the reaction ρ is $\nu_{i\rho}$. Or, more briefly and generally, the foregoing may be written as

$$\sigma = \sum_\alpha J_\alpha X_\alpha, \qquad (2.13a)$$

$$P = \int \left(\sum_\alpha J_\alpha X_\alpha \right) dV, \qquad (2.13b)$$

where the J_α are the rates of irreversible processes and the X_α are the corresponding generalized forces. The J_α and X_α are the conjugate variables associated with the various irreversible processes that are occurring and some of these are given in Table 2.1. Expression (2.13) may be combined with (2.6b) to give

$$\sigma = \sum_\alpha J_\alpha X_\alpha \geq 0 \qquad (2.14a)$$

and

$$P = \int \left(\sum_\alpha J_\alpha X_\alpha \right) dV \geq 0 \qquad (2.14b)$$

as a basic inequality to which the flows J_α and forces X_α must conform

Table 2.1

Process	Flux (J_α)	Generalized force (X_α)	Tensor character
Diffusion	Diffusion flow $\mathbf{j}_i$	Gradient of chemical potential $(-T^{-1} \nabla \mu_i)$†	Vector
Chemical reaction	Reaction rate W_ρ	Affinity $(A_\rho T^{-1})$	Scalar
Heat conduction	Heat flow	Temperature gradient (∇T^{-1})†	Vector

† ∇ is the linear vector operator

$$\left(\mathbf{i} \frac{\partial}{\partial x} + \mathbf{j} \frac{\partial}{\partial y} + \mathbf{k} \frac{\partial}{\partial z} \right)$$

and when operating upon a scalar produces a vector (see Riley 1974, p. 97).

according to the second law—the inequality (>0) denoting irreversible internal entropy production and the equality (=0) denoting a reversible process connecting two equilibrium states. Note that the entropy production is both a *thermodynamic* quantity, through the thermodynamic forces X_α, and a *kinetic* quantity, through the flows J_α. It should also be noted that different sets of generalized flows and forces could be introduced but their products must still give the same sum, σ. How, in principle, the unknown flows J_α are related to the forces X_α, which are known to be functions of composition variables (on the local equilibrium assumption) remains to be examined.

2.4 Stability of the equilibrium state

The question of the stability of states far removed from equilibrium is one of the principal concerns in the thermodynamic study of dissipative structures and their relation to biological systems. However, these considerations have to be viewed against the background of the previous understanding of the stability of the equilibrium state which has been a central theme in the classical developments of Gibbs and Duhem (see Prigogine and Defay 1954, Chapter XVI for a general account, and Glansdorff and Prigogine 1971, Chapters IV, V for an account related to the following).

Using (2.1), (2.7), and the first law written as

$$q = dU + p\, dV, \quad \text{[all external work as pressure work]} \quad (2.15)$$

we obtain

$$T\, d_i S = T\, dS - dU - p\, dV \geq 0, \quad \text{[for irreversible processes]} \quad (2.16)$$

where dS is the change in total entropy S of the system and the inequality represents the second law condition for irreversibility. This may be regarded as the condition for a small fluctuation (when δ would be substituted for the differentials in (2.15)) to be an irreversible, i.e. real, natural, process. Contrariwise

$$\delta U + p\, \delta V - T\, \delta S \geq 0 \quad (2.17)$$

then becomes the condition that a fluctuation (which may be a perturbation resulting from internal molecular fluctuations, as much as from fluctuations at the boundaries from external action) is not irreversible, i.e. it becomes the condition that the equilibrium state will be restored after the fluctuation and that this state is therefore *stable*. For an isolated system (U, V constant) this stability criterion becomes simply $-\delta S_{eq} \geq 0$, that is $\delta S_{eq} \leq 0$, where the subscript shows that the fluctuation is taken as being about the putative equilibrium state. In words, if any fluctuation about an equilibrium state (which is a state of maximum entropy) leads to a negative entropy variation ($\delta S < 0$), then the equilibrium state is stable, for all fluctuations will be damped. Briefly,

$$(\delta S)_{eq} = 0 \text{ for } equilibrium \qquad [\text{constant } U, V] \qquad (2.18a)$$

$$(\Delta S)_{eq} < 0 \text{ for } stability \qquad [\text{constant } U, V] \qquad (2.18b)$$

where the symbol Δ now emphasizes the possibility of finite perturbations.

Other forms of stability criteria for equilibria are possible utilizing other thermodynamic potentials, such as the Gibbs free energy ($G = U + pV - TS$). From (2.16) it follows at once that the stability criterion for equilibrium then takes the form, more familiar to physical chemists:

$$(\delta G)_{eq} = 0 \text{ for } equilibrium \; [\text{constant } T, p] \qquad (2.19a)$$

$$(\Delta G)_{eq} > 0 \text{ for } stability \; [\text{constant } T, p]. \qquad (2.19b)$$

These formulations of stability criteria, and the other parallel ones, in terms of the classical thermodynamic potentials (U, G, enthalpy (H) and Helmholtz free energy (F or A)), suffice for most problems about the stability of equilibrium states. However, with non-equilibrium situations this approach encounters the difficulty that the 'external', system-as-a-whole, properties such as G require a control of fluctuations in variables like p and T not only at the boundaries of the system but also at each point inside it—a requirement not realizable with actual open systems.

For this reason Glansdorff and Prigogine provided a new formulation of equilibrium stability theory applicable to systems with given boundary conditions (e.g. values of T, p) *and* applicable to non-equilibrium cases—always on the 'local equilibrium' assumption, however. The quantity on which they concentrate is the 'excess entropy', $\delta^2 S$, which appears in the equation obtained on expanding the entropy about the equilibrium state in a Taylor series:

$$S = S_{eq} + (\delta S)_{eq} + \tfrac{1}{2}(\delta^2 S)_{eq} \qquad (2.20a)$$

or

$$(\Delta S)_{eq} = S - S_{eq} = (\delta S)_{eq} + \tfrac{1}{2}(\delta^2 S)_{eq} \qquad (2.20b)$$

and so

$$(\Delta S)_{eq} = \tfrac{1}{2}(\delta^2 S)_{eq} \qquad (2.20c)$$

since $(\delta S)_{eq} = 0$ (S a maximum). Intuitively, since at constant internal energy and volume, the entropy is a maximum at stable equilibrium, we could write

$$(\delta^2 S)_{eq} < 0 \qquad [\text{constant } U, V] \qquad (2.21)$$

as characteristic of the region close to equilibrium and therefore as a condition for stability at equilibrium. More explicitly, by differentiating the Gibbs equation (2.8), one obtains

$$\delta^2 S = \delta(T^{-1})\,\delta U + \delta(pT^{-1})\,\delta V - \sum_i (\mu_i T^{-1})\,\delta N_i \qquad (2.22a)$$

or

$$\delta^2 s = \delta(T^{-1})\,\delta u + \delta(pT^{-1})\,\delta v - \sum_i (\mu_i T^{-1})\,\delta N_i. \qquad (2.22b)$$

(N.B. It can be proved that *only* with the set of variables U, V, N is differentiation to this result permissible and can other quantities, such as $\delta^2 U$ and $\delta^2 V$, vanish—see Glansdorff and Prigogine 1971, p. 24 f.) In (2.22a), the macroscopic values of S, U and V have been used; (2.22b) is a form that is useful later and uses the 'per unit *mass*' functions, namely, s, u and v.

If (2.22b) is now combined with the original Gibbs formula (2.8), it may be written as

$$T\delta^2 s = -\delta T \cdot \delta s + \delta p \cdot \delta v - \sum_i \delta\mu_i\,\delta N_i \qquad (2.23)$$

and combination of this with the standard expressions relating the classical potentials μ_i to T, p, N_i, some of the Maxwell relations, the definition of c_v (molar heat capacity at constant volume) and χ (the isothermal compressibility), and standard relationships of the calculus yields (see Glansdorff and Prigogine 1971)

$$\delta^2 s = -T^{-1}\left[c_v T^{-1}(\delta T)^2 + \rho\chi^{-1}(\delta v)^2_{N_i} + \sum_{ij} \mu_{ij}\,\delta N_i \cdot \delta N_j\right], \qquad (2.24)$$

where ρ is the density. The quantities c_v and χ are always positive, being the conditions of assumed thermal and mechanical stabilities, respectively. The condition

$$\sum_{ij} \mu_{ij}\,\delta N_i \cdot \delta N_j > 0 \qquad (2.25)$$

would mean that the response of, say, a binary system, originally at equilibrium, to an inequality of composition is to restore the initial homogeneity, presumably by diffusion. So this condition may be called that of stability with respect to diffusion. Since the thermal, mechanical, and diffusional stability conditions all apply (i.e. the corresponding terms in (2.24) are positive), it follows that

$$(\delta^2 s)_{eq} < 0 \qquad (2.26a)$$

or

$$(\delta^2 S)_{eq} < 0 \qquad (2.26b)$$

STABILITY OF THE EQUILIBRIUM STATE

is a stability condition at equilibrium (since the thermal, mechanical, and diffusional stabilities may be assumed, on other grounds).

Computation of the time derivative of the 'excess entropy' $\delta^2 S$ has been carried out and yields (Glansdorff and Prigogine 1971, pp. 53–4) the elegantly straightforward result that it is equal to the entropy production P of the system:

$$\frac{\partial}{\partial t} \frac{1}{2} (\delta^2 S)_{eq} = P = \int \sigma \cdot dV > 0 \qquad (2.27a)$$

or

$$\frac{\partial}{\partial t} \cdot \frac{1}{2} (\delta^2 s)_{eq} = \sigma > 0 \qquad (2.27b)$$

and this also becomes a stability condition for equilibrium, and the inequality $\partial \frac{1}{2} (\delta^2 S)_{eq} / \partial t > 0$ is true identically by the second law. The evolution of the 'excess entropy' $(\delta^2 S)_{eq}$ with time around equilibrium is illustrated in Fig. 2.3. With increasing time, $(\delta^2 S)_{eq}$ asymptotically approaches zero and vanishes only at the reference (equilibrium) state, which therefore possesses 'asymptotic stability' with respect to small perturbations. For the equilibrium state, $(\delta^2 S)_{eq}$ is what is called a *Lyapunov function* (see Glansdorff and Prigogine 1971, pp. 62–3; Nicolis and Prigogine 1977, pp. 65–6; N. Minorski 1962) which is a function that has the requisite properties to guarantee this kind of stability (roughly, the sign of the slope of the function with time is of the opposite sign to that of the function itself, together with restrictions that ensure the asymptotic nature of the curve as $t \to \infty$).

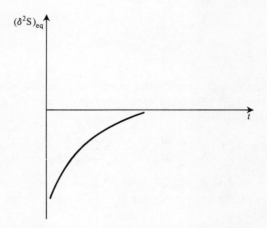

Fig. 2.3. Time evolution of second-order excess entropy $(\delta^2 S)_{eq}$ around equilibrium. (From Nicolis and Prigogine 1977, Fig. 4.2.)

The value of these stability conditions are that in the forms (2.26a) and (2.27a) they can be applied independently of the boundary conditions (e.g. that U and V be kept constant), provided one deals with *small* perturbations (i.e. one can neglect higher terms in (2.20)). It is a more general result than that of the classical Gibbs–Duhem theory and is independent of the existence of any thermodynamic potential. So it is, at least, a candidate for being extended to non-equilibrium systems, and this is in fact the reason why Glansdorff and Prigogine (1971) originally developed it. These results imply, it should be noted, that the equilibrium state cannot be modified if the deviations from equilibrium are only through fluctuations or random disturbances and this is consistent with its lack of spatial or temporal order.

2.5 Linear thermodynamics of irreversible processes

Flows and forces

At the end of section 2.3 it was pointed out that the general formulae (2.14) relating fluxes and forces to entropy production in itself presumed and prescribed no particular relationship between the fluxes and forces, although these latter were, on the local equilibrium assumption, known to be functions of composition variables. In this section, we shall be concerned with the irreversible thermodynamics of processes *close to equilibrium*. At equilibrium itself the generalized forces X become zero (at chemical equilibrium all $A_\rho = 0$, and all $\nabla(\mu_i T^{-1})_{eq} = 0$, the uniform composition of the equilibrium state); moreover there is no macroscopic transport or production of individual constituents, so all

$$(J_\alpha)_{eq} = 0 \tag{2.28a}$$

as well as

$$(X_\alpha)_{eq} = 0, \tag{2.28b}$$

where the subscript 'eq' denotes the equilibrium state. Near equilibrium when the X_α are still relatively weak, the flows can be expanded in a power series of X

$$J_\alpha = (J_\alpha)_{eq} + \sum_\beta \frac{\partial J_\alpha}{\partial J_\beta} X_\beta + \tfrac{1}{2} \sum_{\beta\gamma} \frac{\partial^2 J_\alpha}{\partial X_\beta \, \partial X_\gamma} X_\beta X_\gamma. \tag{2.29}$$

The first term vanishes (by (2.28a)) and the third and subsequent terms can be neglected if the system is near to equilibrium, hence we obtain the 'linear' relation

$$J_\alpha = \sum_\beta L_{\alpha\beta} X_\beta, \tag{2.30}$$

where
$$L_{\alpha\beta} = (\partial J_\alpha/\partial X_\beta)_{eq},$$

and $\alpha, \beta = 1, 2 \ldots \nu$ for ν flows and ν forces. When $\beta = \alpha$, the phenomenological coefficients $L_{\alpha\beta}$ stand for electrical or thermal conductivity, or other coefficients relating flow (e.g. of a substance) to its own specific driving force (e.g. chemical potential gradient). When $\alpha \neq \beta$, the coefficients $L_{\alpha\beta}$ represent the mutual interference between the irreversible processes α and β (e.g. thermal diffusion). The $L_{\alpha\beta}$ are determined by the internal structure of the medium and may depend on the state variables such as p, T, x_i. The rigorous justification of the linearity explicit in (2.30) cannot be effected by thermodynamic considerations alone, but rather by statistical mechanics. It is thereby shown (Prigogine 1949) to be applicable to transport phenomena only when the macroscopic gradients vary on a scale much larger than the mean free path, which is, in fact, one of the features of local thermodynamic equilibrium, already presupposed. However, the linearity of (2.30) is a more restrictive condition for chemical reactions. Thus for the reaction

$$A \underset{k_b}{\overset{k_f}{\rightleftharpoons}} B \tag{2.31}$$

the affinity

$$A = k_B T \ln \frac{k_f x_A}{k_b x_B}, \tag{2.32}$$

where k_B = Boltzmann constant and the rate

$$W = k_f x_A - k_b x_B$$
$$= k_f x_A (1 - \exp(-A/k_B T)). \tag{2.33}$$

This rate can be linearly related to A only if

$$A \ll k_B T, \tag{2.34}$$

when

$$W = \frac{k_f x_A}{k_B T} \cdot A.$$

This is a very restrictive condition limiting the applicability of (2.30) only to chemical reactions in the immediate vicinity of equilibrium (or to reactions with very low A)—and these conditions are not usually satisfied in actual systems, especially biological ones. So there will be a need to extend, if possible, these thermodynamic considerations to the non-linear range (see section 2.6 below). However, when the linear relations (2.30) can be adopted, thermodynamics can yield important information about the coefficients, $L_{\alpha\beta}$. For example, for a system in which two irreversible processes are occurring and to which (2.30) applies, we can write

$$J_1 = L_{11}X_1 + L_{12}X_2$$
$$J_2 = L_{21}X_1 + L_{22}X_2 \qquad (2.35)$$

and the entropy production is given by

$$\sigma = \sum_\alpha J_\alpha X_\alpha = L_{11}X_1^2 + (L_{12}+L_{21})X_1X_2 + L_{22}X_2^2 > 0. \qquad (2.36)$$

The requirement to be positive leads algebraically, to the conditions that

$$L_{11} > 0; \qquad L_{22} > 0, \qquad (2.37a)$$
$$(L_{12}+L_{21})^2 > 4L_{11}L_{22}. \qquad (2.37b)$$

So the *self*-phenomenological coefficients L_{11} and L_{22} are always positive, by (2.37a), whereas the *mutual* ones (L_{12}, L_{21}) have only to obey (2.37b)—and so can be of either sign and may sometimes be negative (as the thermal diffusion coefficient is in certain systems).

The very large number of phenomenological coefficients $L_{\alpha\beta}$, which cannot be obtained from macroscopic thermodynamics, can fortunately be reduced because of both temporal and spatial symmetries. Onsager (1931) showed that the increase in flux J_α per unit increase in force X_β (keeping all other Xs constant) is equal to the increase in flux J_β per unit increase in force X_α (again keeping all other Xs constant), that is,

$$L_{\alpha\beta} = L_{\beta\alpha}, \qquad (\alpha, \beta = 1, 2 \ldots \nu), \qquad (2.38)$$

which is called the *Onsager reciprocity relation*. This relation can be proved (see, for example Prigogine 1955, pp. 46–53) by fluctuation theory joined with the principle of microscopic reversibility (the invariance of all mechanical equations of motion of individual particles with respect to the transformation of time $t \to -t$).

The restrictions on the possible $L_{\alpha\beta}$ imposed by spatial symmetries arise from the properties of the entropy production (p. 23). Table 2.1 divides the irreversible processes of present interest into *vector* phenomena (e.g. diffusion, heat conduction) and a *scalar* phenomenon (e.g. chemical reaction). Now the causes of a phenomena cannot have more symmetry elements than the effects produced, indeed they always have fewer (according to the symmetry principle of Curie (1908); see also Prigogine 1955, pp. 53–4). The close-to-equilibrium state must be isotropic and this condition is usually satisfied by systems at equilibrium, in the absence of external forces such as varying gravitational fields and non-isotropic boundary conditions. If this is so (and this condition is obligatory, Katchalsky and Spangler 1968) a directed flow of (say) heat—a vector—could not be generated solely by a 'scalar' process, such as chemical reaction. The corresponding mutual phenomenological coefficients $L_{\alpha\beta}$ (in this case $L_{\text{chemical, thermal flow}}$) must in these instances therefore be zero. So entropy production can be divided into two uncoupled parts

$$\sigma = \sigma_{\text{chemical}} + \sigma_{\text{diffusion}}, \qquad (2.39)$$

where 'chemical' refers to scalar and 'diffusion' to vectorial processes, given by

$$\sigma_{\text{ch}} = \sum_\rho w_\rho A_\rho T^{-1} \geq 0 \qquad (2.40)$$

and

$$\sigma_{\text{d}} = -\sum_i \mathbf{j}_i \, \nabla \mu_i T^{-1} \geq 0 \qquad (2.41)$$

(shortening the subscripts to 'ch' and 'd' respectively). In these 'linear' circumstances, close to equilibrium, the two groups of processes, 'chemical' and 'diffusion', give *separate* positive contributions to the total positive entropy production. (There is also a third group, in addition to scalar and vector, namely, *tensor* processes, such as viscous dissipation, but these are of less relevance to our present discussion).

Non-equilibrium stationary (steady) states in the linear range

We have already referred (section 2.2) to the concept of *steady states* of open systems in which the state variables, in particular composition (x_i), do not vary with time and in which the total entropy can remain unchanged ($dS = d_i S + d_e S = 0$) by the increase in internal entropy ($d_i S$) being exactly matched by the system as a whole yielding up entropy to its surroundings, so that $d_e S = -d_i S$ (eqn 2.4). Equilibrium is itself a special example of such a steady state for which $d_e S = -d_i S = 0$. At equilibrium, spatial inhomogeneities can only arise by the action of external forces, so that in their absence the systems are homogeneous. Similarly, through the continuity of close-to-equilibrium steady states with the equilibrium state, the steady states are in this region presumed to be homogeneous and the contribution of chemical reactions and diffusion to the evolution of the macrovariables can be separated, and vanish separately. This does not, of course, preclude the possibility that at a critical separation from equilibrium none of these conditions might apply and non-isotropic non-linear features might appear.

The concept of a steady state is very familiar in biochemical kinetics in which a long chain of reactions is regarded as being initiated by a substance received from the outside and is then transformed by a serial chain of linked reactions via intermediate substances to a final product which is returned to the surroundings. This process, usually under enzymatic control, very rapidly reaches a steady state in which the concentrations of the intermediates do not change with time, yet at no stage is *chemical* equilibrium established. Such states are clearly of paramount significance to biology and their thermodynamic study is of key importance. Purely physical examples of such steady states, in the 'linear' range and close to equilibrium, also exist, e.g. the steady gradation of both temperature and concentration (density) in a vertical column of a fluid mixture, gently and steadily heated from below. In this case, the temperature gradient opposes the effects due to gravitational force and heat is transferred, by conduction in a fluid which is at rest with respect to its bulk movement, through a complicated process involving both kinetic terms (movements of molecules) and potential terms (intermolecular energy).

An important result of the thermodynamic analysis of such steady state systems is that, in the linear range denoted by (2.30) and when the Onsager relations (2.38) hold, that is, in this close-to-equilibrium condition, the *entropy production* in the *steady state* can be shown to have a *minimum value* compatible with the prescribed conditions. This may be proved as follows for the case of two pairs of fluxes and forces, such that, by (2.13)

$$\sigma = J_1 X_1 + J_2 X_2, \qquad (2.42)$$

where subscripts 1 could refer to transfer of energy and 2 to transfer of matter, respectively. For a linear situation,

$$J_1 = L_{11} X_1 + L_{12} X_2 \qquad (2.43a)$$
$$J_2 = L_{21} X_1 + L_{22} X_2, \qquad (2.43b)$$

and we shall assume *strict* linearity, namely, that the phenomenological coefficients L_{ij} remain constant, their values depending only on the equilibrium parameters (Nicolis and Prigogine 1977, p. 42). By the Onsager reciprocal relation (2.38)

$$L_{12} = L_{21},$$

so

$$\sigma = L_{11} X_1^2 + 2 L_{21} X_1 X_2 + L_{22} X_2^2. \qquad (2.44)$$

Hence

$$\left(\frac{\partial \sigma}{\partial X_2}\right)_{X_1} = 2 L_{21} X_1 + 2 L_{22} X_2 = 2 J_2. \qquad (2.45)$$

Now, if it *is* a steady state, there is no net transfer of mass, i.e. $J_2 = 0$, so that

$$\left(\frac{\partial \sigma}{\partial X_2}\right)_{X_1} = 0 \qquad (2.46)$$

is also true of the steady state, and likewise (since $J_1 = 0$, there being no net transfer of energy)

$$\left(\frac{\partial \sigma}{\partial X_1}\right)_{X_2} = 0. \qquad (2.47)$$

The conditions $J_1 = 0 = J_2$ and (2.46) and (2.47) are completely equivalent so long as linear conditions (2.43) prevail and the Onsager relation is applicable. The statements (2.46) and (2.47) mean that the entropy production is then minimal with respect to variation in both the forces X_1 and X_2, respectively. The entropy production therefore reaches a minimum value which is constant in time so that, in the steady state,

$$\frac{d\sigma}{dt} = 0 \qquad (2.48a)$$

or, [steady state]

$$\frac{dP}{dt} = 0 \qquad (2.48b)$$

It can furthermore be shown (Prigogine (1945, 1947) and Nicolis and Prigogine (1977, pp. 42–5)) that

$$\frac{d\sigma}{dt} < 0 \qquad (2.49a)$$

or, [away from the steady state]

$$\frac{dP}{dt} < 0. \qquad (2.49b)$$

Hence the changes in internal entropy (S_i) and internal entropy production P ($= d_i S/dt$), or σ, may be represented for the approach to the steady state by Fig. 2.2(b) (middle panel) and by Fig. 2.4(b). Note that the evolution of the internal entropy and internal entropy production in the approach to equilibrium can be regarded as a special case (Fig. 2.2(a), middle panel and Fig. 2.4(a)) of the approach to steady states in the linear, close-to-equilibrium range.

It should be noted, for it will be of significance later (section 2.6), that the time change dP can be decomposed into two parts

$$P = \int \left(\sum_\alpha J_\alpha X_\alpha\right) dV, \quad \text{by (2.13)},$$

$$\therefore dP = \int \left(\sum_\alpha J_\alpha \cdot dX_\alpha\right) dV + \int \left(\sum_\alpha X_\alpha \cdot dJ_\alpha\right) dV \qquad (2.50)$$

$$\equiv d_X P + d_J P \qquad (2.51)$$

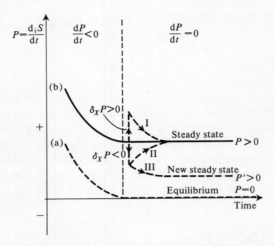

Fig. 2.4. Changes in (internal) entropy production, $P \equiv d_i S/dt$, in evolution to (a) equilibrium, (b) a steady state, in both of which $dP/dt = 0$
Perturbations: (I) a fluctuation for which $\delta_X P > 0$ (depicted by a dotted vertical line, for convenience, on the same ordinate scale as P though, of course, $\delta P \neq \delta_X P$ alone); (II) a fluctuation for which $\delta_X P < 0$, but which reverts to its original steady state; (III) a fluctuation for which $\delta_X P < 0$, but which is amplified and the system passes over into a new steady state and a new regime. The dashed lines indicate a transition and are not meant to be actual plots of P against time. (See text under 'criteria of stability of non-equilibrium states', p. 44.)

where $d_X P$, $d_J P$ represent, respectively, the changes in P due to changes in forces (dX_α) or flux (dJ_α) alone.

Under the 'linear' condition (2.30):

$$\begin{aligned}
d_X P &= \int \left(\sum_\alpha J_\alpha \cdot dX_\alpha \right) dV \\
&= \int \left(\sum_{\alpha\beta} X_\beta L_{\alpha\beta} \cdot dX_\alpha \right) dV \\
&= \int \left(\sum_{\alpha\beta} X_\beta L_{\beta\alpha} \cdot dX_\alpha \right) dV \quad \text{by Onsager (2.38)} \\
&= \int \left(\sum_\beta X_\beta \cdot dJ_\beta \right) dV \\
&= d_J P.
\end{aligned}$$

Hence this result, together with (2.51), yields

$$d_X P = d_J P = \tfrac{1}{2} dP, \quad \text{[linear conditions]} \tag{2.52}$$

where the shorthand 'linear conditions' means the phenomenological laws

LINEAR THERMODYNAMICS OF IRREVERSIBLE PROCESSES

are linear, Onsager's reciprocity relations hold, phenomenological constants (L) are treated as constants—and, as always, there is 'local equilibrium'.

The entropy production σ (or P), with its properties of being itself positive (2.6) and of decreasing with time (2.49) to a zero value (2.48) when it reaches a minimum, suggests that it is a Lyapunov function for the stability of the steady state under equilibrium conditions. This helps to give a picture of what happens to fluctuations about the steady state. The various perturbations—arising either from variation in the external environmental conditions or from within the system itself—act on it (Fig. 2.5) to produce deviations from the set of concentrations x_k^0 that prevail at the steady state (subscript k—any component; superscript zero—steady state) and send it over into time-dependent states ($x_k^0 + \delta x_k$). Because σ^0 is a minimum (2.46, 2.47, 2.48), the entropy production in the new states is larger than in the steady state, and, because of (2.49), it then decreases in time back to the lower, original value σ^0—i.e. the perturbation regresses, as indicated by the arrows in Fig. 2.5. So the theorem of minimum entropy production ensures the asymptotic stability of the steady non-equilibrium state, in the linear range. The entropy production σ is therefore playing a role for this linear range somewhat like that of the thermodynamic potentials in equilibrium theory. It should be stressed that it is entropy *production* ($d_i S/dt$) and not internal entropy itself (and certainly not total entropy, S) that is the concern here, for it is possible that the value of the total entropy S, as such, at a non-equilibrium steady state may actually be less than at the equilibrium state to which it is adjacent. Such a lower entropy cannot reflect the emergence of any

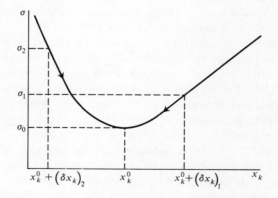

Fig. 2.5. Illustration of theorem of minimum entropy production: x_k^0 = steady-state value of variable x_k; δx_k = perturbation from steady state. (From Nicolis and Prigogine 1977, Fig. 3.2.)

macroscopic order because the requirement of isotropy prevails continuously over the transitions between the equilibrium and the steady states. These latter are indeed essentially uniform in space (in the absence of external constraints, such as a gravitational field, making it otherwise) and their stability in the linear range implies that, in this range, there is no possibility of the spontaneous emergence of order, as spatial or temporal patterns. We must look elsewhere for conditions under which this is possible, namely, to the *non*-linear range.

2.6 Non-linear thermodynamics of irreversible processes

Dissipative structures and 'order through fluctuations'

Although the foregoing reconciles—at least for linear, near-equilibrium systems—the second law of thermodynamics with the existence of biological organisms maintaining a steady, non-equilibrium, state by 'negentropic' absorption from their surroundings, yet the problem still remains of how low entropy, complex systems could have actually come into existence. For the processes occurring in any real system whatever are, by the second law, all irreversible and increase the molecular 'disorder'. Physical and chemical systems ultimately approach a state of maximum homogeneity and minimum order (again see section 6.3 for discussion of 'order') and, according to the second law, they are not expected to move towards the coherent behaviour of biological systems at ordinary temperatures. It is at this critical point that recent studies of the application of thermodynamic and fluctuation theory to dissipative systems have made important conceptual advances. That there *are* indeed systems which become spontaneously more ordered, spatially or temporally or both, under conditions a long way from equilibrium may be illustrated by two famous examples, one chemical, the other physical.

In 1958, Belousov reported an oscillatory chemical reaction in an originally homogeneous solution, namely, the oxidation of citric acid by potassium bromate catalysed by a cerium ion redox system (Ce^{3+}/Ce^{4+}). Zhabotinski (1964) examined this reaction in much greater detail and showed that the same phenomenon also occurred with malonic acid (and other related compounds) and also with other catalysts (e.g. manganese or ferroin). The malonic acid is oxidized by the bromate to carbon dioxide and formic acid. The reaction network includes a number of intermediate stages in the reduction of bromate and there are two reaction pathways, one requiring high concentrations of bromide ions and the other the autocatalytic production of bromous acid with the aid of cerous ions, Ce^{3+}. Both of these ions are products of the final oxidation step so there is positive feedback on to the two alternative pathways, which consequently are operative in turn, the first when bromide ion is abundant, the second as it becomes deplenished. Detailed analyses of this reaction have been made (Noyes and Field 1974) but our present concern is simply to note the changes which occur in the originally homogeneous solution, changes readily observed because of the different absorption spectra of the cerous and ceric ions. Under certain conditions, including constancy of temperature, at first within the vertical reaction tube temporal oscillations are observed, that are randomly distributed spatially; then an inhomogeneity appears at one location which subsequently consolidates into regular spatial patterns in the form of horizontal bands which oscillate temporally and propagate up and down the tube; finally, a spatially static banded pattern is formed in which the Ce^{3+} and Ce^{4+} ions are distributed in alternating horizontal layers (Herschkowitz-Kaufman, 1970). So in this closed system, the reaction network has evolved into an apparently stable state, characterized by spatial inhomogeneity—but not permanently stable in this case, for after several hours the system can change suddenly and catastrophically so that the 'structure' disappears. Such observations suggest that non-linear networks, especially of chemical reactions far from equilibrium, may evolve in unexpected ways, even from a homogeneous phase to a heterogeneous one, which itself may eventually prove unstable and become the starting point of a new development. The need for theoretical analysis of networks evolving so as to include both new elements and new connections is obvious. The example of the Zhabotinsky reaction has served to illustrate the existence of what Prigogine and his colleagues at Brussels have called *dissipative structures*—open systems through which matter and energy flow and which maintain inhomogeneity and order, or rather, organization, at positions far removed from equilibrium. There clearly are phenomena which are not adequately described by the thermodynamic account in terms of entropy which has been outlined so far. There do exist non-organic systems (which in the pre-biotic 'soup' may thereby become biological ones) in which the creation of organized structures may, in fact, occur according to specific non-linear flux–force laws beyond the domain of stability of states on the 'thermodynamic branch' that display the usual thermodynamic behaviour already described.

An actual physical example of this is provided, somewhat surprisingly, by the heated column of a fluid mentioned in section 2.5. When the liquid is at first heated, a continuous relatively small gradation in temperature is produced with the hotter liquid at the bottom near to the source of heat. But as heating continues, the temperature gradient increases and at a critical value (the Bénard instability point) internal convective motion occurs spontaneously in which macroscopic domains of the liquid, containing large numbers of molecules, move in concert. This highly cooperative motion even displays a spatial regularity, in the form of 'cells' of liquid (which may be observed photographically and which take the form of hexagons, rolls, rectangles, etc. according to conditions). The energy of random thermal uncoordinated motion of the molecules has been, at least in part, replaced by the energy of cooperative macroscopically ordered motion. The probability of such cooperation occurring spontaneously through random motion at equilibrium is (on the Boltzmann principle) infinitesimally small; an external constraint, the temperature gradient, has apparently been able to drive the system sufficiently far from equilibrium for ordered, cooperative structures to appear in the liquid. This is an excellent, entirely non-biological, example of what Prigogine has called the 'order through fluctuations' which occurs far away from equilibrium. It seems that new structures may result from an instability, a random stochastic element, namely, a fluctuation which, at the critical point, can be amplified to allow a new structural regime to supersede the old. We have here an intriguing alliance of chance, in the occurrence of the fluctuation, with determinism (necessity)—for once it happens its amplification is inevitable (i.e. it occurs with a probability of unity). The need for a generalized thermodynamics which will include a macroscopic theory of fluctuations is apparent and it is this which Glansdorff and Prigogine (1971) have developed. As already suggested, it is the *non-linear* character of the relation between flows (J_α) and forces (X_α) which is the necessary condition for order to arise through fluctuation. The possibilities of this occurring in purely physical phenomena are somewhat limited (to certain mechanical and hydrodynamic situations), but chemical networks have an unlimited variety and complexity—and moreover, they are often in the non-linear situation (cf. (2.33)), not close to equilibrium, when the reaction rate is

$$J_r \propto \left(\exp\frac{A^f}{RT} - \exp\frac{A^b}{RT}\right), \tag{2.53}$$

that is, not directly proportional to the overall affinity of the reaction $A(=A^f-A^b)$, where superscripts f and b, respectively, refer to the forward and back reactions. It is, therefore, to this non-linear domain, especially with reference to chemical reaction networks, that we must now turn to expound the relevance of thermodynamics to the origin of biological order and complexity.

A general evolution criterion

In the non-linear range of isothermal systems in mechanical equilibrium (e.g. for most chemical reactions for most of which $A \not\ll k_B T$ (cf. 2.34)), the state variables still satisfy (2.11) and the entropy production, on the continued assumption of 'local equilibrium', still takes the form

$$P = \int \left(\sum_\alpha J_\alpha X_\alpha\right) dV \tag{2.13b}$$

in which the various J_α (e.g. $\mathbf{j}_i$ and w_ρ) are related to x_j through non-linear phenomenological laws (such as (2.33)). However, unlike the case when these laws are linear and the Onsager relation is obeyed, the time derivative dP/dt has no special properties and this is the quantity to watch to see how such a non-linear system could evolve to the steady state.

As in (2.51), dP/dt may be split into d_XP/dt and d_JP/dt so that

$$\frac{d_X P}{dt} = \int \left(\sum_\alpha J_\alpha \cdot d_t X_\alpha \right) dV \qquad (2.54a)$$

and

$$\frac{d_X \sigma}{dt} = \sum_\alpha J_\alpha \, d_t X_\alpha. \qquad (2.54b)$$

By an argument (see Glansdorff and Prigogine 1971, pp. 110–12 and Nicolis and Prigogine 1977, pp. 50–1) similar to that which led to the derivation of (2.49), it can be shown that, although dP/dt has no properties of any general validity, yet always

$$\frac{d_X P}{dt} < 0, \quad \text{or} \quad d_X P < 0 \qquad (2.55a)$$

[approach to steady state]

$$\frac{d_X \sigma}{dt} < 0, \quad \text{or} \quad d_X \sigma < 0 \qquad (2.55b)$$

and, as before,

$$\frac{d_X P}{dt} = 0 = \frac{d_X \sigma}{dt} \qquad (2.56a)$$

or

[at the steady state]

$$d_X P = 0 = d_X \sigma. \qquad (2.56b)$$

Thus in the non-linear range, whenever the boundary conditions used are time-independent, this relation for the part of the entropy production due to change in forces X_α operates as a criterion of evolution to the steady state, like the minimum entropy production for the linear range. Moreover, this evolution criterion is independent of any assumption about the phenomenological relations between the J_α and the X_α. It should be noted, however, that nothing of this kind can be affirmed about $d_J P$ and $d_J P/dt$ and so (2.55) cannot prescribe the sign of dP and dP/dt. This evolution criterion, although it is called 'general' or 'universal' by the Brussels school, is confined, as they stress, to conditions under which local equilibrium prevails (but not linearity, or the applicability of the Onsager relations).

In general, the form $d_X P \leq 0$ constitutes a negative definite quadratic form with many variables and non-linear coefficients. It will not be integrable because it contains an antisymmetric part that will not disappear, as it does in the linear range by virtue of the Onsager reciprocity relations (Nicolis 1979). According to Nicolis (1979) the presence of this antisymmetric part leads to two important possibilities: (i) a rotational motion in the space of state variables, corresponding to damped or sustained oscillations of these variables in time (Prigogine and Balescu 1955); and (ii) the absence of any guarantee that the system will return to the stationary state (where $d_X P/dt = 0$), if removed from it. The former possibility will be more the concern of Chapter 4 on kinetic interpretations, but the latter is still within the purview of thermodynamics.

Criteria of stability of non-equilibrium states

Although conformity to the 'general evolution criterion' above (2.55) prescribes permissible directions of change under non-linear non-equilibrium conditions, because the time-variations $d_X P$ and $d_X \sigma$ are not total differentials (leaving out, by definition, the time-variations, $d_J P$ and $d_J \sigma$, in P arising from variation in flows) the evolution criterion does not by itself guarantee the *stability* of any non-linear non-equilibrium state, not even the steady state (for the detailed argument see Nicolis and Prigogine (1977, pp. 51-4); and Stucki (1978, pp. 155-62) for a synopsis of this approach).

Let us take the steady state as a reference state and write all values (X) of flows, forces, entropy production as the sum of two terms: X^0, the value in the steady state, and δX as a virtual variation (compatible with the boundary conditions) arising by a fluctuation from the steady state value of X, viz., X^0, e.g.,

$$X_\alpha = X_\alpha^0 + \delta X_\alpha; \qquad J_\alpha = J_\alpha^0 + \delta J_\alpha; \qquad A_\rho = A_\rho^0 + \delta A_\rho; \qquad (2.57)$$
$$\mu_i = \mu_i^0 + \delta \mu_i, \quad \text{etc.}$$

Then consider first the changes ($\delta_X \ldots$) in entropy production (P or σ) consequent only upon the small but finite changes, δX_α in the forces X_α from their values X^0 in the steady state (i.e. no changes in the J_α). These will be denoted as $\delta_X P$, or $\delta_X \sigma$, and are thereby *defined* as:

$$\delta_X P \equiv \int \left(\sum_\alpha J_\alpha \cdot \delta X_\alpha \right) dV \qquad (2.58a)$$

and

$$\delta_X \sigma \equiv \sum_\alpha J_\alpha \cdot \delta X_\alpha. \qquad (2.58b)$$

It can be shown (e.g. Nicolis and Prigogine 1977, pp. 53, 4) that

$$\int \left(\sum_\alpha J_\alpha^0 \cdot \delta X_\alpha \right) dV = 0, \tag{2.59}$$

and substitution of this in the expansion of J_α according to (2.57) and use of the other definitions (2.58) yields

$$\delta_X P = \int \left(\sum_\alpha \delta J_\alpha \cdot \delta X_\alpha \right) dV, \tag{2.60a}$$

and

$$\delta_X \sigma = \sum_\alpha \delta J_\alpha \cdot \delta X_\alpha. \tag{2.60b}$$

The quantities on both sides of these equations (2.60) are called the *excess entropy production*.

Closely related to the quantity $\delta_X P$, but not identical with it, is the quantity $d_X P$ which we have already met and is defined as the *differential* with respect to time (hence, a small infinitesimal) of the entropy production with infinitesimal time-variations in the forces X_α. By definition

$$d_X P = \int \left(\sum_\alpha J_\alpha \cdot dX_\alpha \right) dV \tag{2.51}$$

and

$$d_X \sigma = \sum_\alpha J_\alpha \cdot dX_\alpha, \tag{cf. (2.51)}$$

and combination of these with the value for X_α relative to the steady state value X_α^0 (see (2.57)) yields (Nicolis and Prigogine 1977, pp 50–2)

$$d_X P = \int \left(\sum_\alpha J_\alpha \cdot d(\delta X_\alpha) \right) dV < 0. \tag{2.61}$$

where the inequality is introduced from (2.55a), and is the 'evolution criterion', i.e. the condition of thermodynamic permissibility of a non-equilibrium process. Note that the quantity $d_X P$ can be regarded as $d_X(\delta P)$, i.e. as the time-variation of that part of a small but finite displacement of P from its steady state reference value, P^0, that is consequent upon the changes δX_α in the forces X_α. For

$$P = P^0 + \delta P$$

so

$$d_X P = 0 + d_X(\delta P),$$

i.e.

$$d_X P = d_X(\delta P), \tag{2.62}$$

where the differentiation ('d') is with respect to time.

It should be noted that the excess entropy production $\delta_X P$, defined by (2.58a) and given by (2.60a), is *not* the same as $d_X P$ given by

$$d_X P \equiv d_X(\delta P) \neq \delta_X P, \tag{2.63}$$

for $d_X P$ refers to a variation in the course of time, whereas $\delta_X P$ refers to a virtual variation arising from a perturbation or a fluctuation, which does not have to follow the laws of dynamics. In fact,

$$d_X P = \left[\begin{array}{c} \text{expression including} \\ d(\tfrac{1}{2}\delta_X P) + \text{antisymmetrical cross} \\ \text{products of } \delta X_\alpha \text{ and } \delta X_{\alpha'} \end{array} \right] \leq 0. \tag{2.64}$$

(If $\delta_X P$ were equal to $d_X P$ then only *positive* fluctuations ($\delta_X P > 0$) would regress because this equality and the subsequent decrease of $\delta_X P$ (putatively $= d_X P < 0$) would together ensure that the $\delta_X P$ would regress to the original unperturbed value; in other words, the criterion of asymptotic stability of steady non-equilibrium states would then be $\delta_X P \geq 0$. However, the presence of the extra terms, involving antisymmetrical cross products, prevents the simple equating of $\delta_X P$ with the $d_X P$ governed by (2.55a) and some other way of establishing this (or another) criterion of stability has to be sought.)

As in section 2.4, when the stability of the equilibrium state was being examined, we write

$$S = S^0 + (\delta S) + \tfrac{1}{2}(\delta^2 S)$$

or

$$\Delta S = S - S^0 = \delta S + \tfrac{1}{2}(\delta^2 S), \tag{2.65}$$

where superscript zero refers to the steady state. Since

$$\delta S = \int dV \sum_i \left(\frac{\partial s}{\partial x_i}\right) \delta x_i = T^{-1} \int dV \sum_i \mu_i^0 \, \delta x_i$$

$$\therefore \quad \delta^2 S = -T^{-1} \int dV \sum_{ij} \left(\frac{\partial \mu_i}{\partial x_j}\right)_0 \delta x_i \, \delta x_j \tag{2.66}$$

and the quantity inside the integral can be shown to be always positive so

$$(\delta^2 S)_0 \leq 0 \quad \text{[at local equilibrium]} \tag{2.67}$$

where the subscript serves to stress that the reference state for $\delta^2 S$ is the steady state. Thus away from the steady state $\delta^2 S$ is negative, indeed as it is in Fig. 2.3 which referred to the equilibrium state; now it is the condition of *local* equilibrium which is ensuring the negativity of $\delta^2 S$. Glansdorff and Prigogine (1971), who deployed this argument, went on to derive the variation with time of $\delta^2 S$ and they were able to show (Chs V and VI; see also Nicolis and Prigogine 1977, pp. 55 ff.) that

$$\frac{\mathrm{d}}{\mathrm{d}t} \cdot \tfrac{1}{2}(\delta^2 S) \equiv \int \left(\sum_\alpha \delta J_\alpha \cdot \delta X_\alpha \right) \mathrm{d}V. \tag{2.68}$$

Given that $\delta^2 S$ is negative by 2.67, it may be regarded, the Brussels school suggest, as a Lyapunov function ensuring stability when the value of its true differential on the l.h.s. of (2.68) is positive. The r.h.s. of (2.68) is, in fact, the *excess entropy production*, $\delta_X P$ (2.60a), so that the stability criterion for non-equilibrium states may now be written as

$$\delta_X P = \frac{\mathrm{d}}{\mathrm{d}t} \cdot \tfrac{1}{2}(\delta^2 S) = \int \left(\sum_\alpha \delta J_\alpha \cdot \delta X_\alpha \right) \mathrm{d}V \geqslant 0 \tag{2.69a}$$

or

$$\delta_X \sigma = \frac{\mathrm{d}}{\mathrm{d}t} \cdot \tfrac{1}{2}(\delta^2 s) = \sum_\alpha \delta J_\alpha \cdot \delta X_\alpha \geqslant 0. \tag{2.69b}$$

(If the steady state in question were the equilibrium state, $\delta J_\alpha \equiv J_\alpha$ and $\delta X_\alpha = X_\alpha$, since $J_{eq} = 0 = X_{eq}$ so (2.68) gives at once

$$\frac{\mathrm{d}}{\mathrm{d}t} \cdot \tfrac{1}{2}(\delta^2 S)_{eq} = \int \sigma \, \mathrm{d}V = P \geqslant 0$$

with $(\delta^2 S)_{eq} \leqslant 0$, as in Fig. 2.3.) It is curious to note that although the conditions involve variation with respect to forces and not to flows, yet forces and flows play a symmetrical role in the stability conditions (2.69).

Prigogine and his colleagues stress that this criterion is only a *sufficient* condition for stability, i.e. if it *is* satisfied (i.e. if $\delta_X P$, or $\delta_X \sigma$, or $\mathrm{d}(\tfrac{1}{2}\delta^2 S)/\mathrm{d}t$, are positive), then the non-linear system is *stable* to fluctuation. However, if it is *not* satisfied, i.e. if $\delta_X P$, $\delta_X \sigma$ or $\mathrm{d}(\tfrac{1}{2}\delta^2 S)/\mathrm{d}t$ are negative, then the system can be *stable or unstable*. In other words, if $\delta_X P$, $\delta_X \sigma$ or $\mathrm{d}(\tfrac{1}{2}\delta^2 S)/\mathrm{d}t$ are negative, there is the *possibility*, but not certainty, of instability and of fluctuations, instead of regressing to zero, becoming increasingly greater and the whole system becoming unstable. Thus under these non-linear conditions, a new structural order *may* (not necessarily will) be established. So (2.69) is only a sufficient but not a necessary condition for stability, because, if it were a necessary one, its violation would *inevitably* lead to disorder and this is not what is being asserted. However one can say that the only route to an instability is through a violation of (2.69), so that an instability is always preceded by $\delta_X \sigma$(etc.) $<$ 0, but a negative fluctuation, $\delta_X \sigma$(etc.) < 0, is not *always* followed by an instability. So, paradoxically, although $\delta_X \sigma$(etc.) > 0, by (2.69), is a *sufficient* condition for *stability*, $\delta_X \sigma$(etc.) < 0 is a *necessary but not sufficient* condition for *instability*. Also note, concerning (2.69), that under these non-linear, far-from-equilibrium circumstances, that it is a positive value of the time derivative of $\delta^2 S$ which is the criterion of stability and not the negativity of $\delta^2 S$ itself, whereas at equilibrium the latter alone is a sufficient stability criterion.

The different sequences of events following a perturbation in the total entropy production (P) are depicted in Fig. 2.4, in an admittedly exaggerated form. The vertical dotted arrow (sequence I, Fig. 2.4) represents a fluctuation obeying the stability condition (2.69), namely $\delta_X P > 0$. This fluctuation is, for convenience, depicted in Fig. 2.4 by a vertical dotted line on the same ordinate scale as the *total* entropy production P but, of course, the sign of δP itself need not be the same as that of $\delta_X P$, since δP also includes first order terms related to $\delta_J P$. Such a fluctuation dies away and the system reverts to its original steady state. A fluctuation for which $\delta_X P < 0$, but which also reverts to the original state, is similarly represented with its consequence (II). But (III) represents an interesting new possibility now manifesting itself under these non-linear conditions, and consequent upon a fluctuation for which $\delta_X P$ is negative, but which is amplified and causes the system to go over to a new regime and a new steady state which may well (though not necessarily) have a different entropy production. The course of the dashed lines in Fig. 2.4 is only schematic and not meant to indicate plots of P against time: indeed the sequence of events indicated by (III) may actually show, in certain circumstances, an intermediate stage in which, after the initial fluctuation, for which $\delta_X P < 0$, the *total* entropy production actually increases before it settles down to a new value characteristic of the new structure. The actual internal entropies before and after such a transition are depicted in Fig. 2.6, and one can see that, after the transition, whether or not the entropy production (the slope of the lines in Fig. 2.6) changes, there could be a net decrease in entropy, indicative of some change in structure— though we have to be careful about asserting *what* changes in view of the problematic connection between organization, especially biological, and entropy (see section 6.3). But it certainly indicates a real change in the system, so we see the thermodynamic possibility that non-equilibrium *might* be the source of order, of new coherently ordered behaviour in space and time, if (2.69) is *not* fulfilled.

More needs to be said about the transition between the 'linear' region where a system is stable to fluctuation and (2.69) is obeyed, to the 'non-linear' region far enough from equilibrium for (2.69) not to be obeyed and so for the possibility of unstable fluctuation. At equilibrium, random fluctuations and disturbances cannot, as we saw in sections 2.4 and 2.5, bring about any ordering and this is also the case for the system as it undergoes a systematic deviation from equilibrium by the gradual change in some constraint (e.g. composition gradient at the boundaries, affinity of

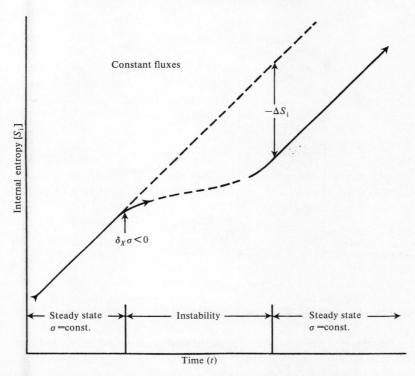

Fig. 2.6. Entropy–time diagram for a transition between two steady states. A fluctuation for which $\delta_X \sigma$ is negative allows, via a region of instability, for a transition to a new steady state, which may have a different value of the slope $\sigma = d_i S/dt$ (actually shown as the same in the diagram) and a generally *lower* value of the internal entropy, S_i.

a reaction) or state parameter. At first, force–flow relationships are linear (2.30) and the steady states close to equilibrium continue to be asymptotically stable (section 2.5). As the deviations increase, the stability in a finite neighbourhood of the equilibrium state can still be maintained along this so-called *thermodynamic branch*. Eventually there must come a point when the linearity (2.30) breaks down and the stability criteria (2.69) become applicable. These allow at least the possibility of fluctuations occurring that will cause the system to break away from this thermodynamic branch, to rupture its continuity with the 'linear' situations including equilibrium, and so to pass to a new stable regime of a different kind that may well be more coherently ordered. There is therefore a region of marginal stability between the 'linear' regions of

50   THE THERMODYNAMICS OF DISSIPATIVE STRUCTURES

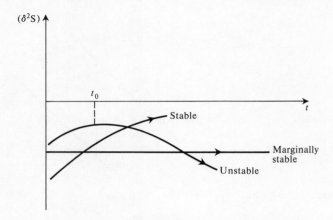

Fig. 2.7. The time evolution of second-order excess entropy ($\delta^2 S$) for the cases of asymptotically stable, marginally stable, and unstable situations. (From Nicolis and Prigogine 1977, Fig. 4.4.)

stability and instability of the non-equilibrium reference state, i.e.

$$\delta_X P = \frac{d}{dt} \cdot \tfrac{1}{2}(\delta^2 S) > 0 \quad \text{[asymptotically stable reference state]}$$
(2.70a)

$$\delta_X P = \frac{d}{dt} \cdot \tfrac{1}{2}(\delta^2 S) = 0 \quad \text{[marginal stability]} \quad (2.70b)$$

$$\delta_X P = \frac{d}{dt} \cdot \tfrac{1}{2}(\delta^2 S) < 0 \quad \text{[possibly unstable reference state]}.$$
(2.70c)

The time course of $\delta^2 S$ for these different stages (corresponding to $t > t_0$ on the diagram) is represented in Fig. 2.7.

Fluctuations and instability

As a result of this development of the stability criteria involving $\delta^2 S$, it is now possible to be more explicit concerning the concept of 'order through fluctuations' mentioned earlier. The classical Einstein formula, based on the Boltzmann principle, for the probability (Pr) of a fluctuation around an equilibrium state is

$$\text{Pr} \propto \exp(\Delta_i S / k_B)$$

where $\Delta_i S$ is the change in internal entropy due to the fluctuation. For an isolated system $\Delta_i S = \Delta S$ and since the entropy is a maximum at equilibrium, $\delta S = 0$ and so (2.65) gives

$$\Delta_i S = \Delta S = \tfrac{1}{2}(\delta^2 S)_{eq}$$

and thus

$$(\text{Pr})_{eq} \propto \exp(\delta^2 S)_{eq}(2k_B)^{-1}.$$

By adopting a phase-space stochastic description, Nicolis and Prigogine (1971) have been able to show that for non-linear systems far from thermodynamic equilibrium and over the domain of validity of the local equilibrium postulate for which (2.13) applies, the excess entropy $\delta^2 S$ also controls the probability of a *small* fluctuation around a non-equilibrium, steady state:

$$\text{Pr} \propto \exp(\delta^2 S)(2k_B)^{-1}, \qquad (2.71)$$

so that short range fluctuations do not, as it were, perceive the overall constraints and behave in the same way as at equilibrium. The values for $\delta^2 S$ and $d(\delta^2 S)/dt$ given in the previous section now allow one to understand the development of a small, thermal fluctuation around such a reference state. If the system is stable $(\delta^2 S)_0 < 0$ and its time derivative is positive (2.69), then according to (2.71) there is a higher probability, as time proceeds, of it approaching the reference state again, following the curve of Fig. 2.3. In other words, such a small fluctuation is damped down and the system regresses to its reference state, namely the non-equilibrium steady state, which on this reckoning is the most probable configuration.

However, if $d(\delta^2 S)/dt$ is negative, as is possible at or slightly beyond an instability, the system, even in a local equilibrium regime, can evolve spontaneously to a new macroscopic state, but only through large thermal fluctuations of macroscopic size that are then amplified to the point where the roles of the basic structure and of such large-scale fluctuations are exchanged. The Einstein formula (2.71) then no longer applies and, as Prigogine puts it (1969), 'What was a fluctuation before now becomes a stable structure maintained through the flow of energy and matter'. In the non-linear range, the relation between order and fluctuation is more complex than at equilibrium (where thermodynamic potentials can determine everything) for not only can fluctuations lead to new structures, different fluctuations may correspond to different structures and more than one non-equilibrium regime might exist—perhaps one should speak rather of 'orders through fluctuation'.

To summarize up to this point, we may say that this thermodynamic treatment affords proof that fluctuations in a steady state can cause either oscillations around or, in favourable circumstances, evolution away from that state to a new steady state via a transition period of instability (for a succinct account see Stucki (1978, pp. 164–73)). More particularly, it provides a condition of stability (2.69) which allows one to predict when instability is *possible* and could lead to a new regime, of a degree of order, and, presumably, also of organization, different from the original system.

In order to understand the application of the criterion proposed by Glansdorff, Prigogine, and others, it is worth examining some chemical systems that might exhibit instability.

Chemical instability

The Glansdorff–Prigogine stability criterion (2.69) is applied to chemical processes in its form,

$$\delta_X \sigma = \sum_\alpha \delta J_\alpha \cdot \delta X_\alpha > 0 \tag{2.69b}$$

where J_α = rate (w) of chemical reaction, $X_\alpha = AT^{-1}$; A is the affinity of the reaction. So, according to the criterion (2.69b) if

$$\delta_X \sigma = \delta w \cdot \delta A \cdot T^{-1} > 0 \tag{2.72}$$

the reaction is stable. How this criterion works is well illustrated by an example considered both by Glansdorff and Prigogne (1971, p. 81) and by Eigen (1971a, Table 6). For the simple autocatalytic reaction

$$X + Y \underset{k^b}{\overset{k^f}{\rightleftharpoons}} 2X \tag{2.73}$$

the rate (w) *near equilibrium* is given by

$$w = [k^f Y - k^b X]X,$$

where X and Y represent the respective concentrations and superscripts f and b, 'forward' and 'backward'. This rate is practically zero since, at equilibrium, $k^f Y = k^b X$. Thus, for *constant Y*

$$\delta w \approx [k^f Y - k^b X]\,dX - k^b X \cdot \delta X$$

and so

$$\delta w \approx -k^b X \cdot \delta X, \tag{2.74}$$

neglecting the second order terms. The affinity, $A(=-\Delta G)$, of the raection is

$$A = k_B T \{\ln K - \ln X/Y\}, \quad \text{where} \quad K = k^f/k^b,$$

so

$$\delta A = -\frac{k_B T}{X} \cdot \delta X, \text{ at constant } Y. \tag{2.75}$$

NON-LINEAR THERMODYNAMICS 53

Hence, from (2.22), (2.74), and (2.75), one obtains

$$\delta_X\sigma = \delta w \cdot \delta A \cdot T^{-1} = RT \cdot k^b(\delta X)^2, \tag{2.76}$$

and this is always positive. Hence fluctuations near to equilibrium are stable. But, *far from equilibrium* and *taking the reverse reaction as negligible*,

$$w = k^f XY$$

and

$$\delta w = (\text{constant})(\delta X), \tag{2.77}$$

if Y is constant. δA is as in (2.75), so now, by (2.77),

$$\delta\sigma = \delta w \cdot \delta A T^{-1} = -(\text{constant})\, k_B T X^{-1}(\delta X)^2 \tag{2.78}$$

and this is always negative (since the constant from (2.77) is positive). So if this reaction is part of a network of reactions maintained far from equilibrium by appropriate constraints, because $\delta\sigma$ is negative it is possible for fluctuations to be amplified, leading to a new regime. Hence this particular reaction could threaten the stability of any large sequence of reactions in which it occurs (by itself, of course, it would simply proceed to equilibrium).

This conclusion should be contrasted with that for the process

$$X \underset{k^b}{\overset{k^f}{\rightleftharpoons}} Y. \tag{2.79}$$

The rate is given *far from equilibrium* by

$$w = k^f X - k^b Y$$

where X, Y represent concentrations and, as in (2.32), the affinity

$$A = k_B T \cdot \ln(k^f X / k^b Y) \tag{2.80}$$

At constant Y and T (and so k^f and k^b)

$$\delta w = k^f \cdot \delta X,$$
$$\delta A = k_B T X^{-1} \cdot \delta X$$

therefore

$$\delta_X\sigma = \delta w \cdot \delta A \cdot T^{-1} = k^f k^b (\delta X)^2 X^{-1}, \tag{2.81}$$

which is always positive, and is stable. So this reaction in a reaction chain will always tend to stabilize the system, unlike the autocatalytic reaction (2.73).

These are, of course, only very simple examples but, even so, one can see how the introduction of an autocatalytic process can possibly make a (negative) destabilizing contribution to the excess entropy production,

$\delta_X\sigma$ or $\delta_X P$. Chemical instabilities, in general, may be unstable with respect either to homogeneous perturbations or with respect to space-dependent inhomogeneous perturbations. If the former, the system can be expected to go from one homogeneous steady state to another (which may or may not prove to be steady); if the latter, then diffusion will play a role, either by making a positive contribution to the excess entropy production (like viscous dissipation in hydrodynamics) thereby stabilizing the steady state, or by increasing the range of perturbations compatible with the macroscopic equation of change (Prigogine 1969, pp. 36–7; see Ch. 4).

The all-pervasive role in biological systems of sequences of chemical reactions that involve feed-back, feed-forward, and autocatalytic processes, and the almost invariable non-linearity of their force–flux relationships (see (2.33)), except very close to equilibrium, constitute one of the principle modes of applicability to biological complexity of the thermodynamic concepts described here—but a mode of application that can only fully be made effective by detailed kinetic analysis of the chemical systems, as we shall see (Ch. 4).

A critique of the Glansdorff–Prigogine stability criterion

The foregoing account has followed fairly faithfully the exposition and arguments of the Brussels school of irreversible thermodynamics about the possibility of order-through-fluctuation with the formation of dissipative structures in non-linear, far-from-equilibrium situations. Although this approach has been widely accepted and it has given much stimulus to others, such as the work of Eigen (see Chapter 5 on selection and evolution at the molecular level) and that of Hess (see Chapter 4) on oscillating biochemical systems, especially the glycolytic pathway, it has not been without criticism at a more fundamental theoretical level. In particular Fox (1979, 1980) and Keizer and Fox (1974) have argued that the excess entropy ($\delta^2 S$) around non-equilibrium steady states is not a Lyapunov function and they have (1974) 'qualms regarding the range of validity of the Glansdorff–Prigogine criterion for stability of non-equilibrium states'. The argument has been brisk and forceful and initially turned on the scope of validity of the stability criterion (2.69), especially in the form

$$\frac{d}{dt}(\tfrac{1}{2}\delta^2 S) > 0. \qquad (2.82)$$

In their application of the criterion to the autocatalytic reaction (2.73) Keizer and Fox (1974) were able to show that when its rate equations are solved exactly (and without assuming the absence of back reaction and constancy of the concentration of the reactant Y), this reaction has a region of stability which could not be determined by the Glansdorff–Prigogine criterion. Another example, the reaction system often referred to colloquially as the 'Brusselator' and described later (section 4.5.2), displays a regime of steady states which this criterion cannot demonstrate as stable. Thus they emphasize that this condition is only a *sufficient* condition for stability and that 'violation of it *does not necessarily* imply the lack of stability'. The criterion, they affirm, only provides a useful condition (by which one presumes they mean a *necessary* as well as a sufficient one) in the neighbourhood of equilibrium.

In their reply, Glansdorff, Nicolis, and Prigogine (1974) agree completely with this last point about the full validity of their criterion in close-to-equilibrium situations, and they have little difficulty in showing that they have always affirmed that the criterion, if satisfied (i.e. $\delta_x P$, etc. >0 in (2.69)) provides only a *sufficient* criterion of *stability*. They state explicitly

> Suppose that one finds a region of values of the parameters for which the excess entropy production [as in (2.69)] ... is a *positive definite* quadratic form ... Then stability (and even asymptotic stability) is assured whatever the solution of the rate equations might be. However, if ... the excess entropy production *may* become negative—one cannot arrive at any conclusion about stability without further information coming from the rate equations. (Glansdorff *et al.* 1974, p. 198.)

and later,

> The stability criterion provides a natural classification of the processes which *could* give rise to instabilities and to a subsequent ordered behavior of the system (Glansdorff *et al.* 1974, p. 199.)

—all of which is consistent with the account developed above based on their earlier published work. To put it another way, their criterion is an *insufficient* criterion of *instability*.

However, the argument continues with the question of whether or not $(\delta^2 S)_{\text{steady state}}$ is a genuine Lyapunov function and Fox (1979) urges that it is not in the strict sense, whereas a function constructed by Keizer (1976) from the covariance matrix for the thermodynamic variables, is such a function for the steady states and so a criterion of stability. Nicolis and Prigogine (1979) reaffirmed the usefulness of their criterion and point out that the use of one Lyapunov function does not exclude the use of another as a stability criterion for the same situation—to which Fox (1980) replied with a counter example. We need not follow this controversy further here—whether $(\delta^2 S)_{\text{steady state}}$ is or is not a Lyapunov function matters less than a clear idea of how it is to be applied, and on that there seems to be less disagreement between the two groups of authors than appears at first sight. How far the alternative stability criterion of Keizer will prove to be useful remains to be seen, but the general approach of the Brussels school has undoubtedly shown by a combination of thermodynamic and fluctuation theory how order can come about through fluctuations and how dissipative structures can thereby result.

2.7 Dissipative structures and evolution

We have seen that linear systems, and in particular those close to equilibrium, evolve to steady states that are stable to fluctuations (as long as equilibrium itself is stable) and are disordered because, in this linear domain, the steady states are continuous extrapolations of equilibrium states and they have the same qualitative properties as these latter (i.e. they are 'disordered' in the sense of Boltzmann). The entropy production in such steady states is a minimum for time-independent boundary conditions (eqns (2.46) and (2.47)), so there is the least possible dissipation consistent with these boundary conditions. However, open systems obeying non-linear force–flux laws can be displaced from equilibrium but can still possess steady states displaying equilibrium-like behaviour. Moreover, along this so-called 'thermodynamic branch' of states, stemming from the equilibrium state, the conditions for stability are not as they are at equilibrium, or for the steady states of linear systems. It now becomes *possible*, if $\delta_X\sigma(\text{etc.}) < 0$ in (2.69), for a fluctuation to be amplified and the system to become unstable and change to a new spatially or temporally more organized state—'order through fluctuation', or non-equilibrium as a source of order. The ordered configurations that emerge beyond such an instability of the thermodynamic branch of non-linear systems have been called *dissipative structures*, because they are created and maintained by the entropy-producing 'dissipative' processes occurring inside the system through which, being open, there is a continuous flux of matter and energy. So far, little consideration has been given to the origin of the fluctuations that give rise to such structures, to the transition process itself, and to the consequences for the open system of changing to a new dissipative structure. The origin of fluctuations is the subject of stochastic methods that cannot be treated here (see Part III of Nicolis and Prigogine 1977, for a relevant introduction), but something further can usefully be developed from thermodynamic considerations of the other two matters.

Dissipation and 'evolutionary feedback'

Before life had evolved, in the prebiotic stage as it is called, one has to presume that a system containing, amongst other species, replicating macromolecules could both maintain itself by means of some simple copying mechanism and have the potentiality of change incorporated into its very structure and function, so as to facilitate a multistage evolution to more complex forms more efficient for survival and reproduction. Some of the kinetic and stochastic problems associated with such prebiotic systems have been very fully investigated by M. Eigen and his colleagues at Göttingen and are expounded in Chapter 5. In the present context, we raise the question asked by Prigogine and his colleagues from a more thermodynamic perspective which may be put thus: 'How can each step in an evolutionary process in which each stage leads to more orderly, or at least more complex, configurations contain within itself the potentiality to change by other, later transitions to yet more ordered, complex configurations?'. Each transition to be part of a *continuous*, multistage evolution cannot become a 'dead-end', that is, it must contain within itself the germ, or possibility, of yet further changes—otherwise the system is simply eliminated in the competition for limited 'food' reserves (of simpler monomers, in this case). What is needed is for change to occur in such a non-linear open system in a way that increases its non-linearity and distance from equilibrium and thereby both ensures and enhances the possibility of further change. This could come about by the change, however brought on, being of such a kind that it increased the interactions between the system and its external environment, such interactions being one of the causes of non-linearity. If this were to happen, this increase of interaction would be reflected in an increase in the entropy production per unit mass, i.e. in the 'dissipation' in the system as an immediate consequence of the transition (since $d_iS = -d_eS$, if total entropy is to remain the same). This increased dissipation would then, temporarily at least, characterize a system which had the potentialities of undergoing further instabilities when appropriate thresholds were passed, and so to form new structures, with again increased dissipation, and so on. A kind of 'feedback' in evolution of such non-linear systems has therefore been postulated by the Brussels group (Prigogine, Nicolis, and Babloyantz, 1972, p. 23) which could be depicted (after Prigogine and Lefever 1975) as:

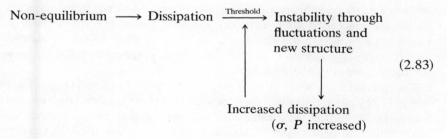

(2.83)

The condition that the 'instability through fluctuations' (2.83) *may* occur is still that $\delta_X\sigma(\text{etc.}) < 0$, by (2.69). Once this instability has been initiated, it is proposed, the system can increase its specific entropy production (s.e.p., entropy produced per unit time per mole, or per unit mass) which they take as a suitable index of the level of dissipation in a system, since variation in the s.e.p. would be 'characteristic of the organizing tendency manifested in the system' (Prigogine and Lefever 1975, p. 521). Each instability would be followed by a higher level of energy dissipation which would mean some of the irreversible processes in the system would be working more intensely, would have therefore departed further from equilibrium, and so would have enhanced the probability that there should exist those kinds of fluctuations that render the processes yet again unstable. They have been able to show that a large number of classes of systems in which biological macromolecules are being synthesized can display this 'evolutionary feedback', principally because the transitions that increase the relative numbers of copies of such macromolecules must have certain autocatalytic properties.

In general they have been able to show (Prigogine *et al.* 1972, p. 38) that if we have a set $\{X_i\}$ of relatively abundant, chemically interacting species X_i ($i = 1, \ldots, n$) then

$$\frac{dX_i}{dt} = F_i^\rho\{X_i\} + \sum_\rho F_i\{X_i\}, \qquad (2.84)$$

where F_i^ρ describes the flow of matter from the surroundings (and is assumed constant) and $\sum_\rho F_i$ is the sum of all reactions of i inside the system. If the existence of at least one steady state is assumed, then it can be shown (Prigogine, *et al.*, pp. 40 ff.) that the formation, by random fluctuation, of even only one new species Y (as an error copy of one of the $\{X_i\}$) can alter the stability properties of the system, even if it were originally stable with respect to perturbations in the $\{X_i\}$. The system may then evolve through unstable transitions to a new state of relatively high concentrations of Y, especially if Y can act as a template for synthesis of a substance which catalyses production of more Y. Prigogine *et al.* (1972, p. 38) have examined this possibility in detail and find that, if the sequence of events concerned with Y occurs as described above, then there is, beyond the transition threshold (during which $\delta_X\sigma(\text{etc.}) < 0$), an increase in the entropy production because the system switches to a fast pathway for the synthesis of the new substance Y.

The general idea behind this model has been substantiated (Prigogine *et al.* 1972, p. 38) by another more specific reaction system involving two polymers, A_1 and B, which mutually act as templates for each other and also diffuse out of the system. Steady polymerization states are predicted and the proportions of A_1 and B can be calculated. It is then assumed that steady-state random fluctuations continuously cause errors in the formation of A_1, catalysed by B, so that a new species A_2 is produced, which further catalyses its own production via intermediaries. It is found (Prigogine *et al.* 1972, Fig. 4) that beyond a particular monomer concentration the fluctuation leading to A_2 is amplified abruptly and the system switches to a new steady state dominated by A_2. Such calculations serve to show again the important role of autocatalytic growth properties on the stability and evolution of such systems.

The entropy production per unit mass has been calculated (Hiernaux and Babloyantz 1976) for another reaction–diffusion system (that of Gierer and Meinhardt 1972) consisting of two substances, 'morphogens', that alter cell development, which diffuse and also undergo non-linear chemical interactions of a specific kind (a short-range activation and a long-range inhibition, see sections 4.2 and 4.5.2 and Fig. 4.22). It was found that the entropy production per unit mass, σ, at first increases abruptly in such a system and then decreases to a lower value. In this respect it is similar to some earlier observations (Zotin and Zotina 1967) that the rate of heat production, a rough measure of entropy production, in chicken eggs is about $6 \, \text{kcal day}^{-1} \, \text{gm}^{-1}$ at the fourth day of development but drops to $1 \, \text{kcal day}^{-1} \, \text{gm}^{-1}$ at the sixteenth day. This heat exchange is paralleled also by specific oxygen consumption which should be an increasing function of specific entropy production; the oxygen consumption increases sharply during early embryonic life and then subsequently decreases.

These calculations on the morphogen system and observations on the chicken embryo show that, as was necessary for coherence and consistency in the thermodynamic treatment, those systems that display an *increase* in dissipation as manifest in the specific entropy production do so only transiently, after the initial threshold fluctuation (for which $\delta_X \sigma(\text{etc.}) < 0$ by (2.69)). For once it is in the new regime that can display this initial enhancement of dissipation, it seems, from the available thermal data, that the system adjusts itself to the constraints and the dissipation then tends to decrease to a new level characteristic of the new non-linear steady state (now further than before from equilibrium). As Prigogine *et al.* (1972, p. 25, box) say: 'one is tempted to argue that only after synthesis of the key substances necessary for its survival (which implies an increase in dissipation) does an organism tend to adjust its entropy production to a low value compatible with the external constraints'.

Stability and evolution
As already mentioned, for a non-linear system to evolve it must undergo changes which are of such a kind that they still, after the change, possess the potentiality of further change. We saw that this was conceivable if,

after the initial fluctuation ($\delta_X\sigma$(etc.) < 0, by (2.69)) leading to instability, the system underwent an increase in dissipation as indicated by the specific entropy production, an increase which would take it non-linearly further from equilibrium. However, we also saw that this state of affairs could not continue indefinitely, and indeed there was experimental and theoretical evidence that eventually the entropy production could decrease to a new value characteristic of the new (dissipative) structure. Briefly, the creating of structure leads to an increase in entropy production, σ, whereas maintenance of the structure once formed follows the theorem of minimum entropy production. The sequence of events postulated is depicted schematically in Fig. 2.8. The time-scale on this figure has been represented as the negative of the logarithm of the time (τ) *backwards* from a reference state of the system (which could be the present) to make clear—purely schematically, it must be stressed—that there is, as is well known to evolutionary biologists, an acceleration of evolution in the course of geological time. It seems that each step in biological evolution increases the chance of the next change and thereby decreases the elapse of time before it will ocur. At the more molecular level of description of living organisms, with which we are concerned here, there is clearly need for something akin to what biologists call an increase in 'selection pressure' if non-equilibrium dissipative systems are going to display an *accelerating* tendency to change, as do the biological organisms they seek to represent and model. This constitutes another reason for proposing 'autocatalytic' processes in the evolution of dissipa-

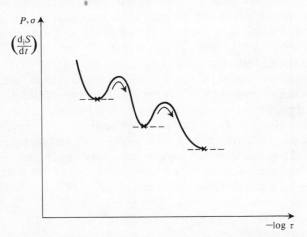

Fig. 2.8. Entropy production during evolution of a series of stabilized steady states. τ measured as time *back* from the (present) reference state of a system. -X- Maintenance at a minimum, steady state. ⌒ Transition to a new steady state with initial increase in P, σ, (d_iS/dt).

tive systems, such as the 'evolutionary feedback' depicted in eqn (2.83)—especially as the actual detailed molecular processes, the atomic exchanges in the chemical reactions, diffusion, and so on, are *not* changing during such evolution. This emphasizes the need for both the macroscopic descriptions of thermodynamics and for the development of kinetic and stochastic theories about organized systems of reactions.

The idea of a succession of temporarily stabilized states as being a useful perspective on evolution was also proposed by J. Bronowski (1970) when he pointed out that 'stratified stability' was the principal feature of the evolutionary sequence, both inorganic and organic. He noted that in the formation of, say, carbon nuclei from helium nuclei, we have an example of that formation of stable complex units from simpler ones which runs not only through atomic physics, but also through chemistry to the formation of the molecules that constitute biological macromolecules, such as the nucleic acids, and also through biology as stable cells, organs, and organisms form. Each more complex form represents a new stratum of stability which has a finite chance of forming under the conditions appropriate to it (very high temperatures for the carbon nucleus; aqueous medium, and other, as yet undetermined, factors for the nucleic acids; etc.). His picture of evolution is: once formed, each stratum of complexity has a stability which allows the possibility of the next stratum appearing by providing the structures (atomic, molecular, organic, or whatever) which form the constituent building blocks of the next level.

> Because stability is stratified, evolution is open and necessarily creates more and more complex forms ... So long as there remains a potential of stability which has not become actual, there is no other way for chance to go. It is as if nature were shuffling a sticky pack of cards, and it is not surprising they hold together in longer and longer runs (Bronowski 1970, p. 32.)

Bronowski speaks here of chance (which is related to entropy) but the strata to which he refers are levels of *energy* minima, successive ones higher in energy than the preceding, and each reached over an energy barrier—rather like ascending and descending a succession of cols into a series of valleys in a mountain range, the level of each valley bottom being higher than its predecessor. This interpretation is given a more sophisticated thermodynamic content in the treatment of the Brussels school. In their exposition, too, it is not instability but a succession of instabilities of dissipative systems, now appearing under the aegis of thermodynamic laws, that bridges the gap between the non-living and the living. This succession of instabilities can be a process of self-organization, and so of evolution, provided there are fulfilled certain conditions that emerge clearly from the foregoing thermodynamic considerations. They are that a process of self-organization can occur in a system if: (i) the system is *open* to the flux of matter and energy; (ii) the system is not at equilibrium and preferably *far from equilibrium*; and (iii) the system must be *non-linear* in its flux–force relationships, i.e. there must be strong coupling between its processes. These simple requirements underlie the ideas developed above, notably by the Brussels school, and are certainly satisfied by all biological systems and organisms and so the following 'theorem', formulated by G. Nicolis (1974) is applicable to them and serves to summarize the whole analysis:

> Consider a single phase system satisfying the above three prerequisites [(i)–(iii)], whose entropy can be defined in terms of macroscopic quantities. Under these conditions, steady states belonging to a finite neighbourhood of the state of thermodynamic equilibrium are *asymptotically stable*. Beyond a *critical distance* from equilibrium they *may* become unstable

The evolution to order through an instability induced by fluctuations, referred to in the last sentence of this quotation is, as we have frequently stressed, only a *possible* not a certain development and in fact requires, along with the subsequent stabilization to a dissipative structure, that some other very stringent conditions be fulfilled. These can only be elucidated in relation to individual systems and usually require the kind of detailed kinetic analyses that are referred to in Part II.

2.8 Dissipative systems in biology

No biological organisms, or indeed the systems within organisms, have ben examined experimentally sufficiently adequately with respect to their thermodynamic parameters to allow any detailed test of the ideas described in the preceding sections. However the general applicability of these concepts finds confirmation in their ability to interpret many different kinds of processes and transitions of biological systems. The concept of dissipative structures, and of the conditions governing the transitions which lead to them, provide a framework that renders intelligible a wide range of situations that are characteristically biological and allows interpretation of them in terms that can be related to the physico-chemical concepts that have been part and parcel of the foregoing thermodynamic treatment. A few typical biological situations to which these ideas have been applied will be described. It should be noted that these applications often depend on the formulation of a model, based on the biological phenomena (cf. (a), Fig. 1.1), which is susceptible to thermodynamic interpretation and from which deductions concerning properties of the system may be made (cf. (b), Fig. 1.1), and interpreted ((c), Fig. 1.1) in relation to actual biological properties.

Enzyme reactions with feedback

(a) *Phosphofructokinase* plays a key regulating role in the glycolytic pathway as an allosteric enzyme whose rate of working is enhanced by the product of its action (ADP) and is inhibited by one of its substrates (ATP)—the other substrate is D-fructose 6-phosphate. The positive feedback effect of ADP is represented by a more general model of the kind

$$\longrightarrow S \longrightarrow E \longrightarrow P \longrightarrow \qquad (2.85)$$
$$\qquad \oplus \uparrow \underline{\qquad\qquad}|$$

where $S = ATP$, $E =$ phosphofructokinase, $P = ADP$. The allosteric enzyme E exists in a number of conformations some of which bind P preferentially and, in accordance with the original ideas of Monod, Wyman, and Changeux (1965), this leads to a sigmoidal binding curve for P on E. So the activation curve, relating rate of working of E and the concentration of P, is also sigmoidal with a steeply rising portion followed by a plateau independent of P. There are three interesting features of this system pertinent to our present concerns.

(1) If one considers the range of concentrations of P and of S, when the rate of working of P is on the rising part of the sigmoidal curve, the stability condition (2.69) can help us to determine if a steady state, with concentrations S_0 and P_0, would be stable with respect to infinitesimal fluctuations in these concentrations. The stability condition (2.69) is used in the form

$$\sum_\alpha \delta J_\alpha \cdot \delta X_\alpha = \sum_\alpha \delta w \, \delta A \cdot T^{-1} > 0, \qquad (2.86)$$

where $w =$ rate of working of E, i.e. of production of P. Writing,

$$\delta w = (\partial w/\partial P)_{P=P_0,\, S=S_0} \cdot \delta P \qquad (2.87a)$$

which is positive, since P is an activator and so $(\partial w/\partial P) > 0$ then

$$\delta A = \delta \ln(SP^{-1}) = -\delta P \cdot P_0^{-1}, \qquad (2.87b)$$

at constant S.

Thus the product $(\delta w \cdot \delta A \cdot T^{-1})$ makes a negative contribution to the sum in (2.69) and the system *could* become unstable if this term were dominant (and the step represented by (2.85) were rate-determining in the glycolytic process—which it often is). Thus thermodynamic considerations indicate the possibility of an instability. The actual conditions under which these can occur, in addition to those given in relation to (2.87), have to be determined by a kinetic analysis, as described in sections 4.5.1 and 4.6.1.

(2) For the moment let us note that such analyses can make explicit what are the characteristics of the dissipative structures to which the instability (now allowed by the application of the thermodynamic stability condition (2.69)) gives rise. Thus Goldbeter (1973) has been able to show that, in the absence of diffusion, this system (2.85) can undergo various kinds of sustained, stable oscillations in the concentrations of S, P and E

that are an example of a *limit cycle*. Moreover, with increasing injection rate of the substrate S, the amplitude of these oscillations goes through a maximum, while their period decreases. These are indeed features of glycolytic oscillations which have in fact been observed and thoroughly examined (see section 4.6.1). The limit cycle is an example of a *temporal* dissipative structure, a coherent state where the concentration of the various constituents oscillate simultaneously with a characteristic phase, different for each constituent but the same throughout the reaction space.

When, however, diffusion is taken into account (Goldbeter and Lefever 1972), the behaviour of the system is related to a dimensionless parameter D^2/L (D = diffusion coefficient and L the length of the system, considering a single dimension initially) that measures the coupling between neighbouring spatial regions. Various combinations of this 'diffusive coupling', and the injection rate of substrate lead to a variety of situations, namely, spatial inhomogeneous steady-state structures, propagating concentration waves, and standing concentration waves—so that both *spatial* and *spatio-temporal dissipative structures* are formed (see section 4.6).

(3) The reaction (2.85) is *in vivo* part of the glycolytic reaction sequence and its function in this sequence in relation to energy dissipation is of interest, in view of the earlier discussions (section 2.7) concerning steady states and entropy production. The minimum energy required to maintain a steady-state flux for such a reaction within a non-equilibrium chain of reactions can be calculated (Hess 1963, 1975). The affinity of the reaction

$$S \underset{k_2}{\overset{k_1}{\rightleftharpoons}} P \qquad (2.88)$$

at the steady state is

$$A = (-\Delta G) = k_B T \cdot \ln k_1 S_0 / k_2 P_0 \qquad (2.89)$$

and the amount of energy per unit time required to maintain the steady state is

$$\frac{d(\Delta G)}{dt} = (k_1 S_0 - k_2 P_0) k_B T \cdot \ln k_1 S_0 / k_2 P_0$$

$$= \text{entropy production rate } \left(\frac{\partial G}{\partial t} = -\frac{\partial S}{\partial t}; S = \text{entropy} \right). \qquad (2.90)$$

Hess (1963) has shown experimentally that the enzyme reactions in the glycolytic pathway are: either, reaction with near-equilibrium enzymes, for which the net fluxes are low relative to the maximal activity of the enzymes in question; or, reactions with enzymes which operate quasi-irreversibly, with a reverse flux that is negligibly small compared with the net flux, and which are regulatory in the pathway. Phosphofructokinase is one of these latter and the free energy of the reaction it catalyses is mostly used for control, along with that of pyruvate kinase. In other words, in the parts of the reaction pathway outside and between these two enzymes there is a kind of 'plateau' of chemical potential with big drops of chemical potential occurring at these two stages. This is consistent with an earlier study of the effects of catalysis on such a reaction chain (Prigogine 1965; see Nicolis and Prigogine 1971, p. 442 ff.) in which A is transformed into a final product B through intermediates X_i

$$A \rightleftharpoons X_1 \rightleftharpoons X_2 \rightleftharpoons X_i \rightleftharpoons X_N \rightleftharpoons B \qquad (2.91)$$

with catalytic feedback from M (product of last stage).

$(i=1,\ldots,N)$ and where M is a product of the last stage and catalyses some (or all) of the earlier reactions. In the absence of M, between A and B there is a continuous degradation of chemical potential

$$\mu_A > \mu_{X_1} > \mu_{X_2} \ldots > \mu_{X_i} \ldots > \mu_{X_N} > \mu_B,$$

with each step $(\mu_{X_i} - \mu_{X_{i-1}})$ displaying only a small drop. The whole process is therefore slow, so that the rate of transformation of A into B is low. The entropy production is the sum of that for each of the steps and at the steady state is given by

$$P_0 = K\left(\frac{A-B}{N+1}\right)\ln(AB^{-1}),$$

(taking all $k_i \equiv K$); or, dividing by the total number of moles in the system,

$$(\sigma_m)_0 = \frac{2K[(A/B)-1]}{(N+1)(N+2)[(A/B)+1]} \cdot \ln\left(\frac{A}{B}\right).$$

So the entropy production per mole at the steady state $(\sigma_m)_0$ depends only on the overall affinity (i.e. on A/B) and is lower the larger the number N of intermediate steps.

However, if there is sufficient catalysis by one of the end-products M, as in (2.91), and if $A/B \to \infty$, then the entropy production, the level of dissipation, is very much increased. Moreover, the effect of the catalysis is to propagate the chemical potential without degradation along the chain, but with a large decrease in the last stage, $X_N \to B$, so

$$\mu_A \simeq \mu_{X_1} \simeq \mu_{X_2} \ldots \simeq \mu_{X_i} \ldots \simeq \mu_{X_N},$$

but
$$\mu_{X_N} \gg \mu_B.$$

By keeping one of the reactions far from equilibrium, the involvement of positive feedback in the system increases its efficiency and, in effect, the system behaves almost as if the chain were shortened to only one step. It is this which seems to be happening with phosphofructokinase and pyruvate kinase in relation, respectively, to the steps that precede them.

(b) *The lac operon–galactosidase–permease system.* Another type of enzyme system that generates dissipative structures that has been much studied both experimentally and theoretically—though not explicitly thermodynamically—is the *lac operon* that regulates the synthesis in *E. coli* of: β-galactosidase needed for the bacteria to utilize lactose as a nutrient instead of glucose; the galactoside permease that allows lactose to enter the cells; and a sequence of enzymes which are not essential for lactose metabolism but which catalyse amongst other reactions certain transformations of the galactose produced by lactose metabolism. One of the distinctive features of these cells is the all-or-none character of their induction, by lactose, allolactose, and certain galactose derivatives, to a state where lactose can be metabolized by the operation of these enzymes. This all-or-none character of the transition is partly the result of the ability of the galactose permease system to be both induced by lactose and at the same time to facilitate its transport. Both *positive feedback*, related to the action of the inducer (lactose) and *catabolite repression* (by the glucose produced from lactose via galactose) are involved. A theoretical model has been worked out for this autocatalytic feedback system (Babloyantz and Sanglier 1972; Sanglier and Nicolis 1976) which also has these and other experimentally confirmable features and shows in detail how dissipative instabilities can arise as transitions in a multiple system of steady states of interlocking sequences of reactions.

(c) *Monod—Jacob type models for induction or repression.* Monod and Jacob (1961) proposed the interaction of several regulatory enzymic pathways as a possible source for alternative steady states important in cell differentiation processes and suggested model systems connecting several controlling systems known to exist in bacteria. Babloyantz and Nicolis (1972) have derived the kinetic equations for a Jacob–Monod system in which the synthesis of proteins is regulated by the correlation between two enzymic pathways (subscripts 1 and 2, respectively in Fig. 2.9). The regulatory gene RG of each enzyme gives a non-active repressor R which upon combination with the product P of the opposite system forms a repressor Re which blocks the formation of enzyme E. It is also assumed that the inactive repressor combines with two product molecules

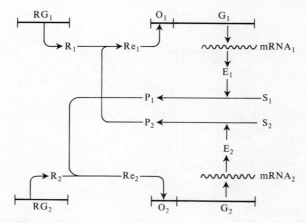

Fig. 2.9. Synthesis of proteins regulated by correlation between two enzymic pathways (subscripts 1 and 2), in a Jacob–Monod-type system. (From Babloyantz and Nicolis 1972, Fig. 1.)

to form Re and that one of the protein products (P_1) can participate in metabolic reactions and be transformed into a decay product F_1. This last step ensures that the system is open to the reservoir of the final decay product F_1 and this adds a further 'degree of freedom' to it. The, rather complicated, kinetic equations were written for each step for variation of the concentrations of the P, Re, E and m-RNA of both systems, assuming that the rate-limiting processes are the synthesis of the messenger (m)-RNA and enzymes and that there is a steady state. The equation for E_1 (which runs to the fifth power of E_1) has been analysed by computer for a large range of values of the different parameters and Fig. 2.10 shows a typical result for the values of E_1 and F_1 consistent with a steady state. It can be seen that, at a particular value of F_1, which is the concentration in a 'reservoir' *external* to the system itself (so F_1 is an external regulator), there is a switch from a regime in which there are three possible steady states (actually only the uppermost and lowest are stable) to a regime of only one steady state. The values of P_1 corresponding to the transition region are quite different from those at equilibrium and are such that one of the pathways may be inhibited to the point of suppression.

Thus the functional behaviour of the system changes radically through the change in an external constraint (in F_1 in the calculation, but it could also have been in substrate concentration or kinetic constants through pH changes) without any alteration in the information carried by the genes. The transition is demonstrated to depend on both the non-linear kinetics of the various processes and on the asymmetry in the system because

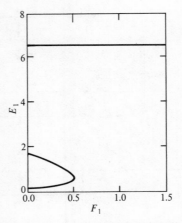

Fig. 2.10. Multiple steady states of enzyme E_1, in Fig. 2.9, for particular values of the rate constants and concentrations of intermediates for m-RNA synthesis. (From Babloyantz and Nicolis 1972, Fig. 2.)

of the non-equilibrium coupling (through F_1) to the environment. These authors conclude that from this point of view

> the phenomenon of switch between different pathways appears to be of the same nature as the transitions leading to dissipative structures (Glansdorff and Prigogine 1971). ... The main point is that since both non-linearity and non-equilibrium conditions are common features of biological systems, the abrupt functional changes (e.g. through Monod–Jacob mechanisms) in various organisms could well be explained, in a large number of cases, as irreversible transitions arising beyond points of chemical instabilities. (Babloyantz and Nicolis 1972, p. 192.)

Cellular systems

Transition to ordered behaviour is not confined to networks of biochemical reactions, for the idea of instabilities leading to dissipative structures seems to be widely applicable in biology. A few examples of these more biological applications, hardly any of which have been analysed in any explicitly thermodynamic manner, will be given as illustrations.

(a) *Depolarization of membranes.* A biological excitable membrane, e.g. the membrane of a nerve cell, may exist in one of two permanent states (approximately speaking). In one of these, it is polarized, with different ionic charges on its two sides and in the other it is depolarized. The transition between the two states is the result of the passage of a pulse concomitant with a change from low ionic permeability to a state of high ionic permeability, closer to thermodynamic equilibrium. It has been shown by Blumenthal, Changeux, and Lefever (1970) that this depolarization may be quantitatively interpreted as a transition arising beyond the point of instability of the polarized state which itself is no longer on the 'thermodynamic branch'. This instability results from the difference in ionic concentrations across the membrane that keeps the system in a far from equilibrium state.

(b) *Morphogenesis.* We shall have cause, in Chapter 4, to recall the original and fundamental work of Turing (1952) on the differentiation of cells in morphogenesis as related to the development of inhomogeneities in the distribution in time and space of form-controlling substances ('morphogens') that were originally distributed uniformly. However, differentiation and growth involve a *succession* of forms and not simply transitions from homogeneity to a final differentiated state, so that one is concerned rather with a succession of states in which instabilities are successively inducible. Martinez (1972) has shown how a system of cells can evolve through distinct states by a series of instabilities. If cell division waits upon a substance (Z) attaining a critical concentration (Z_c), and if the synthesis of Z is controlled by two other substances (X and Y, 'morphogens'), and if the X and Y are formed and destroyed in other reactions and Y can also diffuse between cells, then (depending again on a ratio D_Y^2/L, where D_Y is the diffusion coefficient of Y and L a characteristic length of the system) it can be shown that X, Y, and accordingly Z, will not be distributed homogeneously in a linear array of cells. In some cells Z_c will be reached and so they, but not others, will divide, which destabilizes the pattern first formed and leads to another pattern, with yet more divisions, and so on. Thus instabilities are controlling, through cellular division, the relation between boundary values of parameters (e.g. concentrations) and spatial dimensions of the system.

One could write (cf. Prigogine and Lefever 1975):

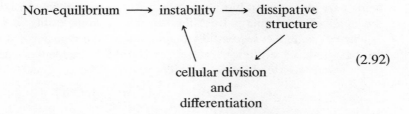

(2.92)

This postulated process of Martinez might justifiably be called 'self-organization' and it should be stressed that it is fundamentally irreversible in character, being a succession of instabilities: indeed this whole interpretation coheres well with those of 'evolutionary feedback' we have already outlined (section 2.7).

(c) *Slime mould aggregation.* Certain unicellular organisms attain during their life cycles a level of organization in which the individual cells aggregate with colonies within which there is a differentiation among the cells. Thus the cellular slime moulds (*Acrasiales*) display remarkable effects of long-range intercellular interaction in their morphogenetic development which are more fully described in section 4.6.2. It has been shown that amongst other things their aggregation is mediated by adenosine 3′,5′-cyclic monophosphate (cAMP) and that this species also produces an enzyme that converts the cAMP into 5′-AMP which is inactive in the whole process. The initiation of the aggregation can be interpreted as an instability in the uniform distribution of a homogeneous population of cells consequent upon starvation and the production of 3′,5′-cyclic AMP by the cells in a process that is far from equilibrium, i.e. practically irreversible. Aggregation may thus be viewed as a breakdown of stability caused by intrinsic changes in the basic parameters which characterize the system. In other words, the first stages, at least, of the aggregation of these cells may be interpreted as a spatial dissipative structure.

2.9 The role of irreversible thermodynamics in biological interpretation

In this chapter, we have been describing a very substantial and sustained endeavour over the last three to four decades to extend the thermodynamics applicable to equilibrium and near-equilibrium systems, for which fluxes and forces are linearly related, to the non-linear range dominated by irreversible processes. This endeavour has been significant for theoretical biology, because biological organisms are *ex hypothesi* not at equilibrium, which is equivalent to the death of an organism; are highly internally and externally coupled and so are non-linear; and are open systems containing many sub-systems in steady states, as well as being part of a general biosphere which is in a steady state in relation to the steady flow of energy from the sun that sustains all life.

The extension has successfully been made to the non-linear domain to produce both an 'evolution criterion', controlling the likely succession of states towards a steady state, namely

$$\frac{d_X P}{dt} < 0; \qquad \frac{d_X \sigma}{dt} < 0, \qquad (2.55)$$

and also a 'stability criterion', namely

$$\delta_X P \geq 0 \quad \text{and} \quad \delta_X \sigma \geq 0. \qquad (2.69)$$

These relationships are significant discoveries but, because they cover only the variations with forces ($d_X P$, $d_X \sigma$, $\delta_X P$, $\delta_X \sigma$) and not fluxes, the differentials are not total and satisfaction of the stability criterion is sufficient for stability, but its contravention is *in*sufficient for *in*stability. Hence we are in fact unable to *predict* instability through violation of (2.69), and we are able only to say if it is possible. Clearly supplementary information is needed for assessing stability, once it can be seen to be allowed, and this can only come from direct analysis of the rate equations for the processes occurring in the biological system, or in the model seeking to represent some aspect of it.

Thermodynamic analysis links naturally to fluctuation theory because of the Einstein relationship (2.71) between the probability of a fluctuation, at or near equilibrium, and the exponential of the second order differential of the thermodynamic potentials. This is one of its strengths relative to kinetic treatments. In the non-linear range, the Einstein relation (2.71) applies only to *small* thermal fluctuations about steady states far from equilibrium, under conditions of local equilibrium. Such states are thereby shown to be stable only to small thermal fluctuations. However, large-scale fluctuations, for which local equilibrium does not apply, can lead to a change of macroscopic state in non-linear systems far from equilibrium (Nicolis and Prigogine 1971) at or slightly beyond an instability. When this happens the Einstein relationship is no longer applicable and there is a fundamental limitation to the ability of thermodynamics to interpret evolution in the non-linear range of irreversible processes. For such evolution is in general *not* reducible (unlike the equilibrium and near-equilibrium cases) to control by the properties of a thermodynamic potential. So again, the problem of large-scale fluctuations in the non-linear range, well away from equilibrium, turns out to be a dynamical problem, involving analysis of the rate equations governing the probability distributions.

So, one may well ask, what *is* to be gained from the application of irreversible thermodynamic concepts and criteria to biological systems? The primary and overriding gain is undoubtedly in the ability of thermodynamics to provide, as it were, an architectonic skeletal framework which limits but does not in detail prescribe. One can then build on this framework by using other resources of dynamical theory, of kinetics, of fluctuation theory—and of precise experimental information and new knowledge of modes of control and regulation at all levels in biology. Structural order comes from the existence of constraints and the macroscopic and phenomenological approach of thermodynamics is uniquely fitted to handle such factors. Thus it is that the thermodynamic analyses we have been describing can serve to eliminate some putative models of biological situations as being incompatible with macroscopic physical laws, while permitting others, if not actually determining the choice between them. This can be a useful role, since so much of theoretical biology is concerned with formulating and testing mathematical models of the phenomena in question (cf. Fig. 1.1) and thermodynamics can be a help, for example, in restricting the rate laws that might be relevant. Indeed for some areas of theoretical enquiry, the thermodynamic constraints on the building of models are often the most reliable knowledge of the situation available—for example, in any modelling of prebiotic evolution.

So thermodynamics can never work in isolation from other approaches based on the theory of fluctuations, of stability, of stochastic processes and of non-linear differential equations. However it has its own unique insights which serve to link reflection on biological systems with the whole corpus of physico-chemical theory. For the new concepts in irreversible thermodynamics of non-equilibrium as the source of order, of 'order through fluctuations', of the decisive role of non-equilibrium constraints and of dissipative structures (spatial, temporal, and both) in open systems, broaden and deepen immeasurably our perspective on biological systems and whole organisms and, indeed, have already proved to be a stimulus for the kind of detailed work (see Chapters 4 and 5) that is required to give them a 'local habitation and a name'.

3 Network thermodynamics

3.1 Thermodynamics and electrical circuits

IT is clear from what has already been said that biological structures are extremely complex, hierarchical systems and that they are heterogeneous and non-linear. They involve a subtle interplay between the energetic processes of transport, reaction, and conformational change and cybernetic flows of information whose regulatory effects are out of all proportion to their energy content. It is also clear that, if we are to have theories which exploit the hierarchical functions of a biological system, they must be able to represent the interconnections that exist between the simpler sub-systems in the hierarchies. As soon as one attempts to make any sort of plan or drawing representing such a system, one very quickly sees the analogy between this attempt and modelling complex electrical circuits, which are often simply nested sequences of black boxes. The way the sub-units are connected up is absolutely crucial for electrical circuits and indeed electrical circuit theory is about these interconnections and their topology. There is a sense in which a radio or television set is more than the sum of the characteristics of its components and no one would attempt to describe a television set by integrating Maxwell's equations! It is no use saying that this is possible 'in principle', for the complexity of, for example, a mitochondrion is more of the order of that of a television set (in fact very much greater) than that of an anisotropic medium. In 'network thermodynamics' electrical circuit theory has been brought into play to model very complex systems of the kind which arise in living organisms (for introductory expositions see Oster, Perelson, and Katchalsky, 1971, 1973; Oster and Auslander 1971; Oster and Perelson 1973; for the most recent full account, see Schnakenberg 1977). But how is it that network theory, which is just another name for electrical circuit theory in our present context, comes to be linked with thermodynamics?

First, a few elementary points about electrical circuits which are certainly irreversible thermodynamic systems. For our present purposes, electrical circuits may be regarded as consisting of three kinds of physical elements, capacitances (C), inductances (L), and resistances (R). The first two, the capacitances and inductances, store energy and the resistances dissipate energy. Each of these elements is characterized by a 'constitutive relation' which is a relationship between two of the parameters of the circuit at the point where the element is present. These relationships are

Table 3.1. Physical elements, constitutive relations, and incremental quantities in electrical circuit theory

Physical elements	Constitutive relation	Incremental quantity
C: capacitances	$\phi_C(\text{voltage }(v), \text{charge }(q)) = 0$	$C \equiv \partial q/\partial v$
L: inductances	$\phi_L(\text{flux }(\phi), \text{current }(i)) = 0$	$L \equiv \partial \phi/\partial i$
R: resistances	$\phi_R(\text{voltage }(v), \text{current }(i)) = 0$	$R \equiv \partial v/\partial i$

given in Table 3.1 which also lists the incremental capacitances, inductances, and resistances which are simply the differential relationships between the two parameters that are involved in the constitutive relations.

In electrical circuit theory there are certain topological constraints, the boundary conditions that one element imposes on another. These are: (i) Kirchhoff's current law (KCL), which is a law of conservation and says that the total current flowing into any one point must equal the total current flowing out of it; and (ii) Kirchhoff's voltage law (KVL) which is the law of continuity and says the electrical potential is unique at each point at a given time, and changes continuously as one goes through the electrical network, so that the algebraic sum of the potential differences in a closed mesh is zero. These two laws, together with the topological description of the network, are enough to solve all questions about the properties of an electrical circuit, in which the wires have no resistance and are just communicators between elements, by the usual conventions of electrical circuit diagram-making. The reason why these relationships and this approach to a complex circuitry is of more general importance than simply solving circuit problems is that the parameters which we use to characterize the electrical circuit are examples of a much wider class of parameters which can in fact be used to characterize any thermodynamic and dynamic system. Thus:

the current, i, is a 'through-variable' (measured at *one* point) and is an example of a *flow*, f;

the voltage, v, is an 'across-variable' (it needs *two* points to measure it) and is an example of an effort, e;

the charge, q, is an integrated through-variable ($q(t) = q(0) + \int i \cdot dt$) and is an example of a generalized displacement, q;

the flux, ϕ, is an integrated across-variable ($\phi(t) = \phi(0) + \int v \cdot dt$) and is an example of a generalized momentum, p.

3.2 Energy storage and dissipation: bond graphs

Ideal elements

The possibility of extending these concepts and intellectual techniques to thermodynamic systems, notably biological hierarchies, arises because at each point in any such system there is both energy storage and dissipation of energy, just as there is storage of energy in capacitances and inductances and dissipation of energy in resistances in electrical circuits. In both systems energy is processed and sometimes it is processed in two or three different ways at the same physical point in space. In order to extend electrical network theory to thermodynamic systems one examines the system (e.g. some part of a biological organism or organelle) and notes what energy transactions (reversible or irreversible) are occurring, each characterized by a distinctive constitutive relationship. It must be emphasized that more than one of these conceptually separate transactions, which will constitute the *ideal elements* and are the analogue to the physical elements of the electrical network, may be transpiring at any one point in space. For example, both reversible and irreversible processes take place in a membrane and, in a homogeneous medium in which several chemical reactions are occurring, several different transformations occur in any given small volume element. The basis for generalizing electrical network theory to other (e.g. biological) systems is simply that the through- and across-variables, both in their ordinary and integrated forms (f, e, q, p), can be exemplified in many different ways (see Table 3.2). Each ideal element in the network representing the thermodynamic system, like each physical element in an electrical network, is distinguished by its constitutive relation between two of these variables. It is therefore legitimate to speak of a generalized capacitance (C), a generalized inductance (L), and a generalized resistance (R). As before, the generalized capacitance and inductance represent storage of energy and the generalized resistance represents irreversible dissipation of energy. Table 3.3 lists, in parallel with Table 3.1, the constitutive relationships in the generalized case and the definitions of the corresponding incremental quantities. It also gives some typical examples of the incremental capacitances, etc. which have been obtained by substituting for the general notation (for the variables (e, f, p, q)) the more familiar quantities given in Table 3.2. Table 3.3 also includes both an effort source (E) which is simply an infinite capacitance, one in which the effort is always constant whatever the flow, and a flow source (F), an infinite inductance in which the flow is always constant whatever the effort. The set of ideal elements listed in Table 3.3 (together with one other element, a transductance (TD)) are sufficient to construct a representation of most thermodynamic

Table 3.2. Common through- and across-variables.

Energy domain	f Flow (KCL) through- variable	e Effort (KVL) across- variable	q Integrated through- variable	p Integrated across- variable
Electrical	Current i	Voltage v	Charge q	Flux ϕ
Diffusion	Mass flow J_i	Chemical potential μ_i	Mass m_i or no. of moles n_i	—
Thermal	Entropy flow $\dot{S}$	Temperature T	Entropy S as thermal change	
Chemical reaction	Reaction rate J_r	Affinity $A = -\Delta G$ $= -\sum \nu_i \mu_i$	Advancement (extent) of reaction ξ	—

$$\underbrace{}_{L}$$
$$\underbrace{}_{R} \quad \underbrace{}_{C}$$

systems. It should be noted that the letters R, L, etc. are used as a notation both for the ideal elements and for the constitutive relationships themselves.

'Ports'

Each ideal element (or 'sub-system' or 'sub-process') has a finite number of ways of interacting with its surroundings and with other processes and

Table 3.3. Ideal elements, constitutive relations, and incremental quantities in network thermodynamic representations.

	Generalized quantity	Constitutive relation	Incremental quantity	Example
Energy storage (reversible)	Capacitance C	$\phi_C(e, q) = 0$	$C \equiv \partial q/\partial e$	$\partial n_i/\partial \mu_i$ $= n_i/RT$ (Diffusion)
Energy storage (reversible)	Inductance L	$\phi_L(f, p) = 0$	$L \equiv \partial p/\partial f$	—
Energy dissipation (irreversible)	Resistance R	$\phi_R(f, e) = 0$	$R \equiv \partial e/\partial f$	$\partial(\Delta \mu_i)/\partial J_i$ (diffusion) $\partial A/\partial J_i$ (chemical reaction)
	Effort source E	$e(f) = $ constant (E) for all f		
	Flow source F	$f(e) = $ constant (F) for all e		

ENERGY STORAGE AND DISSIPATION: BOND GRAPHS 77

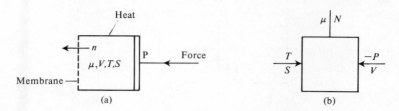

Fig. 3.1. (a) Thermomechanical three-port. (b) Three-port (schematic, see text).

these are called its number of *ports*. The word 'port' is a picturesque nomenclature for the external interaction itself; it is not necessarily an object. Each mode of interaction, each port, has two independent variables and so one characteristic constitutive relation. Thus if we had a gas enclosed in a cylinder, equipped with a piston at one end and a membrane permeable to the gas at the other, then it would interact in three ways with its surroundings: mechanically, thermally, and by diffusion (Fig. 3.1(a)). Each of these modes of interaction would have their specific constitutive relation (between pressure (p) and volume (V); entropy (S) and temperature (T); chemical potential (μ) and amount of gas (n)) and so the system could be represented as a three-port system by some such symbol as Fig. 3.1(b), in which each single line represents a mode of external interaction, a 'port'.

More relevant to our purposes would be the case of a membrane (m) across which a substance is diffusing between constant chemical potential reservoirs I and II, with flux J (Fig. 3.2(a)). If we consider a thin lamina adjacent to the side of the membrane (m) facing I, two processes can be distinguished: (i) a flow (J_I) of the substance with dissipation of energy across a chemical potential difference, $\Delta\mu_I(=\mu_I-\mu_m)$, where μ_m is the chemical potential in the lamina; and (ii) a storage of the substance, and so of energy, within the membrane (before it then flows out into II). In the network representation the first will be shown as a one-port (generalized) resistance, R, with a constitutive relation between a 'flow' (the molar flux inward, J_I) and an 'effort' ($\Delta\mu_I$) at side I; the second will be represented as a one-port (generalized) capacitance, C, with a constitutive relation between an integrated through-variable (the number of moles of the substance stored per unit time which is the difference, $\Delta J_I = J_I - J_m$ between an inward and an outward flow with respect to the lamina in the membrane) and an effort variable (the chemical potential, μ_m of the substance inside the membrane). The symbols for these would then be

$$\frac{\Delta\mu_I}{J_I} R_I$$

78 NETWORK THERMODYNAMICS

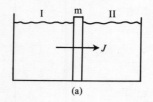

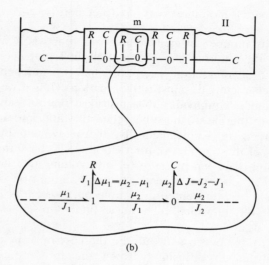

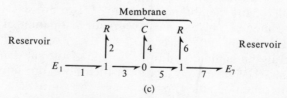

Fig. 3.2. Bond graph for diffusion through a membrane. (Oster *et al.* 1971, Fig. 1.)

(subscript to R showing side I) and

$$\frac{\mu_m}{n_m (\text{or } \Delta J_1)} C_m$$

(subscript m to C showing inside the membrane), where the lines have been used simply to represent interactions, i.e. exchanges of energy, according to a particular constitutive relation (resistive and capacitative, respectively).

Basic procedure: state variables

The basic procedure of network thermodynamics is best summarized in the words of the publication which has been most influential in disseminating its ideas, namely, that of Oster *et al.* (1973, p. 11):

> If the overall system is continuous, it is first subdivided mentally into homogeneous subsystems; and ... each subsystem is further separated conceptually into reversible and irreversible parts. The reversible subsystems are assumed to store energy without loss, while the irreversible subsystems are assumed to dissipate energy without storage. Each of these subsystems will be identified with an n-port. The fundamental thermodynamic quantity in this treatment is energy rate, or 'power', instead of energy. Systems amenable to a network representation frequently share one common property: the energy-rate processes may be expressed as a product of an *effort* or force variable e and a *flow* variable f such that the power $P = ef$. (In electrical networks these are, of course, voltage and current.) Most thermodynamic processes can be so characterized. Note that we are *defining* effort and flow as the primitive state variables, rather than constructing them from the Gibbs equation, as is the usual practice; consequently, it is not necessary that their product have the units of power.

Energy bonds

The meaning of the lines radiating out from the symbol for an ideal element (the symbol which, as mentioned above, can be thought of as also representing the constitutive relation proper to that element) can be extended naturally to be a symbol to represent connections between elements. They are then called ideal *energy bonds*, 'ideal' because they represent an instantaneous transmission of energy without loss between elements—so they are entirely analogous to the idealized, perfectly conducting wires of electrical circuitry. The energy bonds stand for the connections between elements and so for the boundary conditions which one energy transaction (with its own constitutive relation) exerts upon another; they are therefore closely related to how the whole system interlocks and interacts, so that the topology of the network manifests these interactions (*not*, necessarily, the spatial organization). The ideal elements R, L, C, E, and F are all one-ports, that is they are elements described by one constitutive relation between a single 'through' and a single 'across' variable (or their corresponding integrated forms). So that they always appear in the networks as

$$\frac{e}{f}\text{R}, \quad \frac{e}{q}\text{C}, \quad \frac{f}{p}\text{L}, \quad \frac{e}{f}\text{E}, \quad \frac{f}{e}\text{F},$$

from which an arrow convention to represent reference directions of flows relative to efforts has been omitted.

This convention takes the form of a half-arrow, showing the direction of power flow, positive towards an element when the arrow points into the element symbol, thus

$$\frac{e}{f} \rightharpoonup R \text{ implies } e = Rf, \text{ with power delivered} = +ef,$$

and

$$\frac{e}{f} \leftharpoonup R \text{ implies } e = -Rf, \text{ with power delivered} = -ef.$$

Ideal junctions

For physical systems which comprise many types of energy flow and couplings, as do the biological, the topological representation of 'bond graphs' (Paynter 1961; Karnopp and Rosenberg 1968; Thoma 1975a) is especially useful (the alternative linear-graph representation is somewhat unwieldy when the systems are complex). The bond-graph representation usefully depicts two *ideal junctions* which allows one to join the ports of energy-processing elements through energy bonds. They are really a graphical notation for the linear constraints that the two Kirchhoff laws (now involving generalized across- and through-variables) place upon the interaction between elements. Thus these junctions represent the ways in which the influence of the boundary conditions of one constitutive relation of a particular energy transaction may be exerted upon the constitutive relation of another energy transaction. These ways are limited by the laws of conservation and continuity, KCL and KVL, respectively, and so the two kinds of junction are:

(1) The *zero-junction* (cf. a parallel connection) at which flows split when incident upon it. It is represented by

$$\begin{array}{c} 2\diagdown \overset{3}{\mid}\diagup i \\ 1\text{—}0\text{—}n \end{array} \quad [\textit{or by:} \text{—}p\text{—}] \tag{3.1}$$

and around it

$$e_1 = e_2 = \cdots e_i \cdots = e_n.$$

Since the junction is ideal in the sense that it neither stores nor dissipates power, $\sum e_i f_i = 0$ and so $\sum f_i = 0$; it obeys KCL.

(2) The *one-junction* (cf. series connection) around which efforts redistribute. It is represented by

$$\begin{array}{c} 2\diagdown \overset{3}{\mid}\diagup i \\ 1\text{—}1\text{—}n \end{array} \quad [\textit{or by:} \text{—}s\text{—}] \tag{3.2}$$

and around it

$$f_1 = f_2 = \cdots f_i \cdots = f_n.$$

Since it, too, is an ideal junction $\sum e_i f_i = 0$, so $\sum e_i = 0$; it obeys KVL.

3.3 Diffusion through a membrane

We are now in a position to develop further the network representation of diffusion through a membrane. The ideal elements R_I and C_I which we distinguished in the process occurring on side I of the membrane depicted in Fig. 3.2(a) can now be represented as

$$E_I \xrightarrow{\;\;J_I/\mu_I\;\;} 1 \begin{array}{c} R \\ \bigg| J_I \; \big| \Delta\mu_I \end{array} \xrightarrow{\;\;J_I/\mu_m\;\;} 0 \begin{array}{c} C_m \\ \Delta J_I \big| \mu_m \end{array} \xrightarrow{\;\;J_m/\mu_m\;\;} \tag{3.3}$$

(with reference directions, arrows, omitted). In practice, this notation will only be a fair representation of an infinitesimally thin element in a membrane. Any finite membrane must be regarded as a series of such laminae $(1, 2 \cdots)$, each of which has a capacity for storing energy by taking in matter at a given chemical potential, so that the true representation of a finite membrane would be as in Fig. 3.2(b). The enlargement shown in the lower part of Fig. 3.2(b) shows how the flows (J_1 in the figure) are equal around a one-junction and the efforts (μ_2 in the figure) around a zero-junction.

A bond graph, such as Fig. 3.2(c) depicts the conceptual separation of the two processes or subsystems (resistive and capacitative elements) occurring in each lamina. Consider a membrane (m) with only a single, inner compartment in which the diffusing substance is stored at a particular chemical potential (μ_m) and the bounding compartments, I and II, have infinite capacitance, that is, are constant effort (chemical potential) sources, denoted by E_1, then Fig. 3.2(c) (now with arrows depicting the correct reference directions) is the network representation of this 'one-lump' membrane. Since the elements are constitutive relations and the junction represents their mutual interaction through the boundary conditions they make for each other, this diagram is simply an algorithm, a kind of pictorial shorthand, of the various constitutive relations and of the mutual constraints which each exerts on the other through the conceptual topology depicted by the network (related to but not the same as the spatial topology). By writing out in detail, for the capacitance C, the linear constitutive relations (bond 4, Fig. 3.2(c) between the effort (e_4), i.e. chemical potential ($\mu_m \equiv e_4$), and reversible flow, f_4 ($=f_3-f_5$), of the diffusing material, and by taking account of the equality of the various flows around the one-junction as well as the equality of the efforts (chemical potentials) around the zero-junction, the relation between the relaxation time (τ_m) for the membrane process and the diffusion coefficient (D_m) and membrane thickness (Δx) can be obtained as $\tau_m = (\Delta x)^2/2D_m$ (Oster et al., 1971).

The significant feature of this procedure is that the dynamical equation can be 'read off' from the bond graph for arbitrarily complex systems; for the graph is perfectly equivalent to the differential equations themselves, but with the added advantage that it reveals the system *topology*, often obscured in a catena of equations. So computer programs can be devised which will accept the bond graph as its input and then compute the dynamical behaviour directly, without dealing with the differential equations explicitly—a great advantage when the complexity of the system precludes analytical solution.

The algorithmic derivation of the differential equations from a bond graph has been described by Oster and Auslander (1971) in the following terms:

> To derive the system equations in the form of n first-order, ordinary differential equations, n elements are chosen which fix the system state and determine a set of state variables. In most cases the state variables are the output variables of L's or C's and the first step of the derivation is to write the constitutive relations for each. This step yields a set of first-order equations with the derivative of a state variable on the left-hand side of each equation and algebraic terms involving some variables that are state variables and some that are not on the right. Following the directions indicated by the causal strokes, the equations are extended by substitution through the junctions and other elements until the right-hand side of the equation contains nothing but state variables and input (forcing) variables. The process is repeated on each of the original equations so the final result is a set of equations of the form
>
> $$(d/dt)\mathbf{x} = \mathbf{F}(\mathbf{x}, \mathbf{u}),$$
>
> where $\mathbf{x}$ is the state vector and $\mathbf{u}$ is the forcing vector. (Oster and Auslander 1971, p. 15; see pp. 15 and 16 for a worked example.)

3.4 Causality

The bond-graph representation can be enhanced by including a representation of experimental causality. Thus in the membrane diffusion process discussed in the preceding section, the filling of membrane (m) is represented by the flow

$$J_m = \frac{dn_m}{dt} = \frac{dn_m}{d\mu_m} \cdot \frac{d\mu_m}{dt} = C_m \cdot \frac{d\mu_m}{dt}, \qquad (3.4)$$

where n_m = number of moles of permeant in the membrane; and C_m =

generalized capacitance (see Table 3.3, where $C = \partial q/\partial e$, and Table 3.2, where e, q are, for diffusion, chemical potential μ and the number of moles of diffusing substance, respectively). Now μ_m cannot change instantaneously ($d\mu_m/dt < \infty$) unless the flow J_m becomes infinite. So the flow is the 'natural' input, the independent variable, the 'cause', for a capacitative element. The converse is true for inductive elements, for which the basic relation (Table 3.3) is

$$L(\partial f/\partial t) = e, \qquad (3.5)$$

so inductances have the effort e (e.g. voltage) as their natural input. For resistances, however, the relation between flow and effort includes no time derivatives, so there is no independent variable of choice.

It is often important to know which is the independent variable, the 'cause', and the *causal stroke*, proposed by Paynter (1961), represents this in a bond graph. The causality in a bond is indicated by a single vertical stroke (Fig. 3.3). In Fig. 3.3(a), the vertical stroke, as viewed from *inside* the system, indicates that e is the independent input variable (the 'cause') and f is the dependent, output, variable. In Fig. 3.3(b), the position of the stroke implies that f is the input variable (the 'cause') and e the output. (The convention is a natural one for R ⊢, meaning *effort*-controlled, can be thought of as a plunger exerting effort on R, while R ⊣, meaning flow-controlled can be thought of as an arrow R ← directing flow into R.)

When applied to the junctions, this leads to some simple rules: for a 1-junction (Fig. 3.3(c)) the fs on all the bonds are equal so, viewed from the junction, only one f can be an independent variable (the 'cause'); for a 0-junction, (Fig. 3.3(d)), since all es are equal, only one e is an

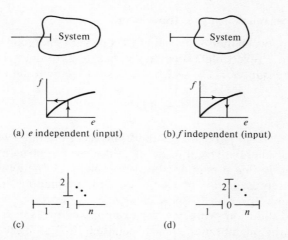

Fig. 3.3. System causality. (Oster and Auslander 1971, Fig. 5.)

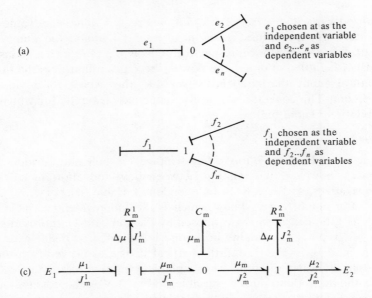

Fig. 3.4. The causality convention around ideal junctions in bond-graph representations (Oster *et al.* 1973, Fig. 3.4.)

independent variable (see also Fig. 3.4). By this convention, Fig. 3.2(c) for the membrane becomes Fig. 3.4(c) (note that the placing of a causality stroke is not related to the flow-sign conventions represented by the half arrows).

3.5 Chemical reactions: the 'transducer'

In order for network thermodynamics to cope with energetic coupling and conversion in irreversible processes, it has to introduce another ideal element, a *transducer* (TD), which is a 2-port element denoted by

$$\frac{e_1}{f_1} \text{TD} \frac{e_2}{f_2}, \tag{3.6}$$

where $e_1 = re_2$; $f_1 = -f_2/r$; and r is the 'transducer modulus' governing the conversion of energy. The transducer, although an 'element', is like an ideal junction in that it neither stores nor dissipates energy and in that it is a pictorial representation of a set of constraint equations on e and f. An important application of this new element is to represent hydrodynamically coupled diffusion processes, for example, of two substances, A and B, through the membrane already denoted in Fig. 3.2(c). Two such diagrams, one each for A and B, are coupled now through an element TD

(Fig. 3.5(a); the elaboration and proof of this is given by Oster *et al.* (1971)). The network bond graph can also represent the coupling of a chemical reaction with the diffusional flows. The transducer is then used to represent the stoichiometric conversion of chemical species in reactions, where the 'flow' of the reaction (J_r) is

$$f \equiv J_r = \frac{d\xi}{dt} = \frac{1}{\nu_i}\frac{dN_i}{dt} \tag{3.7}$$

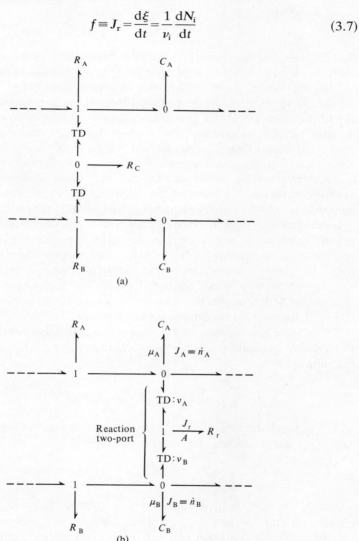

Fig. 3.5. Bond graph of diffusion coupling (a), and of reaction coupling (b). (From Oster *et al.* 1971, Fig. 1.)

where ξ = extent of the reaction; N_i = molar amount of species i; ν_i = stoichiometric coefficient of species i in the reaction equation; and

$e \equiv A$, the affinity of the reaction,

$= -\Delta G$ of the reaction,

$= -\sum \nu_i \mu_i$.

Conversion in a chemical process is a 'reaction 2-port', shown, as an example, by the bracketed portion in Fig. 3.5(b), for near-equilibrium conditions. When there is linearity between J_r and the affinity, this transducer relation can be represented by a one-port resistance (for proof see Oster *et al.* 1971). By procedures similar to that outlined previously for the simpler, uncoupled, membrane diffusion process, the correct dynamical description can be derived from the bond-graph representation, which has the added advantage of clearly distinguishing the reversible and irreversible processes. In this and other complex cases, this network technique has allowed the calculation of experimentally testable conclusions, which in many cases have been verified (Oster *et al.* 1971).

In biochemistry, one is often dealing with interlocked reacting systems, which are naturally represented as networks, representing stoichiometric relations, on the metabolic 'maps' which adorn biochemistry laboratory walls. The network representation of the thermodynamic processes occurring in such systems would seem to be a natural application in which the stoichiometry determines the topology of the resulting network representation of the various thermodynamic processes.

As we have already mentioned, near equilibrium the reaction process is represented by a one-port resistance and the rate, J_r, is linearly related to the affinity, A by

$$J_r = \frac{\bar{v}^f}{RT} \cdot A \quad \text{[at or close to equilibrium]} \quad (3.8)$$

where $\bar{v}^f$ represents the forward reaction rate at equilibrium at temperature T. If the reaction is far from equilibrium

$$\frac{\partial J_r}{\partial A} = L_r,$$

which is not well-defined (see (2.31) to (2.34)), since both A and J_r, depend on both A^f and A^b (f, b stand for 'forward' and 'back' reaction, respectively). So the ideal element representing the reaction is then a 2-port resistance denoted by

$$\cdots \frac{J_r^f}{A^f} \mathrm{R}_r \frac{J_r^b}{A^b}. \quad (3.9)$$

CHEMICAL REACTIONS: THE TRANSDUCER

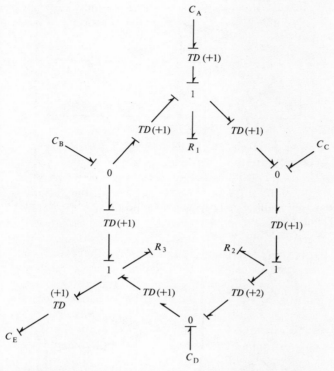

Fig. 3.6. Bond-graph representation of a system of near-equilibrium reactions: $A+B \rightleftharpoons C$, $C \rightleftharpoons 2D$, $D+B \rightleftharpoons E$. (From Oster *et al.* 1973, Fig. 5.4.)

Figure 3.6 shows a representation of the reaction system $A+B \rightleftharpoons C \rightleftharpoons 2D$, $D+B \rightleftharpoons E$, for the near-equilibrium condition. Figure 3.7 depicts the bond graph for $A+B \rightleftharpoons C+D$, in which the dissipation in the reaction is represented by a 2-port resistance, for the conditions not close to equilibrium. A biologically significant process is represented in Fig. 3.8, namely,

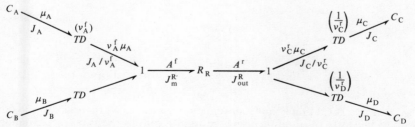

Fig. 3.7. Bond graphs for the reaction $A+B \rightleftharpoons C+D$, far from equilibrium. (From Oster *et al.* 1973, Fig. 5.6.)

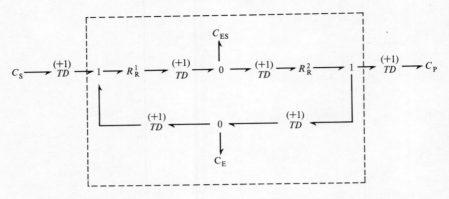

Fig. 3.8. Representation of an enzyme reaction $S+E \underset{\rightleftharpoons}{\overset{1}{}} ES \underset{\rightleftharpoons}{\overset{2}{}} E+P$. (From Oster *et al.* 1973, Fig. 5.9.)

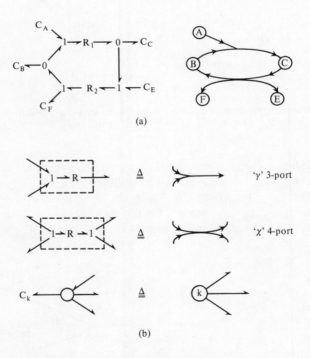

Fig. 3.9. (a) Bond-graph and conventional biochemical representations of a reaction network. (b) Equivalences used to interconvert the two representations. (From Perelson 1975, Fig. 10—based on Morowitz 1973.)

CHEMICAL REACTIONS: THE TRANSDUCER

Table 3.4

Element	Symbol	Defining equations	Remarks
0-junction	$\underset{1}{\overset{2}{\mid}}0\underset{n}{\cdot\cdot\cdot}$	$\sum \sigma_i f_i = 0; e_1 = e_2 = \ldots e_n$	Generalized 'parallel' connexion
1-junction	$\underset{1}{\overset{2}{\mid}}1\underset{n}{\cdot\cdot\cdot}$	$\sum \sigma_i e_i = 0; f_1 = f_2 = \ldots f_n$	Generalized 'series' connexion
Resistance†	$\rightarrow$R	$\phi_R = (e, f) = 0$	Ideal dissipative element
Capacitance‡	$\vdash$C	$\phi_C(e, q) = 0$	Capacitive (displacement) energy storage
Inductance§	$\dashv$L	$\phi_L(p, f) = 0$	Inductive (kinetic) energy storage
Transducer	$\rightarrow$TD$\leftarrow$	$\begin{bmatrix} e_1 \\ f_1 \end{bmatrix} = \begin{bmatrix} r & 0 \\ 0 & y_{-1/r} \end{bmatrix}\begin{bmatrix} e_2 \\ f_2 \end{bmatrix}$	Energy conversion and signal modulation
Effort source	E$\dashv$	e = constant	Ideal energy source (effort)
Flow source	F$\vdash$	f = constant	Ideal energy source (flow)

† The half-arrow is the sign convention: power is considered positive *into* all elements.
‡ The causal stroke $\vdash$C indicates that the natural input (independent) variable is the flow variable, since for C, the dynamic equation is $C(de/dt) = f$, and the physical restriction $P = ef < \infty$ prohibits step inputs of effort.
§ The natural input (independent) variable for L is e. All other elements are causally neutral.

the enzyme–substrate interaction to form a complex ES, and then product. In the bond graph, the dotted line encloses the portion to which the enzyme is confined. These diagrams look complex but they can be generated algorithmically from conventional biochemical reaction flow diagrams, as shown in Fig. 3.9.

The notation of bond graphs, in the form of a list of ideal elements is summarized in Table 3.4 and Fig. 3.10 is a diagram summarizing the constitutive relations.

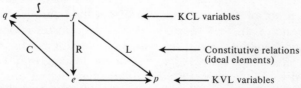

Fig. 3.10. The relation between the variables of network thermodynamics, constitutive relations, and ideal elements.

3.6 Network properties

Tellegen's theorem

The systems represented by the networks of bond graphs (and indeed of any other graphical system) are assumed to be operating irreversibly, like electrical circuits. The energy dissipated is therefore equal to the overall decrease in the free energy of the energy and matter reservoirs which act as the constraints on the system maintaining it in its non-equilibrium state. So for given constraints, the total power flow through the system *and* the reservoirs, taken together, must be zero. In a closed network (or, open network and reservoirs) bond graphs include connections with these reservoirs of energy and matter and it can be shown that (see Oster *et al.* 1973, pp. 44, 5), provided the elements are linked only through 0- or 1-junctions and provided the effort variables are unique and flows are conserved, that the products of the efforts (e_i) acting on the element i and the conjugated flows (f_i) through the same element then sum to zero, i.e.

$$\sum_{\substack{i \\ \text{elements}}} e_i f_i = 0, \qquad (3.10)$$

where the sum is taken over *all* elements (reversible and irreversible, i.e. capacitative as well as resistive and inductive). This result, which is known as *Tellegen's theorem* (Tellegen 1952), is purely topological and is completely independent of the nature of the elements in the network and of the particular relations between efforts and flows, linear or non-linear. It is *inter alia* a power-conservation theorem, meaning conservation of total power both stored in generalized capacitances or dissipated in generalized resistances and inductances. Its meaning is in fact more general than that of conservation of power and this is of significance in connection with evolving networks (Atlan and Katzir-Katchalsky 1973; and see below). There is a logical link between this theorem and Kirchhoff's laws, for these latter are also based on effort variables being unique and flows being conserved. The proof (see Oster *et al.* 1973; Atlan and Katzir-Katchalsky 1973) depends on all bonds being directed into elements, i.e. that efforts and flows do not have independent directions (i.e. they are 'associated'; see Oster *et al.* (1973, p. 33) and the remarks under *Energy bonds*, above). The proof of Tellegen's theorem from the bond-graph representation of an energy dissipating and storing system makes it clear that the 'elements' and bonds in (3.10), for which the sum of products of efforts and flows is made, are restricted to the bonds leading only to one-port elements and to non-conservative multiport elements, in order to avoid counting some products more than once (Atlan and Katzir-Katchalsky 1973). Indeed Kirchhoff's laws can be regarded as a direct result of the topological constraints imposed by Tellegen's theorem on the permissible arrangements of efforts and flows in an electrical network (Chen 1976).

Stability

The sum in (3.10) could be split into its resistive (energy dissipating and irreversible) and its capacitative (energy storage and reversible) elements, so

$$\sum_R e_i f_i + \sum_C e_i f_i = 0, \qquad (3.11)$$

where R and C denote, respectively, sums over these two kinds of elements. Differentiating (3.11) with respect to a real time evolution of the system and taking into account the Kirchhoff laws (in particular KCL, i.e. at each element $d(\sum f_i)/dt = 0$) yields

$$\sum_R \left(\frac{de_i}{dt}\right) f_i = -\sum_C \left(\frac{de_i}{dt}\right) f_i. \qquad (3.12)$$

But for the (reversible) capacitative elements (see Tables 3.2, 3.3),

$$\frac{de_i}{dt} = \frac{\partial e_i}{\partial q_i} \cdot \frac{\partial q_i}{\partial t} = \frac{1}{C} \cdot f_i, \quad \text{[capacitative elements]} \qquad (3.13)$$

where q_i are integrated through variables (Table 3.2). So the sum on the r.h.s. of (3.12) becomes $\sum f_i^2/C$, and this must always be positive since C is always positive, being one of the conditions that *local* equilibrium prevails (see section 2.3), as it still must do in the network approach as much as in that described in Chapter 2. So the r.h.s. of (3.12) is negative, and

$$\sum_R \left(\frac{de_i}{dt}\right) f_i \leq 0, \qquad (3.14)$$

and this also applies to non-linear and non-stationary systems. Now f_i and (de_i/dt) are equivalent, in the network notation of this chapter, to J_α and (dX_α/dt), respectively, of the previous one. So (3.14) is the network counterpart of the general evolution criterion of Glansdorff and Prigogine (eqn (2.54), (2.55)), written as

$$\frac{d_X \sigma}{dt} = \sum \left(\frac{dX_\alpha}{dt}\right) J_\alpha \leq 0, \qquad (3.15)$$

and both (3.14) and (3.15) represent a condition imposed on the dissipation of energy during the evolution of a system. Equation (3.14), for a discrete system, states that $\sum f_i(de_i/dt)$ will decrease until a steady state is reached. So any perturbation will increase this function and the criterion for stability may be written as

$$\sum f_i \cdot \delta e_i \geq 0. \tag{3.16}$$

On introducing into (3.16) the expansion of f_i in a Taylor series about the steady-state value $(f_i)_0$, namely,

$$f_i = (f_i)_0 + \sum_{ij} \overline{(\delta f_i/\delta e_j)} \delta e_j, \tag{3.17}$$

one obtains

$$\sum (f_i)_0 \delta e_i + \sum_{ij} \overline{(\delta f_i/\delta e_j)} \delta e_j \cdot \delta e_i \geq 0, \tag{3.18}$$

(ignoring higher-order terms).

At the steady state $\sum (f_i)_0 \cdot \delta e_i = 0$ and

$$\sum (\delta f_i/\delta e_j) \delta e_j = \sum \delta f_i,$$

hence (3.18) becomes

$$\sum \delta f_i \cdot \delta e_i \geq 0 \tag{3.19}$$

as the condition for stability. This result of network thermodynamics for a discrete system is the equivalent of the Glansdorff–Prigogine stability condition (2.69b)

$$\delta_X \sigma = \sum \delta X_\alpha \cdot \delta J_\alpha \geq 0,$$

given in the previous chapter for a continuous system (see Oster *et al.* 1973, pp. 126–7).

Evolving networks

We have already mentioned that Tellegen's theorem does not depend on the topology of the closed network to which it applies, namely, the linked set of one-port elements that constitute it. (Note that any non-conservative multiport element—such as the 2-port resistance for a chemical process far from equilibrium—can be reticulated into an equivalent set, from the point of view of flow of power, of disconnected one-port elements.) This has the important consequence that Tellegen's theorem is applicable even if the topology of the network itself is changing, so long as the one-port elements remain the same, even if they vary with time and are non-linear. Transformations of the 0- into 1-junctions, and vice versa, and even the appearance of new junctions leave invariant the sum of products of efforts and flows, provided the number of one-port elements is unchanged (Atlan and Katzir-Katchalsky 1973).

Hence the Tellegen theorem is an invariance relation that could be applied not only to non-linear, time-variable networks but also to *evolving* networks whose junctions are changing, i.e. whose actual organization alters with time. Living systems are chemico-diffusional networks (chemical reactions plus diffusions) that evolve with time, so the representation of this aspect of their behaviour by network thermodynamics could be promising (Atlan 1975). Two different ways of developing this approach, especially in relation to the creation of biological 'information' (related, roughly, to the complexity of biological systems—but see Chapter 6) have been proposed (Atlan 1978).

(1) If one compares the initial homogeneity of an unstable chemico-diffusional system with its subsequent inhomogeneity beyond the instability point (as, for example, in the Belousov–Zhabotinsky reaction), then it is clear that two different bond graphs will have to be written to represent the two networks of energy-dissipation storage and flow. The later state possesses inhomogeneities in concentration and transport that can be quite sharp, so that they almost act as membranes—and so new generalized capacitances and resistances as new elements, with new junctions, would have to be introduced. A trivial way to treat such a network changing as a result of its own functioning would be to represent the final state from the beginning with the relevant elements as zero. The evolution of these particular elements would be reduced to that of their time-dependent coefficient, changing from zero to their final values (Atlan 1975). Clearly in such cases the final state needs to be already known.

(2) A different, more sophisticated, approach would assume a system structure to be stochastic, which would imply, for network thermodynamics, probability distributions for the values of the constitutive elements and their connections. Atlan concludes his (1978) discussion of these possibilities as follows:

> It is our opinion that together with invariance laws, a combination of deterministic theories of the kind of network thermodynamics with a probabilistic theory of organization mentioned in this paper should be fruitful in the analysis of natural information processing systems where something—but not all—is known about their structure and function.

These procedures also have a bearing on stability, for application of the second, probabilistic approach led to the conclusion that increasing the connectivity in a system decreases its probability of being stable (see Atlan (1978) for an account of this, with references). Since a great connectivity means both a kind of redundancy and rigidity in a structure, these latter must be associated with a low probability of finding conditions favourable to stability. Contrariwise, a more flexible system would be more able to keep its overall functioning structure in spite of the random perturbations arising from varying circumstances (cf. the precision of conditions required for the stability of crystals compared with the flexibility for stability of basic macromolecular structure, with changing configurations).

3.7 Applications and assessments

Although it is true that 'network thermodynamics has produced no new physical results' (Nicolis 1979, p. 248)—see, for example, its cross-connections with irreversible thermodynamics (section 3.6)—it has nevertheless provided new ways of representing, for example, the complexity of biological systems by considering them from the viewpoint of power flow, and of energy dissipation and storage. The network of connections thus made explicit is not the same as, though of course it is related to, the spatially organized complex of components (molecules, or cells, or organs) that is normally expressed by biologists, in, for example, metabolic maps, physiological flow diagrams, or neuronal nets. So although not constituting any physically fundamental theoretical advance, network thermodynamics could find its justification in its practical usefulness for handling and articulating biological complexity. Some interesting biological applications of network thermodynamics were advanced in the seminal paper by Oster *et al.* (1973) and heat flows and production, so important in relation to chemical and biochemical reactions, were formulated in terms of bond graphs by J. Thoma (1971, 1975a,b, 1976) who developed the concept of entropy as a kind of thermal 'charge' in relation to thermal energy flow—just as rate of change of electric charge, 'electric current', is related to flow of electrical energy (see Table 3.2). However there was a relatively long lag period before there were further applications. Papers applying bond graphs to dynamical systems have approximately doubled every three to four years since 1972 (180 in 1979) but those on network thermodynamics of biological systems have been more sparse. However, 1979 saw the publication of some significant attempts to apply these ideas. (For an important collection of papers on bond graphs in relation to biology see the special issue of the *Journal of the Franklin Institute* on 'Bond graph techniques for dynamic systems in engineering and biology' Volume 308 (1979), pp. 173 ff., which concludes with a useful bibliography, by V. D. Gebben, of bond-graph publications.) A few selected applications will now be described before making an assessment of its usefulness.

Generation of heat in biological systems

Since cellular processes involve conversion of chemical energy from one molecular form to another with concomitant generation of heat, any bond-graph representation of biological systems must include this partial transformation of the free energy of reactants into heat. Following the treatment of Thoma (1971, 1975b, 1976), as developed by Plant and Horowitz (1979), let a (biological) reaction be assumed to be open, i.e. to be able to transfer heat to or from a reservoir in the environment at temperature T_e and to be able to exchange reactants with reservoirs in the environment at a temperature T, the temperature of the reaction itself (this avoids convection effects). Flows of heat can be represented as the products of conjugate variables: the temperature (T), as effort, and entropy flow ($\dot{S} = dS/dt$). The contributions to the heat flows and exchanges are: (i) the generation of heat in the reaction itself; (ii) the uptake of heat by the reactant reservoirs, consequent upon the reactant (j) being formed or removed; (iii) the uptake of heat associated with the thermal capacity of the whole reaction system; (iv) flow of heat out of the reaction system across the temperature difference ($T - T_e$). The conservation of the total heat flow dependent on (i)–(iv) can be represented on the bond graph for the reaction by an additional 0-junction, around which the flows (f_i) of heat split, with always $\sum f_i = 0$ by KCL. Let R_r represent the generalized resistance for the reaction (cf. Figs. 3.6–3.8), C_j the generalized capacitances of the reactant reservoirs, and C_T the thermal capacitance of the whole reaction system; the respective conjugate variables and the energy bonds joining these elements to the new 0-junction on the bond graph are given in Table 3.5. The 0-junction of this table is *one* particular junction which, with the connecting energy bonds given there, must now supplement any bond graphs of a chemical reaction if heat exchanges are to be represented. The bond graph for the chemical

Table 3.5. Bond-graph representation of heat flows associated with a chemical reaction. (See Plant and Horowitz 1979.)

Process (see text)	Conjugate variables	Bond-graph representation
(i)	T and $\dot{S}_i$ = internal entropy production in the reaction	$R_r \xrightarrow{\frac{T}{\dot{S}_i}} 0$
(ii)	T and $\dot{S}_j$ = time variation of entropy because of formation or removal of reactant j	$C_j \xrightarrow{\frac{T}{\dot{S}_j}} 0$
(iii)	T and $\dot{S}$ = rate of change of entropy of whole reaction system	$C_T \xleftarrow{\frac{T}{\dot{S}}} 0$
(iv)	$(T - T_e)$ and J_S^e = entropy flux to the environment (e) due to heat flow (Nature of this flow, e.g. if a reactant, would determine choice of element connection to the 0-junction)	$\xleftarrow{\frac{T}{J_S^e}} 0$

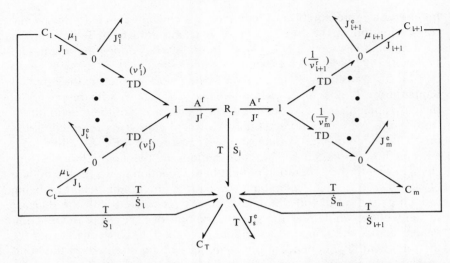

Fig. 3.11. Bond graph for a chemical reaction with heat production and flow. For symbols, see text and Table 3.5; other symbols as for other bond-graph representations of reactions (section 3.5 and Figs. 3.6–3.8). (From Plant and Horowitz 1979, Fig. 2.)

reaction involving m chemical components (j), l of them reactants and $(m-l)$ products, with stoichiometric coefficients ν^f and ν^b for forward and back reactions, respectively (with subscripts indicating the component), is then given by Fig. 3.11.

The constitutive relations corresponding to processes (i)–(iii) have been derived, for restricted conditions, and make it worthwhile applying the bond-graph representation to typical energy conversions in biological systems. Plant and Horowitz (1979) devised, for example, the bond graph for the coupled transport of sodium ions out of a cell, and across the membrane, to compensate for the leak of sodium ions back into the cell. The graph also included the heat exchanges involved and, through its constitutive equations, satisfactorily represented various known relationships and their mutual coherence and consistency with the laws of thermodynamics.

A further application of the bond-graph representation of heat exchanges in biological reaction systems has been made by Horowitz, Giacchino, and Horowitz (1979) to represent the power flow in 'brown fat' cells—the cells in the brown adipose tissue of certain mammals that can generate heat, without shivering, when the mammal is exposed to the cold (reviewed by Smith and Horowitz (1969) and Horowitz (1978)). Once a brown fat cell has been activated by the appropriate neural stimulus, the basic chemical process producing heat is

$$\mathrm{FFA} + O_2 \rightarrow CO_2 + H_2O + T \cdot J_S^e, \tag{3.20}$$

where FFA = free fatty acids, stored as triglycerides in fat globules within the brown fat cell, and $T \cdot J_S^e$ is the heat flow (J_S^e as above). For the intriguing details of this activating process the review and the paper by Horowitz *et al.* (1979) should be consulted. Here we are concerned with the use of the bond graphs of network thermodynamics in the analysis of the power flow within a brown fat cell, for which two possible mechanisms have been proposed differing with respect to the pathways postulated for the flow of protons in the mitochondria of the brown fat cell.

Most of the cells contain mitochondria which are surrounded by two membranes, an inner and an outer. There are three possible pathways (labelled (a), (b), (c) in Fig. 3.12(a)) for the flow of protons across the inner membrane (between the intermembrane space and the 'matrix' within this membrane).

(a) Protons are 'pumped uphill' against the electrochemical potential gradient from the matrix to the intermembrane space by drawing on chemical energy available from oxidation (and so involving $NAD^+/NADH$). This process is essential to enable the two others to operate for any length of time; these latter are movements of protons down the electrochemical gradient, (b) and (c) as follows.

(b) In brown fat mitochondria protons are able to move 'downhill' across the inner membrane without being coupled to chemical reactions (a protein, P in Fig. 3.12, may control this flow). This process represents a dissipation of electrochemical potential without, say, any ATP synthesis.

(c) Protons move down the electrochemical gradient through the inner membrane and their fall in potential drives ATP synthesis.

Figure 3.12(b) shows the bond graph of Horowitz *et al.* (1979) for these three kinds of proton flow across the inner mitochondrial membrane.

The two mechanisms postulated for proton flow in brown fat cells are (Horowitz *et al.* 1979) either, (I) = (a) + (c), or (II) = (a) + (b). In mechanism (II), protons would dissipate all electrochemical potential as heat by moving 'down' (b), whereas in (I), by going 'down' (c), some of this potential would be conserved as chemical energy in ATP, though heat would again also be produced. Horowitz *et al.* devised the bond graph for both of these mechanisms (which need not eventually be mutually exclusive)—that for mechanism (I) is shown in Fig. 3.13(b) (Fig. 3.13(a) represents schematically the spatial features of the proposed mechanism). It contains an additional port for the power associated with the flow of ATP and ADP + inorganic phosphate. Whether over pathway (b) or (c), proton flow is driven by the difference in electrochemical potential

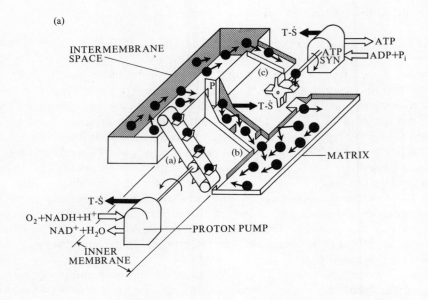

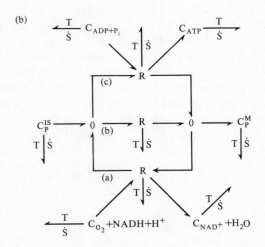

Fig. 3.12. (a) Diagram of proton flow across the inner membrane of a mitochondrion. (b) Bond graph for the proton flow shown in (a). Superscript M denotes matrix and IS denotes intermembrane space. Subscript P denotes proton flux. (From Horowitz *et al.*, 1979, Fig. 5.)

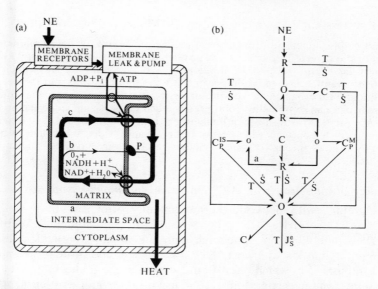

Fig. 3.13. Bond graphs for power flow in brown fat cells. A diagram of the pathway involving the plasma membrane pump is depicted in (a) and the associated bond graph is shown in (b). There are many mitochondria as well as other organelles within each cell, and this diagram is not to scale. Moreover, the outer membrane is a smooth continuous membrane while the inner membrane, although continuous, has many folds. NE = norepinephrine, a neurotransmitter. (From Horowitz *et al.* 1979, Fig. 5.)

between protons in capacitances C_p^{IS} and C_p^M (IS = intermembrane space; M = matrix). Combination of the bond graph for either mechanism (I) or (II) with those for the Na^+/K^+ pump and for the overall energy conversion in the nerve and brown fat cell would be very complex. The problem was justifiably simplified by Horowitz *et al.* (1979) by assuming that the release and uptake of the neurotransmitter, norepinephrine (NE), involves relatively little energy; that all of the energy lost to the system through chemical reaction and ion translocation is converted directly to heat (i.e. no turbulence, etc.); and by making certain other plausible assumptions (cf. Horowitz and Plant 1978). This simplified bond graph for the whole system, incorporating mechanism (I), allowed its simulation by the ENPORT computer program. The heat developed in the system could be calculated by summing the power over the ports of the resistances which dissipate power as heat. The calculation showed enhanced rates (a six-fold increase) of energy conversion to heat as membrane permeability was lowered and the ion leak increased. The bond-graph model also simulated other aspects of the regulation of heat production and of potassium and sodium ion levels. These authors conclude:

> Bond graphs appear to provide an attractive procedure for describing interaction of the diverse elements—from protons to chemical reactions—involved in brown fat thermogenesis. In developing a bond graph for a physiological system, attention is focused on power flows through networks and the need for additional experiments in specific areas becomes clear... For the complex system one can draw many bond graphs, each emphasizing some facet of the whole system. An advantage of bond graphs is that one does not have to remain committed to a particular model, but rather as new evidence becomes available on power flows through a system, models can be easily modified. (Horowitz *et al.* 1979, p. 294)

Coupling of ionic transport, ATP reactions, and protein synthesis

Bond graphs, with network thermodynamic formalism, have similarly been found useful by Atlan, Panet, Sidoroff, Salomon, and Weisbuch (1979) in modelling the functional coupling in rabbit reticulocytes between potassium ion transport, ATP metabolism, and protein (haemoglobin) synthesis, for which they found evidence from the effects of two known potassium carriers on ATP concentration and protein synthesis (^{14}C-leucine incorporation), as well as on the rate of efflux of potassium ions. The bond graphs were built up stage by stage. First the coupling of ATP metabolism and protein synthesis was depicted by the bond graph in Fig. 3.14, which represents ATP as the energy source of protein synthesis in the vertical energy bonds linking the horizontal top sequence (for ATP breakdown, R_T) through a transducer to the lower sequence (for amino acids ($aa_1 \ldots aa_{20}$) being incorporated, R_P, into protein (prot)). Since the ATP that is directly linked into protein synthesis becomes AMP and inorganic phosphate (P_i), these may also be regarded as products of the reaction (R_T) producing the protein—hence the diagonal energy bonds from the lower sequence to the capacitances representing these products. The large stoichiometric coefficients written alongside the transducers

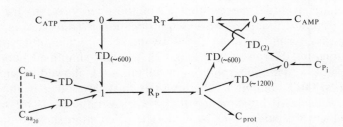

Fig. 3.14. Bond-graph representation of a direct (stoichiometric) coupling between a simplified representation of ATP synthesis and breakdown to AMP and inorganic phosphate, P_i, and protein synthesis. (From Atlan *et al.* 1979, Fig. 3.)

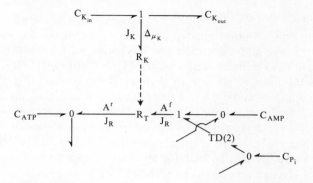

Fig. 3.15. Coupling via a signal bond between K^+ efflux through the membrane (represented in a simplified way as a diffusion process) and ATP synthesis and breakdown (represented in a simplified way as a global reaction of double phosphorylation of AMP (AMP + $2P_i$ ⇌ ATP)). (From Atlan et al. 1979, Fig. 5.)

take account of the fact that 600 molecules of ATP and of AMP, and 1200 of P_i, are involved in the formation of each haemoglobin molecule. The coupling of potassium transport with ATP synthesis was represented, in a second stage of building up the bond graph, Fig. 3.15, as an informational coupling (indicated by a vertical dashed 'signal bond')—an indirect effect of changes in the membrane permeability to K^+ on ATP synthesis. The resulting bond graph, Fig. 3.15, can then be fitted on to the previous one, Fig. 3.14, since the lower part of the former forms the upper part of the latter. All of the processes depicted in these two bond graphs, taken together, have been grossly simplified (e.g. no intermediate steps have been given in the linking of membrane permeability to K^+, the possible role of a phosphorylated enzyme in protein synthesis has been omitted, K^+ transport through the membrane has been represented, in Fig. 3.15, as if it were a diffusion process, etc.). Even so, this representation helped the authors to visualize at a phenomenological level the indirect effects of the two K^+ carriers on protein synthesis and to formulate kinetic equations for their action. The bond graphs thus helped them to analyse the experimental observations, to raise additional hypotheses and to suggest experimental verification of their proposed links through new correlations. One of the advantages of the bond graph method that they report is the finding that it can be used at different levels of integration. The oversimplifications, mentioned above, can be eliminated, and then additional parameters and the relationships between them would be implicit in the new elaborated bond graph—and this again would suggest new experiments.

Skeletal muscle glucose metabolism

Perhaps the most ambitious application of network thermodynamics allied to the bond-graph notation that has so far been attempted is that of Hunter, Peura, Crushberg, and Harvey (1979) to model and then simulate the metabolism of glucose in skeletal muscle—including the insulin-modulated, carrier-mediated transport of glucose into skeletal muscle, the biochemical reactions of the Embden–Meyerhof pathway, the synthesis and degradation of glycogen controlled by protein kinase, itself dependent on cyclic $3',5'$-adenosine monophosphate, and the diffusion of lactate from muscle. So the chosen field for analysis comprised the whole cycle of reactions of glycogen metabolism initiated by adenyl cyclase and proceeding to the formation of glucose 1-phosphate and the conversion of glucose to lactate during glycolysis. The model, depicted as a bond graph, was based on a thermodynamic analysis of the energy transformations in the biochemical reactions coupled with the dynamics associated with an energy-storage capacity. The resistive, or energy-dissipating, elements correspond to the biochemical reactions, and membrane phenomena, and the capacitative, or energy-storing, elements result from the potential energy associated with storage of metabolites. These elements were identified in relation to the three open compartments of the system (muscle, membrane, and extracellular fluid) and the energy flows between them. An energy flow graph was first devised that was virtually the bond graph for these three 'compartments' and their energy exchanges and this was expanded on the assumption that the electrostatic potential of each compartment remains constant and the capacitative storage element was replaced by an effort source (for the other simplifying assumptions see Hunter *et al.* 1979, pp. 502, 503, 508). The resulting bond graph (Hunter *et al.* 1979, Fig. 3) is of formidable complexity, displaying as it does all the kinds of transformations and displacements of energy in two major interlocking metabolic networks (those of glycolysis, with ancillary reactions, and of glycogen synthesis and degradation). The constitutive equation for each biochemical reaction, as a dissipating resistance, was developed from experimental information in the literature (dissociation constants, Michaelis coefficients, forward and reverse rate constants). These non-linear algebraic equations were derived using equilibrium and steady-state enzyme kinetics. Constitutive equations for the capacitative storage elements were based upon the assumption of thermodynamically 'ideal' metabolites, in order to give values to the pertinent chemical potentials needed to calculate the capacitances (and also affinities for reactions). By the procedures already outlined, the information about the constitutive relations and the actual numerical values of the parameters were incorporated into a computer program (Digital PDP-10 system) and the stepped response of the whole system to plasma epinephrine was

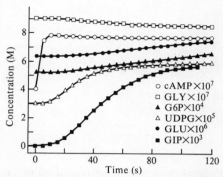

Fig. 3.16. Transient response of the theoretical model of Hunter *et al.* (1979) for skeletal muscle glucose metabolism to a change in plasma epinephrine concentration. The initial epinephrine concentration was 273 nM. cAMP: cyclic adenosine monophosphate. GLY: glycogen. G6P: glucose 6-phosphate. UDPG: uridine diphosphoglucose. GLU: glucose. G1P: glucose 1-phosphate. (From Hunter *et al.* 1979, Fig. 4.)

simulated. The simulated output included the concentration and chemical potential of each metabolite, the velocity and entropy production of each chemical reaction and the total entropy production of the system. The effects of changes in epinephrine on selected metabolite concentrations and on the 3′,5′-adenosine monophosphate-dependent protein kinase system are shown in Figs. 3.16 and 3.17. The simulated resulted results

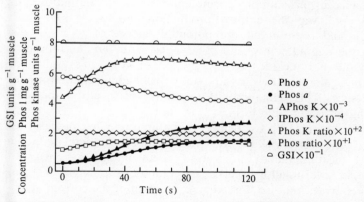

Fig. 3.17. Simulated response to a change in epinephrine concentration of the cAMP-dependent control system in the model of Hunter *et al.* (1979) for skeletal muscle glucose metabolism. Epinephrine concentrations as in Fig. 3.16. Phos *b*: phosphorylase *b*. Phos *a*: phosphorylase *a*. A Phos K: Active phosphorylase kinase. I Phos K: Inactive phosphorylase kinase. Phos ratio: (Phos *a*)/(Phos *b*). Phos K ratio: (A Phos K)/(I Phos K). GSI: glycogen synthase-independent. (From Hunter *et al.* 1979, Fig. 5.)

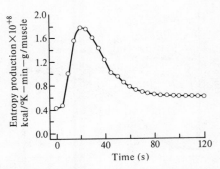

Fig. 3.18. Calculated transient entropy production resulting from the disturbance in plasma epinephrine concentration given in Fig. 3.16 relating to the theoretical model of Hunter *et al.* (1979) for skeletal muscle glucose metabolism. (From Hunter *et al.* 1979, Fig. 6.)

agreed well both qualitatively and quantitatively with *in vivo* measurements where these were available in the literature—and at the same time predicted many changes not previously examined properly and raised new questions concerning the location and effective concentration of many of the enzymes.

One particularly interesting result, in relation to the ideas of the Brussels school concerning enhanced dissipation immediately following an instability in a dissipative system (section 2.7), is illustrated in Fig. 3.18 which shows the changes in entropy production as a result of the disturbance in plasma epinephrine concentration. The major component of this peak was the entropy production due to the phosphorylase *a* reaction and represented the potential energy of glycogen dissipated during the degradation of glucose 1-phosphate. The entropy production (which did not include the protein kinase system) eventually became constant again in a new steady state that had resulted from the input disturbance. This dying out of the transient is also indicated by the levelling off observable in Figs. 3.16 and 3.17.

Hunter *et al.* (1979) conclude that this modelling process, using the bond graphs of network thermodynamics, not only stimulated the acquisition of more detailed information about unknown constants of the system, in order to allow a successful computer simulation of the response to epinephrine, but also 'forced an evaluation of the current-state-of-the-art and an extension of existing theory'.

Starch gelatinization

A quite different application of network thermodynamics has been made by Blanshard (1978, 1979) to the problem of unravelling the interplay of the various processes that occur when water diffuses into a starch granule.

The starch granule is a semicrystalline polymer spherulite exhibiting both X-ray crystallinity and birefringence (which reflect intra- and intermolecular order). On heating the granules in water a temperature is reached at which the molecular ordering is lost and the granules swell. The process is termed gelatinization and in practice occurs over a temperature range.

Earlier research had largely concentrated on investigating this phase transition under essentially equilibrium conditions. These processes, however, are almost universally non-equilibrium in character, so it became desirable to formulate a model of starch gelatinization using a non-equilibrium approach and network thermodynamics appeared appropriate for this purpose.

By carefully considering the energy transactions that were occurring, Blanshard was able to identify three constituent, concomitant processes and to represent them in a bond graph (Fig. 3.19). They are: (i) diffusion

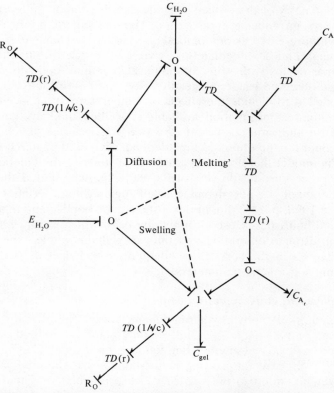

Fig. 3.19. Bond-graph representation of Blanshard (1978) of the process of starch gelatinization.

of the water into the starch granule or crystallites; (ii) a helix–coil transition that is facilitated by hydration and is accompanied by disaggregation of the helices in the crystallites, amounting to an apparent 'melting'; and (iii) swelling of the granule, a kind of diffusion. The swelling may clearly also modify the diffusion of water into the granule crystallites; so any model must permit such feedback. The bond graph shown in Fig. 3.19 was devised to represent these mutually interlocking processes.

The diffusion of water into the granule is described by a generalized resistance and the effects of temperature and concentration upon the rate of diffusion are included by incorporating appropriate transducers. E_{H_2O} represents the chemical potential of the water which is the essential driving force in the system. The actual loss of order in the crystallites within the granule is formulated as a chemical reaction represented as the discharge of two generalized capacitances (the water content of the granule and its chemical potential (C_{H_2O}), and the starch crystallites (C_x)). The stoichiometry of the process and the effect of temperature are again modelled by introducing transducers. The immediate product of the reaction is represented as randomized polysaccharide chains (C_r), but this charged capacitance immediately discharges through the interaction of the polymer chains with water at chemical potential E_{H_2O}, to yield the swollen gel observed inside the gelatinized granules. This latter process is again modelled to include the effects of temperature and concentration.

On the limited information available it seemed unlikely that known values of the diffusion coefficient of water in gelatinized starch systems could account for the observed time dependency noted in starch gelatinization. The model therefore suggested that a study of the kinetics of the 'melting' process would prove worthwhile. This necessitated the design and development of a new dynamic, small angle light scattering technique (Blanshard 1979, p. 145). So, in this instance, the bond-graph representation proved heuristically useful in making plain which variables it would be most fruitful to determine and thus shaped the whole experimental approach, even in the absence of adequate knowledge of all the thermodynamic and kinetic parameters.

Other topological graph representations

The originators of network thermodynamics were quite aware that the bond-graph representation was not the only graphical method of representing power flows and energy transactions in a system. Thus in the last lecture he ever gave (at Göttingen in May 1972), Aharon Katchalsky-Katzir drew attention to the way in which a reticulated system may be comprehended by topological analysis and cited the famous eighteenth century *bridge problem* of Leonhard Euler (see Neumann (1979) for an account of this lecture). In the city of Konigsberg, the river divides into two arms with an island placed at the centre of the bifurcation zone; various bridges interconnect the three banks of the river at the branching point and the island. It was a popular puzzle to ask whether it is possible to cross, in a *continuous* walk, all the bridges without using a bridge twice. In Euler's time, there were seven bridges and a topological representation of Euler (letting the three distinctive *areas* of land involved be represented by points, joined by lines representing the bridges) easily showed that such a walk was impossible. But, later, when an eighth (and later still, when a ninth) bridge was added, the topological representation easily shows it is possible to make such an 'Euler walk' (see Katchalsky's lecture, Fig. 2, reported by Neumann). This power of topological representation is quite general for reticulated systems, and *linear graph theory*, for example, can be usefully applied (see Oster *et al.* 1973, Section II) to both mechanical systems and to the membrane problems we discussed earlier (section 3.3). However this particular kind of representation became cumbersome when applied to systems involving coupled flows, i.e. energy transductions (Oster *et al.* 1973, p 32) and bond graphs have become the preferred method of representation in this context. However, other methods are used: in particular, Mikulecky and his colleagues have used a representation of Peusner (1970) who developed a means by which electrical network diagrams could be used to represent the flow–force equations of non-equilibrium thermodynamics (Mikulecky, Wiegand, and Shiner 1977; Mikulecky 1977; Mikulecky and Thomas 1979). They have developed his methods and extended them to non-linear ranges and in fact their technique resembles the equivalent circuit approach widely used to represent ion flows in cells and membranes (Finkelstein and Mauro 1963), although it is also applicable to coupling in a multi-component system. Mikulecky and his colleagues have applied their topological technique to sodium ion uptake in epidermis (Mikulecky, Huf, and Thomas 1979) and to volume flows through epithelial membranes (Mikulecky and Thomas 1978). They claim that their representation is easier for biologists to understand being closely related to electrical circuitry which is familiar to most investigators, but do not deny the attractive features and power of the bond-graph representation once one has mastered the symbolism.

3.8 Conclusion

A. S. Perelson concluded his 'overview' of network thermodynamics (presented in the special issue of the *Biophysical Journal* of July 1975, on 'Thermodynamics of Living Systems' and dedicated to Aharon Katchalsky) somewhat prophetically as follows:

> Let me end by mentioning an area in which I foresee future growth of network thermodynamics. The goal of biophysics is to understand how biological systems work. Traditionally we have approached this problem through reductionist analyses. However, when we have isolated every enzyme and catalogued every reaction that occurs in a cell will we understand how the system works? I think not, for there are complex dynamic interactions that impart to matter the property that we call life. However, if we can design and synthesize systems which have these dynamic characteristics we will have made significant progress towards understanding them. Engineers have enormous experience in synthesis and design, and it is my hope that through network thermodynamics, these techniques can be applied to synthesize chemical networks with prescribed behaviors.
>
> Network thermodynamics is based upon an idea of great simplicity—that the logical foundation of finite-dimensional thermodynamic models is formally identical to that of network theory. It was Aharon's hope that this similarity could be exploited to enrich thermodynamics and to provide a practical tool for the experimental biophysicist. (Perelson 1975, pp. 683–4.)

Yet, four years later, Mikulecky and Thomas (1979, p. 322) could express the view that 'the use of network simulation in biology is only beginning'. For, they argue, the network approach (with bond graphs or Peusner's method) is more than merely another novel kind of simulation, for in it the relation of the whole to its parts is the central consideration. If such an approach were linked with appropriate computer programs incorporating a more biochemically and physiologically pertinent 'language', then these workers think 'the possibilities are awesome' for 'as the possibility of simple, fast simulation becomes more widely known, the impact of network thermodynamics will be immense'.

It cannot be affirmed of the decade or so since it was first made explicitly relevant to biological problems that the impact of network thermodynamics can be said to have been very great, but the spate of papers in 1979 suggests that perhaps it may become so, especially as computerization becomes more accessible. Let the last words be those of Aharon Katchalsky, the chief founder of network thermodynamics in the bond-graph representation:

> The main importance of the network approach to bio-thermodynamics is of a conceptual nature. Network thermodynamics serves as a bridge between classical thermodynamics and the general dynamic theory of modern physics, and allows the introduction of thermodynamic concepts into the systems approach devoted to the fundamental problems of biological organization. (Aharon Katchalsky in 1972, quoted by Neumann 1979.)

Part II Kinetic interpretations of living systems

4 The kinetics of biological self-organization

4.1 Chemical kinetics and living systems

IT is both the strength and the weakness of thermodynamic interpretations of physico-chemical processes and systems that they are phenomenological and macroscopic, pointing directions, circumscribing possibilities, and indicating potentialities of evolutionary change but unable to provide the basis for postulating the detailed nature of these changes in terms of molecular organization and transformations. Chemical kinetics, on the other hand, however abstract and formal its mathematical formulations may become, always develops its proposals and predictions in very close relation to the stability and transformation of actual molecules. In even the most formalized treatments, the principle parameters are *concentrations of molecules* as a function of *time*, and for this reason it has a more direct and immediate rapport with the way of thinking of the biochemist and biologist concerned with the evolution in time of actual biological structures, whether at the molecular or a higher level. This is not to belittle the power and the scope, usually underestimated by experimental biological scientists, of thermodynamic concepts—as I hope Part I has, however inadequately, served to demonstrate—but it does mean that any attempt at physico-chemical interpretations of biological complexity must explore the possibilities afforded by our present knowledge of the kinetics of chemical change. This knowledge was established in its broad outlines in the first four decades of this century in that great 'classical' period of the development of physical chemistry into one of the most intellectually elegant and at the same time most practically relevant aspects of chemistry as a whole (never better outlined as an intellectual adventure than in *The structure of physical chemistry* by C. N. Hinshelwood (1951)). The development of an understanding of the kinetics of chemical processes was a major component in this emergence of physical chemistry as a distinctive area of science characterized by its special fusion of the fundamental concepts and formulations of physics with the knowledge of molecular structures and their transformations that organic and inorganic chemists were achieving.

In the second part of this volume, we shall therefore seek to discern the broad outline of how the concepts of chemical kinetics, and ways of thinking about biological systems derived from such concepts, have contributed to our understanding of living systems and are still doing so. This account will be confined mainly to picking out salient features in the publications of the last decade or so, a period that has witnessed an almost explosive expansion in this area. Behind the more recent applications of chemical kinetics to biological systems lie important theoretical developments in the physical understanding of fluctuations in physical systems—a development closely linked to the progress in the thermodynamics of systems far from equilibrium that is described in Chapter 2—and even in developments in mathematics itself, in bifurcation theory, stability analysis, and stochastic theory. Comprehensive accounts of these theoretical developments are increasingly available (see references in section 4.3 below), so that, although they are highly pertinent to our theme, only a broadly descriptive account of the relevance of their results will be attempted here.

As I have already intimated, the application of the ideas of chemical kinetics to biological systems is pre-eminently a twentieth-century development and confined to the last few decades, after a basis had been established for the kinetics of chemical change as such. However, significant pioneering work in the formulation of ideas about kinetic processes of significance to biology has occurred in every decade of this century. So, before an attempt is made to assess the present situation, it is pertinent, and it may be didactically helpful, to recall the pioneering efforts of those who blazed this particular trail—frequently either misunderstood by their contemporaries in the biological sciences, or ignored, if not actually disdained, by those in the physical sciences. Those efforts will be the main theme of the immediately following section (4.2). However to obtain an accurate perspective, it is worth noting that an understanding of the rates and organization of biological processes at the molecular level has been a long-cherished dream amongst biological scientists, even those who have had no overly reductionist intentions. For it is clear, even to the most holistic biologist, that the atoms and molecules in biological organisms must obey the laws of physics and chemistry, whatever else they may wish to affirm—and since antivitalists do not necessarily have to be reductionists, there will no doubt be much else to be affirmed. Even so, the possibility of life must be at least based on the properties of the molecules that occur in living systems, and that includes their potentialities for chemical transformation and the rates thereof.

It is intriguing to note how, in each century preceding the twentieth, there have been those who, in the language of their day, have been concerned with this issue. Thus, at the end of the seventeenth century John Ray, the greatest English biologist before Darwin, in developing a 'natural theology' in his influential work of 1691, felt it necessary nevertheless to recognize a 'plastic virtue' inherent in material bodies whereby they became living, growing, and adapting. Admittedly in Ray's thought this plastic nature was a kind of impersonal force upholding the natural order and devoted to the rule of law (and itself a secondary instrument of divine agency). His hypothesis did not solve the problem but its invention illustrates the recognition by one of the founders of biology of *both* the material basis of living organisms *and* their distinctive properties. More than a century later Erasmus Darwin, the grandfather of Charles, could write in his *Zoonomia* (1794–6), somewhat prophetically we would now have to admit, of the 'one living filament' from which all warm-blooded animals had derived:

> From thus meditating on the great similarity of the structure of the warm-blooded animals, and at the same time of the great changes they undergo both before and after their nativity; and by considering in how minute proportion of time many of the changes of animals described have been produced; would it be too bold to imagine, that in the great length of time, since the earth began to exist, perhaps millions of ages before the commencement of the history of mankind, would it be too bold to imagine, that all warm-blooded animals have arisen from one living filament, which THE GREAT FIRST CAUSE endued with animality, with the power of acquiring new parts attended with new propensities, directed by irritations, sensations, volitions, and associations; and thus possessing the faculty of continuing to improve by its own inherent activity, and of delivering down those improvements by generation to its posterity, world without end? (*Zoonomia*, 4th edn, Philadelphia, 1818, Vol. I, 397)

He then went on to suggest that all living forms, and not only warm-blooded animals, had been derived from this single, original 'filament'. For Erasmus Darwin 'generation was a process of organic transformation produced by the interaction between matter possessing certain propensities and the forces which acted upon it from within and without' (Greene 1959, p. 167).

Some sixty years later the German pathologist Virchow to whose insights we have already had cause to refer, foresaw the paths that would lead from the biology of his day to the molecular biology of ours in a lecture on *Die Mechanische Auffassung des Lebens* delivered in 1858:

> *Life is cell activity; its uniqueness is the uniqueness of the cell.* The cell is a concrete structure composed of definite chemical substances and built up according to a definite law. Its activity varies with the material which forms it and which it contains; its function changes, waxes and wanes, originates and disappears, with the alteration, accumulation, and diminution of this material. However, in its elements this material is no different from that of the inorganic, the non-living, world from which, rather, it continually replenishes itself, and into which it sinks back again after it has fulfilled its special purposes. As particular, as peculiar, and as much interiorized as life is, so little is it withdrawn from the rule of chemical and physical law. Rather does every new step on the path of knowledge lead us nearer to an understanding of the chemical and physical processes on whose course life rests. Every peculiarity of life finds its explanation in peculiar arrangements of an anatomical or chemical character, in peculiar configurations of material whose universally occurring

properties and powers are expressed in these configurations, though apparently quite otherwise than in the inorganic world. (Virchow, pp. 106 and 107 in the 1959 English translation.)

Virchow's apparently reductionist faith that '*every* peculiarity of life' (my italics) will eventually 'find its explanation in peculiar arrangements of an anatomical or chemical character' may need some considerable qualification. Nevertheless the motive, that Virchow expresses, of pressing forward with inquiry into the molecular 'arrangements' in living organisms has undoubtedly been a most fruitful prescription, provided the 'peculiar' features of living organisms are not lost sight of. The implementation of such a strategy of research had, however, to wait upon the development of a more sophisticated structural and physical chemistry, as much as on a more precise biochemical and physiological understanding of living systems.

4.2 Pioneers of the kinetic approach

Living matter

One of the most sustained attempts to interpret the organization of living matter as an integrated and interlocking *kinetic* system of chemical processes was undoubtedly that of Sir Cyril Hinshelwood and a long succession of collaborators in the Physical Chemistry Laboratory at Oxford over the period from the late 1930s until his death in 1967 (the first papers on 'Physico-chemical aspects of bacterial growth' are those of Dagley and Hinshelwood 1938). The full scope of his approach is developed in *Growth, function, and regulation in bacterial cells* (Dean and Hinshelwood 1966) which he published in 1966 in collaboration with A. C. R. Dean as successor to an earlier volume of his on *The chemical kinetics of the bacterial cell* (Hinshelwood 1946). Hinshelwood was, of course, himself one of the pioneers in the study of chemical kinetics, particularly of reactions in the gaseous phase, for which he received the Nobel Prize in 1956 (see for example *Kinetics of chemical change* (Hinshelwood 1940)). His physico-chemical description of living matter, notable for its use of the terminology of chemical kinetics, has already been quoted (Section 1.2) but the style of his approach is perhaps better represented in his Faraday Lecture on 'Autosynthesis' to the Chemical Society of London in March, 1952:

> In the reproduction of a cell there is indeed copying, but degradation of energy also occurs, and highly complex series of reactions combine, as it were symphonically, to give among the total products certain substances of very low entropy. It is the interplay of all these processes which must make autosynthesis possible, not the replication of individual genes as such. ... The interplay presents kinetic problems of great interest, and the biological analogies which the results indicate, ... are highly suggestive. They are certainly of significance

though in just what ways is perhaps hardly yet agreed. In the meantime we may at least say that we are studying a new type of chemical reaction system. (Hinshelwood 1953, p. 1948.)

In that same lecture, he recapitulates a fundamental principle concerning autosynthesis by referring to a quite simple system of kinetic equations that represent the interlocking nature of the chemical processes that occur in living cells—the fact that the synthesis of one enzyme (X_1) is linked, through suitable intermediates, with the working of another substance (X_2) which may also, though not necessarily, be called 'enzymes'. It is straightforward to show (Hinshelwood and Lewis 1948) that, if the intermediate (c) between X_1 and X_2 remains at a steady concentration, then the rate of formation of X_2 is proportional to the concentration of X_1 (which determines the steady-state concentration of the intermediate between them).

Thus if we have

'Enzyme'   Intermediate   'Enzyme'

$$X_1 \xrightarrow{k_{x_1}} c \xrightarrow{k_{x_2}} X_2 \qquad (4.1)$$

where the symbols X_1, X_2, c (italic) represent also their respective concentrations,† then

$$\frac{dX_2}{dt} = v_{x_2} \cdot k_{x_2} \cdot c, \qquad (4.2)$$

where v_{x_2} is a stoichiometric factor. Now the steady-state constancy of the intermediate concentration gives

$$\frac{dc}{dt} = v_{x_1} \cdot k_{x_1} X_1 - k_{x_2} c = 0, \qquad (4.3)$$

if there are no losses of c by diffusion etc. (where v_{x_1} is also a stoichiometric factor). So

$$\frac{dX_2}{dt} = \beta X_1,$$

where

$$\beta = v_{x_1} v_{x_2} k_{x_1}.$$

Now if the 'enzyme' X_2 also contributes to the formation of X_1, whether directly or indirectly, so that (if the former) we can write

'Enzyme'   Intermediate   'enzyme'

$$X_2 \longrightarrow b \longrightarrow X_1$$

† Here and elsewhere, unless otherwise stated, roman letters represent substances, or molecules, whereas italic letters represent concentrations of substances or molecules.

and $db/dt = 0$ in the steady state, we have correspondingly

$$\frac{dX_1}{dt} = \alpha X_2. \qquad (4.4)$$

The concentrations of these two substances X_1 and X_2 each of which is formed at a rate determined by the other, can be deduced (Hinshelwood and Lewis 1947; Hinshelwood 1952) as functions of the initial concentrations of X_1 and X_2, i.e. $(X_1)_0$ and $(X_2)_0$ and of α and β and the time (t). When t is large,

$$\frac{X_1}{X_2} = \left(\frac{\alpha}{\beta}\right)^{\frac{1}{2}} = \frac{\alpha}{k} = \frac{k}{\beta} \qquad (4.5)$$

writing $k^2 = \alpha\beta$ (see Hinshelwood and Lewis 1947, p. 321; Caldwell and Hinshelwood 1950). The rates of formation of both X_1 and X_2 now appear as autocatalytic, for

$$\frac{dX_1}{dt} = kX_1; \qquad \frac{dX_2}{dt} = kX_2. \qquad (4.6)$$

Or:

$$X_1 = (X_1)_0 \exp(kt); \qquad X_2 = (X_2)_0 \exp(kt). \qquad (4.7)$$

Thus each mutually linked constituent, X_1 and X_2, appears to be 'autosynthetic', or self-duplicating. Of course, one of X_1 and X_2 could be ribonucleic (RNA) and the other a protein, in which case the mutual regulation of protein synthesis by RNA and of RNA by enzymes (proteins) would be seen to lead inevitably to an autocatalytic exponential rate law for the growth of the whole system (cf. Caldwell and Hinshelwood 1950). This result was shown to be generalizable to any number of constituents formed in such a way that the increase of each depends on the catalytic action of the others, with the cycle closing at the last member back to the first. Thus, if

$$\frac{dX_1}{dt} = \alpha_1 X_2; \quad \frac{dX_2}{dt} = \alpha_2 X_3; \quad \ldots \frac{dX_i}{dt} = \alpha_i X_{i+1} \ldots \frac{dX_n}{dt} = \alpha_n X_1, \quad (4.8)$$

it can be proved (Hinshelwood 1952) that the autocatalytic law (4.6) prevails as t becomes large, with an autocatalytic constant $k = (\pi_i \alpha_i)^{\frac{1}{n}}$, and the relative proportions of the X_i settling to fixed ratios which are functions of the α_i.

If now, a portion of this system in which such constant ratios are established (e.g. eqn (4.5) for the two-constituent case) is separated and becomes the starting point of a new autosynthetic system (as when bacteria are subcultured) then the new initial concentrations $(X_1)_0$ and

$(X_2)_0$ are automatically in the ratio given by (4.5), i.e.

$$(X_1)_0/(X_2)_0 = \alpha/k, \tag{4.9}$$

and the growth rates of X_1 and X_2 would at once settle to the 'autocatalytic' values given again by (4.6).

Usually, however, the ratios of X_1 and X_2 (and other interlinked substances in the more-than-two-constituent case) are far removed from the values satisfying (4.5) (or the corresponding equations involving the α_i for the multiple case), and there will be periodic variations in the relative concentration of X_1 and X_2, or of the X_i in the multiple case (Hinshelwood 1952). When the number of interlocking constituents is large, as it is in actual cases of, say, the subculturing of bacteria, the amplitude of these fluctuations will be small compared with other purely exponential terms and so may not be readily detectable in the growth rate of the *whole* system (i.e. the bacterial cell). Nevertheless this would account for the long lags and irregular bursts of growth and arrests that are frequently observed when a bacterial culture is transferred to a completely different growth medium (Hinshelwood 1952, p. 748).

More significantly, Hinshelwood (1953) was able to show how an autosynthetic system would automatically adjust the proportion of its components in the presence of agents, or under conditions, that interfere with the interlinking of the X_i. Thus suppose we have a system such as (4.1) in which an enzyme X_1 produces a substance used in its formation by a second enzyme X_2 but with the additional feature that the intermediate (c) is diffusable. Let c still represent the concentration of the intermediate, but let $\mathbf{X}_1$ and $\mathbf{X}_2$ represent the *total amounts* of X_1 and X_2, respectively, contained in n cells. Furthermore, let X_2 be the component upon the formation of which cell division awaits, so that $n = \beta \mathbf{X}_2$, where β is a constant. From the conditions for the steady state, it follows that

$$\frac{dc}{dt} = A\left(\frac{\mathbf{X}_1}{n}\right) - B\left(\frac{\mathbf{X}_2}{n}\right)c - Cc = 0, \tag{4.10}$$

where A, B, and C are constants—the first term representing formation of intermediate by X_1, the second its consumption by X_2, and the third loss by other means, e.g. diffusion. From (4.10),

$$c = \alpha \mathbf{X}_1/\mathbf{X}_2 = \alpha\phi, \tag{4.11}$$

where α is another constant, and $\phi = \mathbf{X}_1/\mathbf{X}_2$.

If X_1 and X_2 are autosynthetic systems (no doubt for the reasons already outlined above) then we can write

$$d\mathbf{X}_1/dt = k\mathbf{X}_1 \tag{4.12a}$$

$$\frac{d\mathbf{X}_2}{dt} = k_2\mathbf{X}_2 c = k_2\alpha\mathbf{X}_1 = k_2\alpha\phi\mathbf{X}_2. \tag{4.12b}$$

From these equations, it follows, by differentiating $\mathbf{X}_1/\mathbf{X}_2$, that

$$\frac{d\phi}{dt} = \phi(k - k_2\alpha\phi), \tag{4.13}$$

and, at the steady state, when $d\phi/dt = 0$, ϕ has reached the value given by

$$k = k_2\alpha\phi. \quad \text{[steady state]} \tag{4.14}$$

Under these conditions, the cell growth rate is, by (4.12b), (4.14) and the definition of ϕ,

$$\frac{d(\ln n)}{dt} = \frac{d(\ln \mathbf{X}_2)}{dt} = k. \tag{4.15}$$

If now, c decreases, owing to the action of some toxic drug, or other cause, then $d\mathbf{X}_2/dt$, which is proportional to c (4.12b), decreases. The cell growth rate $dn/dt = \beta\, d\mathbf{X}_2/dt$ will also decrease. The immediate effect of the drop in c is also to decrease α to a new lower value α' so that $k > (k_2\alpha'\phi)$ in (4.13), and $d\phi/dt$ becomes positive instead of zero, i.e. $\mathbf{X}_1$ begins to increase relative to $\mathbf{X}_2$. It will do so until a new $\mathbf{X}_1'$ is reached for which again $k = k_2\alpha'(\mathbf{X}_1'/\mathbf{X}_2)$ so that (as for (4.14)), $d\phi'/dt$ is again zero, where $\phi' = \mathbf{X}_1'/\mathbf{X}_2$. Thus a new steady state $(d\phi'/dt = 0)$ establishes itself and the growth rate is then, using the last equality of (4.12b)

$$\frac{d(\ln n)}{dt} = \frac{d(\ln \mathbf{X}_2)}{dt} = \frac{1}{\mathbf{X}_2}\frac{d\mathbf{X}_2}{dt} = k_2\alpha'\phi' = k. \tag{4.16}$$

In other words, the amount of $\mathbf{X}_1$ has automatically increased, as a consequence of the reduction in c, to restore the cell growth rate to its original value.

This principle of automatic kinetic adjustment was much employed by Hinshelwood and his colleagues in the interpretation of the effects of bacteriostatic agents on the growth rates of bacteria and on their lag phases (see Hinshelwood (1946) and, particularly, Dean and Hinshelwood (1966) where the relation of such ideas to alternative interpretations, available in some cases, that are based on selection of pre-existing mutants, is discussed in detail). We need not follow these elaborations here but perhaps the foregoing will serve to illustrate the 'kinetic principles' of cell growth that Hinshelwood expounded thus (1953, p. 1950):

(1) Autosynthetic function is not inherent in any one cell component, but emerges from the harmonious cooperation of many, so the growth of the cell and the reproduction of its matter are very intimately linked with their own working, since it is this which in turn enables others to work and provide the synthetic materials for them. One could say

'enzyme' + metabolite 1 → more 'enzyme' + metabolite 2.

(2) As a consequence of (1), on continual growth the proportions of the various kinds of material present in a system will settle down to stable values—and will change if conditions of growth change to alter the velocity constants.

(3) Because living matter is organized in cells, and because cell division must come about at some critical value of the area/volume changes that accompany change of size, then it is a natural hypothesis that at the moment of division some important cell component(s) plays a key role and the division process begins when this key substance reaches some critical limit (*cf.* constancy of DNA per cell).

(4) From (1)–(3) it follows, by mathematical calculations with simple models, that in such systems those proportions of constituents will finally be established which lead to a maximum rate of growth in the actual environment.

This pioneering attempt at a kinetic interpretation of biological complexity has been summed up by Dean and Hinshelwood in what they refer to as the *network theorem*, thus:

> The growth of living matter depends upon closed cycles of mutually dependent interactions, as a result of which, in a constant environment, a steady state is reached where the proportions of the various constituents have settled down to constant values. These correspond to an optimum rate of increase, since any lack of the perfect balance achieved in the steady state leads to bottle-necks and rate-limiting steps. In a new environment the proportions modulate to new values again consistent with optimum growth in the changed circumstances.
>
> The reaction pattern is a branched one, with the formation of some components dependent on more than one other, and some themselves serving the needs of more than one other. There may also be alternative routes from given initial substances to the final cell components, and the relative extent to which the alternatives are used depends upon the medium in which growth occurs. Components such as enzymes on which large demands are made tend to increase in amount: others which are no longer useful diminish. (Dean and Hinshelwood 1966, p. 113.)

The network theorem, they suggest, reflects much of the plasticity and adaptability of living matter but by itself is incomplete insofar as it oversimplifies the nature of the mutual dependences, it has dealt only with total masses of components under supposedly constant conditions, it assumes a condition for cell division, and takes no account of the spatial relationships (at a number of levels) of the macromolecules in the cell. Their book should be consulted for the full working out of this 'network theorem' in the light of modern molecular biology.

Another major contribution to interpretations of biological organization in terms of chemical kinetics was that of H. Kacser (1957, 1960). He started from an awareness that living organisms have 'systemic properties' which depend on organization and functional relationships and so are not simply the result of an 'additive' combination of their constituent parts.

120 THE KINETICS OF BIOLOGICAL SELF-ORGANIZATION

> There are no concepts in chemistry or physics equivalent to genes, regulation, epigenesis, pleiotropy, pheno-copy, acquired character, etc., precisely because these are properties only possible in systems of greater complexity than have been subjected to detailed analysis by those sciences. (Kacser 1957, pp. 192–3.)

Nevertheless he sought to show that systems composed of 'orthodox inanimate components' could display properties normally associated with living organisms and considered that the appropriate method of description of the 'molecular calculus' of the processes occurring, say, in gene action, was that of chemical kinetics. However, even when a particular kinetic model is equatable with a genetic or physiological phenomenon, he would not have this taken as anything other than a *possible* mechanism, or working hypothesis, to be refuted or confirmed by subsequent experiment (for his general understanding of 'analogies' and 'models' see his 1960 article). Perhaps because of his location in a biological laboratory, the scope of Kacser's interpretative ambitions extends in his 1957 and 1960 articles to quite general properties of biological organisms, whereas Hinshelwood and his colleagues, working upwards as it were from physical chemistry, concerned themselves rather with the kinetic organization of the individual cell, as such. However, the characteristic features of specifically *living* matter which Kacser aimed to interpret were much the same as those discussed by Hinshelwood and his colleagues—and there is considerable convergence in their respective chosen models and the conclusions they draw—though, it would seem, there is no evidence of any direct exchange of ideas between them.

Kacser pointed out that a closed system of two molecular species that can be transformed into each other would, by the ordinary laws of chemical kinetics and equilibrium, have the properties both of approaching its equilibrium state, from an arbitrary initial composition, at rates depending on individual rate constants (and so on the presence of catalysts), and of attaining a final equilibrium composition independent of these same rate constants (and catalysts). This is summarized in Table 4.1.

However, the situation is quite different for an *open* system—open to the influx and efflux of matter. Kacser (1957) considered the system:

$$X \xrightarrow[\text{Constant flow in}]{k_X} A \underset{k_{BA}}{\overset{k_{AB}}{\rightleftarrows}} B \underset{k_{CB}}{\overset{k_{BC}}{\rightleftarrows}} C \underset{k_{DC}}{\overset{k_{CD}}{\rightleftarrows}} D \xrightarrow{k_Y} Y.$$

with $B \underset{k_{BE}}{\overset{k_{EB}}{\rightleftarrows}} E$

Boundary 1   Boundary 2   (4.17)

Again let symbols for components denote concentrations (when italicized) and assume X is sufficiently large that its transformation into A makes

Table 4.1. Closed system

$$\bar{N} = f(K, N_0)$$
$$\bar{N}/\sum \bar{N} = f(K)$$
$$\frac{dN}{dt} = f(K, N_0, k)$$

N = amount of each component (bar—at equilibrium; subscript zero—initially); k = rate constants; K = equilibrium constant.

only a negligible difference to its concentration (or its concentration is maintained steady by some other mechanism). So the rate of transformation of X into A has the constant value Q_X $(= k_X X)$. The rates of formation of A, etc. will be given by

$$\frac{dA}{dt} = k_X X - k_{AB} A + k_{BA} B$$
$$\frac{dB}{dt} = k_{AB} A + \ldots \text{etc.}$$
(4.18)

At the steady state this constant inflow of X will be equal to the flow out as Y, and the following conclusions could be drawn (summarized in Table 4.2).

(1) The expressions for $\bar{A}$, $\bar{B}$, etc. (where the bar denotes the steady state when dA/dt, (etc.) = 0) include rate constants (k) as well as equilibrium constants, (K), e.g.

$$\bar{A} = Q_X \left\{ \frac{K_{AB} K_{BC} K_{CD}}{k_Y} + \frac{K_{AB} K_{BC}}{k_{CD}} + \frac{K_{AB}}{k_{BC}} + \frac{1}{k_{AB}} \right\}$$
$$\bar{B} = Q_X \left\{ \frac{K_{BC} K_{CD}}{k_Y} + \frac{K_{BC}}{k_{CD}} + \frac{1}{k_{BC}} \right\}$$
$$\bar{C} = Q_X \left\{ \frac{K_{CD}}{k_Y} + \frac{1}{k_{CD}} \right\}$$
$$\bar{D} = Q_X \cdot \frac{1}{k_Y}$$
$$\bar{E} = K_{EB} \cdot \bar{B}.$$
(4.19)

Table 4.2. Open system

$$\bar{N} = f(Q, K, k)$$
$$\bar{N}/\bar{M} = f(K, k)$$
$$dN/dt = f(K, N_0, k)$$

So the absolute values of the rate constants determine the final proportions of the components. The influence of a change in the amount of any one catalyst (at constant Q_X) on the composition of the 'adult' system is of genetic interest. Consideration of the set of equations (4.19) shows that change in a given k has different effects on the various intermediates in the reaction sequence. Thus if it is the rate of interchange of B and C (i.e. k_{BC} and k_{CB}) that is affected by change of amount of catalyst, then $\bar{A}$, $\bar{B}$, and $\bar{E}$, all of which contain k_{BC}, but not $\bar{C}$ and $\bar{D}$ will be affected (K_{BC} is unchanged). So the distribution and number of effects consequent upon the change in a single enzyme depends on the location of that change—and on other kinetic factors relevant to the position of each effect. So the relation of these 'substitution' effects to the primary site of *gene* action may be even more complex and no simple conclusions concerning the relative proximity of effects and gene action are possible.

(2) The steady-state values, $\bar{A}$, $\bar{B}$, etc. are independent of the initial concentrations not only relatively but absolutely. The time taken for this steady state to be attained will vary but, as long as the ks are the same, the final state will be the same, and this is consistent with the steadiness of living systems which are also open. This property of 'equifinality' is, Kacser points out, paralleled in many biological regulation phenomena, e.g. the amount of cytoplasm relative to the nucleus does not affect the ability of early cleavage cells, of echinoderms and amphibia, to produce normal cells. Moreover, these kinetic considerations also show that if some part of any reactant is removed, the system will in due course revert to the *same* steady state—and this, Kacser suggests (1957, p. 204), must be compared with the power of regeneration of certain species (e.g. the planarian worm and hydroids) after amputation, and this may also be parallel to embryonic regulation.

(3) The flux rate (Q_X), which is regarded as constant in the system in question, enters all the expressions for $\bar{A}$, etc. only as a factor, so its value does not alter the *ratios* of $\bar{A}$ to $\bar{B}$ etc.: this is consistent with the invariance of composition of organisms, even when starved, when (within limits) they show remarkable changes in size, depending on the food supply, but not in composition. Moreover, although the absolute size of the adult system will depend on the flux rate, its composition, and all other properties dependent thereon, will be unaffected by its size.

Kacser (1957) instances a number of other features of biological systems which parallel the properties of the kinetic model (4.17) and (4.19), for example, *temperature effects* (smaller organisms and more rapid development at high temperatures, for non-thermostatic organisms, cf. rise of temperature increases the ks of (4.17) and (4.19) more than the Ks and has relatively little effect on the diffusion coefficients k_X and k_Y and thereby increases the rate of approach to a steady state in which all levels, except $\bar{D}$, are reduced); '*pleiotropy*' (multiple phenotypic variation with the same underlying genetic constitution, cf. the 'substitution' discussed under (1) above); *development abnormalities* yet with the same final, adult form (cf. in (4.17) and (4.19), alteration of k_{EB} and k_{BE} affects the intermediate values of E and the rate of formation of, say, C, but the final steady-state values of all components are unaltered); the passage of an organism through a *larval stage* as a temporary organization that is eventually 'sucked back' into the main channel of adult formation (cf. effects of changes of k_{EB} and k_{BE} just described, thinking of E now as a sub-system rather than as a single substance); *size effects*; *irreversible* (and so 'maternal') *effects*; and the *buffering capabilities* of both kinetic models and living organisms.

Like Hinshelwood and co-workers, Kacser also considers the kind of interlocking of reactions whereby the *formation* of one macromolecular enzyme, that acts on smaller molecules, is itself catalysed by another, whose formation may likewise be catalysed by yet another in a kind of 'hierarchy of catalysts', though he does not develop any detailed kinetic analysis of such a system. However, he does elaborate (Kacser 1960) an interesting kinetic model of a 'switch' mechanism that might operate in the emergence of functionally different parts which is such an outstanding aspect of development in biological organisms. He takes the sequence of catalysed reactions depicted in Fig. 4.1. Note that in these cycles C and N

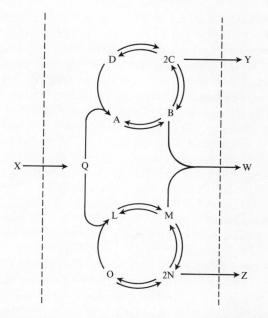

Fig. 4.1. System having two stable steady-state solutions. (From Kacser 1960, Fig. 6.)

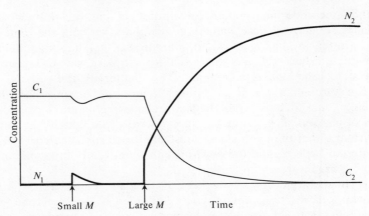

Fig. 4.2. Generation of alternative patterns indicated in Fig. 4.1. (From Kacser 1960, Fig. 7.)

are fed back (anticlockwise and clockwise directions to D and 0, respectively) to combine with Q to form A or L. Such loops, or cycles, are common in metabolism. Such a system can be shown to have two stable steady-state solutions: either (1) C is positive at a value C_1 and $N_1 = 0$; or (2) N is positive at a value of N_2, and $C_2 = 0$, i.e., *either* the upper *or* the lower loop (Fig. 4.1) proceeds, but not both. So either of the two reaction patterns can be established according to circumstances. Thus, in Fig. 4.2, the steady state with C finite and N zero is stable to addition of small quantities of M (or any other member of the lower sequence) but not to a larger amount of M above a critical value which, once it brings the N-producing cycle into operation, inhibits the other one. The system can switch from one synthetic pathway to another. This could provide the beginnings of an explanation of the formation of diverse cell patterns in embryonic development, for many reaction cycles have only one molecular species in common (e.g. Q in Fig. 4.1). If, in an egg cell, there was a diffusion gradient of, say component M, then the boundary at which it passes the critical concentration will be a boundary between two regions, one producing C and the other N (Fig. 4.3)—and it is not difficult to envisage how subsequent cell division might then lead to cell differentiation. It should be noted that this 'switch', between C- and N-regions, has been produced by continuously varying functions, namely reaction rates and diffusion.

Kacser did not develop further (in 1960) this theme of the interplay of chemical reaction rates and diffusion: this must be the concern of the next sub-section for it has subsequently transpired that this combination in particular provides intriguing clues for physico-chemical interpretations of biological complexity.

Oscillations and symmetry-breaking

The observation of periodicities, whether spatial or temporal or both, in chemical processes tends to come as a surprise and novelty to physical chemists who have become accustomed to thinking of the systems with which they deal evolving to steady states, usually equilibrium, of maximum disorder. The second law was thought to preclude any systematic and regular deviations from homogeneity, for example, in the form of sustained oscillations—although we have seen (Chapter 2) that more recent developments in thermodynamics lead to a qualification of this prohibition. One of the earliest purely chemical periodicities to be observed was the variegated colours of the famous Liesegang rings (1905) that are formed when a crystal of silver nitrate is placed in the centre of a thin layer of gelatin gel that contains potassium chromate. Within a few

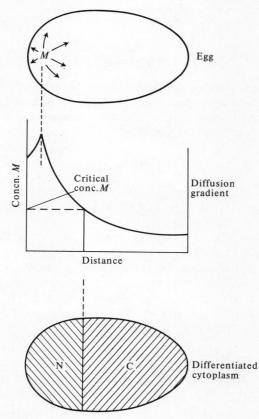

Fig. 4.3. Formation of diverse cell patterns at embryonic stage. (From Kacser 1960, Fig. 8.)

days, as a result of the outward diffusion of the silver nitrate, rings of silver chromate encircle the crystal. For a long time this remained a chemical curiosity but now, as we shall see in the next section, chemical and biochemical oscillations, periodicities in both space and time, have been increasingly observed and analysed, as well as biological oscillations. The founders of the study of such systems are, interestingly from the point of view of the historians of science, primarily those who investigated the theoretical *possibility* of such oscillations, long before experimentalists came across them by chance or, later, began to look explicitly for them.

The first theoretical model that predicts sustained oscillations appears to be that of A. J. Lotka (1910) who considered the series of consecutive reactions

$$A \xrightarrow{k_1} X$$

$$X + Y \xrightarrow{k_2} 2Y$$

$$Y \xrightarrow{k_3} E, \tag{4.20}$$

of which the second is autocatalytic. He assumed that the first process was slow compared with the establishment of equilibrium and that A† was constant (e.g. representing, say, a saturated vapour), so that the rate of the first process was also constant. Then

$$\frac{dX}{dt} = k_1 A - k_2 XY$$
$$\frac{dY}{dt} = k_2 XY - k_3 Y. \tag{4.21}$$

Although solutions of these equations present difficulties, the final stages of the process, near to a final steady state (in which $dX/dt = 0 = dY/dt$), can be deduced. It transpires that the differential equations for dx/dt and dy/dt (where x, y represent the deviations of the concentrations of X and Y, respectively, from the final steady state values $\bar{X} = k_3 k_2^{-1}$ and $\bar{Y} = k_1 A k_3^{-1}$) are second order and of the form associated with 'damped vibrations'. The reaction is periodic if

$$(k_1 A) k_2 < 4 k_3^2.$$

The sequence is: X increases; but increase in X causes a rapid rise in Y, which causes a decline in X; depletion of X ultimately causes a decline in Y (which is being removed all the time) and meanwhile A has recovered and increases and the whole cycle repeats itself. The whole sequence can

---

† See footnote on page 115 about the use of italics.

be strikingly demonstrated by a bead game, 'Struggle', devised by Eigen and Winkler (1981a, pp. 91–5) in which the cycle of concentration, or population, changes in A, X, and Y are represented by different coloured beads which are placed on an 8×8 squared playing board and which are added or removed from particular squares, according to rules simulating the reactions (4.23) below, and these particular squares are located by the simultaneous throw of two octahedral dice with numbered faces.

It is intriguing to record that it was the autocatalytic growth rate of *living* matter that suggested the above case to Lotka (1910, p. 274, *n.*1). Later in his classic work on the *Elements of physical biology* (1924, also in 1920a,b), he described in more detail the solution of similar equations in a form that also include 'autocatalytic' production of the first element (X) in the chain:

$$\frac{dX}{dt} = k_1 AX - k_2 XY$$

$$\frac{dY}{dt} = k_2 XY - k_3 Y, \qquad (4.22)$$

where A, k_1, k_2, k_3 are constants.

Lotka used a notation relevant to a problem in parasitology of relative host and parasite populations (1924, pp. 88 ff.) and to the increase in masses of a plant and of a herbivore which ate the plant (1920a). His results are shown in Fig. 4.4 for the case in which all the ks are constant (in Fig. 4.4, x_1 and x_2 represent the deviations of X and Y ($\equiv X_1$ and X_2 in Lotka's notation), from their steady-state values, namely k_3/k_2 and $(k_1 A)/k_2$, respectively, which are represented by the x_1, x_2 origin). Lotka's solution (1920a,b, 1924) is constituted by an infinite family of closed curves. When terms of higher degree in X and Y are added to equations (4.22), the solutions, plotted as in Fig. 4.4, are then a spiral winding into the steady state point, and so represent damped oscillations in X and Y. V. Volterra (1931) subsequently studied this predator–prey system in greater detail and so it is usually called the Lotka–Volterra model. For our present purposes it is pertinent to refer (like Glansdorff and Prigogine (1971, p. 225); cf. Prigogine and Balescu 1955) the equations (4.22) of this model, not to a predator–prey system but, as in Lotka's original 1910 paper, to a consecutive series of reactions involving now *two* autocatalytic steps ((a) and (b), below; not one as in (4.20)), thus:

$$A + X \xrightarrow{k_1} 2X \qquad (4.23a)$$

$$X + Y \xrightarrow{k_2} 2Y \qquad (4.23b)$$

$$Y \xrightarrow{k_3} E. \qquad (4.23c)$$

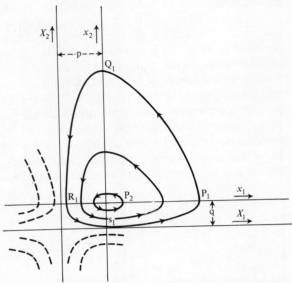

Fig. 4.4. Course of parasitic invasion of insect species, according to Lotka. (From Lotka 1924, Fig. 13 and Lotka 1920a, Fig. 1.)

The system is an open one in contact with reservoirs of A, B, and E so that their concentrations remain constant over the time-scale considered and the two variables are the concentrations X and Y. All reactions are written as if the affinities are very large so that the processes (a), (b), and (c) are virtually irreversible. The rate equations for such a reaction system are then indeed those of (4.22), with solutions as represented in Fig. 4.4. Each member of the (infinite) family of closed curves corresponds to a particular frequency and amplitude of oscillation about the *steady state*, not about equilibrium. Which trajectory is adopted depends on the initial conditions. The Lotka–Volterra model has a continuous spectrum of frequencies associated with this infinity of periodic trajectories and therefore fluctuations do not decay with restoration of the *same* trajectory (i.e. a particular cycle of variation of X and Y with the same relevant parameters). For consequent upon any infinitesimal fluctuation, the system will continuously change to trajectories with another frequency and there will be no 'preferred' trajectory—although the *average* concentration of X and Y over an arbitrary trajectory is always equal to the steady-state values (Volterra 1931). Such systems are not strictly 'structurally stable' (when its trajectories in concentration-space would *not* be affected by small disturbances) and so cannot be models for stable chemical oscillations observed in the laboratory, or stable biological oscillations. Because the system has a 'constant of motion', given by

$$k_2(X+Y) - k_3 \ln X - k_1 A \ln Y = \text{constant}, \tag{4.24}$$

it is called *conservative*. (For a fuller analysis of this system including its stability in terms of $\delta_X P$ (cf. Chapter 2), see also Glansdorff and Prigogine (1971, pp. 225 ff.), Nicolis and Prigogine (1977, pp. 160 ff., 264 ff.), Lefever, Nicolis and Prigogine (1967), and Nicolis (1971, pp. 247 ff.)). Similar 'conservative oscillations' have been postulated to describe the control of protein synthesis (Goodwin 1963, 1965) and the time-dependent properties of neural networks (Cowan 1968).

The kinetics of open reaction systems, into and from which material flows, were subsequently investigated by K. Denbigh and his colleagues, first in connection with continuous processes in the chemical industry, but then explicitly (Denbigh, Hicks, and Page 1948) in relation to the possibilities of interpreting biological processes in terms of such kinetics. They concluded that the properties of an open reaction system allow the output of energy to take place continuously 'without departing from its own internal constancy' (Denbigh *et al.* 1948, p. 491) and to do so over a time-scale greatly extended beyond that available to a closed system (of the same content and structure presumably). Such a system is not 'isolated', in the strict thermodynamic sense, so can increase its entropy at the expense of degradation of substances flowing through it (cf. section 2.1); moreover it can adapt itself to changes in its surroundings. They recognized autocatalytic reactions, such as those of the Lotka–Volterra type, as anomalous in exhibiting the possibility of having more than one steady state and discussed the mechanism and character of transitions from one steady state to another. They also recognized the 'anomaly' of the possibility of oscillations in open reaction systems involving autocatalysis and Moore (1949) in the same group extended Lotka's case of one or two autocatalytic reactions to a much longer sequence of such reactions, modifying his rather artificial condition that the precursor concentration (A in (4.23a)) is constant, to the more natural condition that it enters the system by a first-order diffusion process—and she also allowed for other reactants to leave the system by diffusion. She proved that, if a non-zero steady state exists for the system, the concentrations of the substances in the reaction sequence do not settle down to the steady-state values but oscillate about them, with a slight degree of damping, provided (a) that the rate of entry of precursor is great enough and (b) the rate of removal of substances is small enough. If (a) does not obtain there is a higher degree of damping, even to the point of making the oscillations not detectable. If the reactants leave the system too rapidly the later reactants will vanish, leaving a shorter chain with oscillating concentrations. Special conditions for oscillations apply when the number of autocatalytic reactions (other than $A \to X$) is odd, e.g. the concentration of A must exceed a certain critical value.

One of the earliest applications of physico-chemical ideas to understanding the development of biological complexity was that of A. M. Turing who, in what is now widely regarded as a remarkable paper entitled 'The chemical basis of morphogenesis' (1952), suggested that a system of chemical substances, 'morphogens', *reacting* together and *diffusing* through a tissue, is adequate to account for the main phenomena of morphogenesis. He proved that such a system, although it may originally be quite homogeneous, can later develop a pattern or structure due to an instability of the homogeneous equilibrium, which is triggered off by random disturbances. In spite of these latter considerations, this paper was quite independent of the mainstream of stability analysis in the physical sciences and stood alone in the biological literature. Its ideas were subsequently applied suggestively about a decade later by Maynard Smith and Sondhi (1961; see also Maynard Smith 1960) to certain morphogenetic problems, but it took almost another decade before the importance and originality of Turing's work became fully recognized. Although, with hindsight, it is now possible to extend (e.g. Gmitro and Scriven 1966; Prigogine and Nicolis 1967; Othmer and Scriven 1969) his ideas and to discern more precisely the conditions under which his equations apply (Bard and Lauder 1974; Erneux and Hiernaux 1980), we shall here confine ourselves to picking out some of the main conclusions of this historic paper.

Turing conceived of 'morphogens' (= form producer) as substances which, diffusing into a tissue or cell, somehow induce it to develop in ways otherwise than if it had been absent. He then proceeds to show that there can be a breakdown of symmetry or homogeneity in a system originally homogeneous with respect to distribution of a pair of morphogens (X and Y), as a result of mutual reaction and independent diffusion. This, of course, is counterintuitive for one would expect that, since the functions relating chemical reaction rates and concentration are the same in all cells of a given kind, then the effect of diffusion would be simply to smooth out any differences of concentration that might arise at different points, i.e. as between different cells, which would then be expected to behave in the same way. In effect, Turing asked 'Given a group of cells coupled by diffusion, are there any particular chemical kinetics that will induce *differences* of concentrations over space in a region in which they were originally the same?'. He showed that there *were* such kinetics and that they were, biologically speaking, sufficiently plausible to merit serious consideration as the basis of morphogenesis. Thus, although it is true that in almost all cases diffusion does even out spatial inhomogeneities of concentration and each cell behaves identically, yet under a particular range of conditions, it is possible for the chemical reactions to exploit, as it were, any small diffusion fluxes and to depart from this uniform behaviour to generate non-homogeneous distributions of X and Y as between cells in a group, between which X and Y diffuse. The kinetic constants have to satisfy certain criteria both in relation to each other and to the diffusion coefficients for this to happen. This is what Turing was able to demonstrate was possible.

Initially, to illustrate his meaning, he took a numerical example, which is worth reproducing. Suppose two living cells, (1) and (2) each contain two morphogens X and Y and suppose the rate of formation of each depends on the other according to the not implausible equation

$$\frac{dX}{dt} = 5X - 6Y + 1 \qquad (4.23)$$
$$\frac{dY}{dt} = 6X - 7Y + 1.$$

(Turing justifies the plausibility of these rate equations.) X and Y can diffuse between the two cells thus

$$\begin{array}{ll} \text{Cell 1} & \text{Cell 2} \\ X_1 \leftrightarrow X_2 & \\ Y_1 \leftrightarrow Y_2 & \end{array} \qquad (4.24)$$

where subscripts denote location in a particular cell. Let the rate of diffusion of X per unit time into a cell be 0.5 for unit concentration difference of X ($\Delta X = 1$), and 4.5 for Y ($\Delta Y = 1$). Now if $X_1 = Y_1 = X_2 = Y_2 = 1$, the rates in (4.23) all become zero and no concentration difference (ΔX or ΔY) for either morphogen is generated between the cells, so there is no net flow of X or Y in either direction. But suppose that, for whatever cause, the values of X and Y in each cell deviated from unity by a small amount, e.g. $X_1 = 1.06$, $Y_1 = 1.02$, $X_2 = 0.94$, $Y_2 = 0.98$, with the same total amount of X and Y present. Then, by (4.23),

$$dX_1/dt = +0.18; \quad dY_1/dt = +0.22,$$
$$dX_2/dt = -0.18; \quad dY_2/dt = -0.22$$

and the net flows per unit time of X and Y would be

diffusion of X from (1) to (2): $0.5(\Delta X) = 0.5 \times 0.12 = +0.06$
diffusion of Y from (1) to (2): $4.5(\Delta Y) = 4.5 \times 0.04 = +0.18$.

So the net gain of X per unit time, reaction and diffusion together, is

+0.12 in (1) and −0.12 in (2); and the net gain of Y is +0.04 in (1) and −0.04 in (2). In other words there is a flow from the second cell to the first of both X and Y that accentuates the already existing (small) differences between the cell. More generally, if, at some moment

$$X_1 = 1+3\xi; \quad X_2 = 1-3\xi$$
$$Y_1 = 1+\xi; \quad Y_2 = 1-\xi$$

then the four concentrations *continue* to be so expressible with ξ increasing with time. ξ may be regarded as an arbitrary fluctuation, but only certain kinetic relationships ((4.23) in this case) will give this result that the drift away from equality of distribution occurs with almost any small displacement. So such a two-cell system would be unstable, with inhomogeneity succeeding homogeneity—i.e. symmetry is broken and heterogeneity created.

In order to examine mathematically with more rigour such a breakdown of homogeneity through instability, Turing considered the system depicted in Fig. 4.5, a ring of N cells between which two morphogens X and Y diffuse, and within which they can react, as in (4.24). Concentrations of X and Y in cell r are written X_r and Y_r, respectively. Let μ represent the cell-to-cell diffusion coefficient of X and ν of Y and let cell r exchange only with adjacent cells, $r-1$ and $r+1$. so there are $2N$ differential equations of the form

$$\frac{dX_r}{dt} = f(X_r, Y_r) + \mu(X_{r+1} - 2X_r + X_{r-1})$$
$$\frac{dY_r}{dt} = g(X_r, Y_r) + \nu(Y_{r+1} - 2Y_r + Y_{r-1}),$$
(4.25)

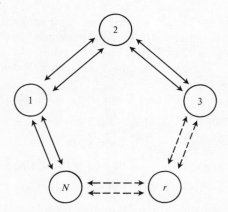

Fig. 4.5. Turing's ring of cells around which two morphogens can diffuse and react (see text).

where $r = 1, \ldots, N$. The system has an equilibrium state in which X and Y are in chemical equilibrium within each cell, at the equilibrium concentrations ($X = h$ and $Y = k$) that would prevail in isolated cells. The actual departures of the concentrations, in the rth cell from the equilibrium values can be written as

$$x_r = X_r - h \quad \text{and} \quad y_r = Y_r - k. \tag{4.26}$$

Assuming the system is not very far from equilibrium, at which f and g are both equal to zero, one may ignore higher powers of x and y, and write the linear relationships

$$f(X, Y) = ax + by$$
$$g(X, Y) = cx + dy \quad \text{[for a given } r\text{]}. \tag{4.27}$$

Equations (4.25) then become

$$\frac{dx_r}{dt} = ax_r + by_r + \mu(x_{r+1} - 2x_r + x_{r-1})$$
$$\frac{dy_r}{dt} = cx_r + dy_r + \nu(y_{r+1} - 2y_r + y_{r-1}). \tag{4.28}$$

Solutions of these equations were obtained by Turing for x_r and y_r, and thence X_r and Y_r, respectively, as sums of exponential terms containing in their indices the time and coefficients which are functions of a, b, c, d, μ, and ν (as well as, naturally, numbers also dependent on r). The movements of X_r and Y_r with time, and so the stability, depend on the relative contributions of the various exponential terms and Turing deduced the mathematical relationships between a, b, c, d (the 'marginal reaction rates'), μ, and ν (the diffusion coefficients) that prevail for different kinds of solution to exist, and also the conditions for their stability. His paper (and that of Bard and Lauder 1974, who set it out particularly clearly) should be consulted for the details. Note that his analysis assumes that initially at $t = 0$ the system is not quite homogeneous having suffered a (small) disturbance from this state, but that after $t = 0$ further disturbances are presumed absent.

From Turing's examination of the solutions of (4.28) two kinds of state of the system transpired to be possible; they were the *stationary* and the *oscillatory*. In the first case there are stationary waves of concentrations of X and Y around the ring and they have a particular number of lobes or crests so that the wavelength is the circumference of the ring divided by this number. The pattern for one morphogen determines that for the other. In the oscillatory case, the waves in the concentration of X and Y travel around the ring, with a certain velocity, and so frequency; there are two wave trains moving round the ring in opposite directions. The wavelengths of the patterns are submultiples of the circumference of the ring, so depend on it as well as on the chemical parameters. There is however a 'chemical wavelength' independent of the ring dimensions, which is the limit to which the wavelengths tend as the rings are made larger. Within this classification into stationary and oscillatory cases, six subcases may be distinguished (each with their own specific mathematical conditions).

(a) *Stationary with extreme long wavelength.* The contents of all the cells are the same, and there is no resultant flow from one to the other by diffusion. Each is isolated, but is in unstable equilibrium and easily slips out of it into synchronism with the others. This case, being rather featureless, is of no great biological import unless Fourier components other than the most unstable are also allowed to contribute to the pattern, when 'dappled' patterns become possible—and are perhaps relevant, Turing suggests, to early stages of foetal development that are small enough for the morphogens to traverse by diffusion.

(b) *Oscillations with extreme long wavelengths.* As in (a), each cell behaves as if it were isolated, but now the departure from equilibrium is oscillatory.

(c) *Stationary waves of extreme short wavelengths.* The concentrations in each cell remain stationary and are *different* from those in its nearest neighbours, (contrast (a)). If there is a drift from equilibrium, it will be in different directions in contiguous cells.

(d) *Stationary waves of finite wavelength*—the case of most biological import. There is a stationary wave pattern on the ring, with no time variation, apart from a slow increase in amplitude, the pattern becoming more marked. The peaks of concentration are uniformly distributed and, if it had been a ring of continuous tissue rather than discrete cells in which these reactions and diffusion had been recurring, the variation of concentration of a morphogen with respect to position around the ring would be sinusoidal. The number of peaks is given approximately by dividing the 'chemical wavelength' (defined above) into the circumference of the ring and taking the nearest whole number. Turing considered two sets of chemical reactions that would, with appropriate rate constants, fulfill the conditions (d). One of these is depicted schematically in Fig. 4.6 (for the particular constants chosen see Turing 1952, p. 65). Bard and Lauder (1974) have used the same constants as Turing (and, additionally assumed $h = k = 4$) and calculated (1974, p. 526) that the 'chemical wavelength' according to Turing's equations should be 6 to 7 cells long. They were able to simulate by computer—a technique not available to Turing—the development with time of X and Y for this system, using Turing's equations. Their results are shown in Fig. 4.7—which nicely illustrates how, as Bard and Lauder put it, 'from an initially bland piece

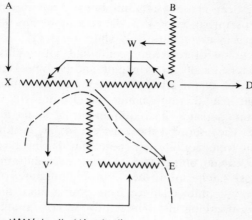

⋎⋎⋎⋎ implies '+' or 'and'

Fig. 4.6. Turing's second set of chemical kinetics: a diagrammatic flow diagram. X and Y: morphogens. V and V': two forms of an enzyme. The reactions represented are: $A \rightarrow X$; $X+Y \rightarrow C$; $C \rightarrow X+Y$; $C \rightarrow D$; $B+C \rightarrow W$; $W \rightarrow Y+C$; $Y \rightarrow E$; $Y+V \rightarrow V'$; $V' \rightarrow E+V$. (From Bard and Lauder 1974, Fig. 1.)

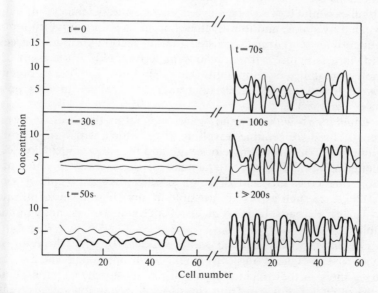

Fig. 4.7. Turing's second set of chemical kinetics; the development with time of the pattern of morphogens X and Y over a line of 60 cells following evocator (V) production in each cell. Fine line: concentration of morphogen X. Heavy line: concentration of morphogen Y (From Bard and Lauder 1974, Fig. 3.)

of tissue we have been able, by creating in each cell an enzyme evocator [i.e. by allowing perturbation of V in the scheme of Fig. 4.6], to produce a chemical pattern in a comparatively short time' (1974, p. 500). As it happens this same set of reactions, combined with diffusion, has been analysed thermodynamically by Prigogine and Nicolis (1967) in terms of the general approach of the Brussels school (see Chapter 2). They were able to show that near thermodynamic equilibrium, the steady state of this reaction system satisfies the requirements of linear non-equilibrium thermodynamics. They showed the existence of an instability for situations very distant from equilibrium (note that Turing had taken all the reactions in his scheme of Fig. 4.6 to be irreversible and so far from equilibrium). Most significantly, they also demonstrated that for steady states, at a large but finite distance from equilibrium, there exist well-defined symmetry-breaking instabilities beyond critical affinities. Beyond this transition point the stable steady state is inhomogeneous, the diffusion compensating the differences in reaction rates.

The biological significance of these possibilities have become increasingly apparent in the three decades since Turing's work, not only in connection with morphogenesis but also with the origin of life. Turing himself suggested that this hypothetical situation (d) is relevant to the development of the tentacles from the open (head) end of *Hydra* which has circular symmetry. At one stage it develops a 'patchiness' in its cells, as revealed by stains, and these patches occur at points at which tentacles subsequently arise. Turing suggested that a reaction–diffusion process of type (d) in a structure with circular symmetry could be the basis of this symmetry-breaking development. He also cites the development of whorls of leaves on Woodruff (*Asperula odorata*) as another possible instance.

(e) *Oscillations with a finite wavelength* (for three or more morphogens). These are genuine travelling waves which, with a ring, would be two in number, one travelling clockwise, and the other anticlockwise. The wavelengths, velocities, and frequencies of these waves depend on the reaction rate constants and diffusion coefficients—and, indeed, on the geometry, since such waves are possible in any anatomical form, according to Turing. Perhaps the movements of spermatozoa tails are a biological example.

(f) *Oscillations with extreme short wavelengths* (for three or more morphogens). Neighbouring cells would be in opposite phases.

Turing then, in the concluding part of this most original paper, extends the idea of chemical waves from rings to spheres and is able to treat the breakdown of homogeneity through the theory of spherical harmonics and concludes that, under certain not very restrictive conditions (small sphere, increasing in size), the pattern of the breakdown is axially symmetrical about a *new* axis that is determined by the disturbing influences and is not the original axis of the spherical polar coordinates. So if one of the morphogens is (or encourages production of) a growth hormone, this could explain how a blastula would develop in an axially symmetrical manner, but at a greater rate at one end of the axis than at the other—and this, under many circumstances, might lead to gastrulation.

The biological application of his ideas were, of course, extremely speculative at the time, but with hindsight they have a prophetic character and have remained a stimulus chiefly because of the fundamental physico-chemical basis for pattern formation that Turing demonstrated was inherent in reaction–diffusion systems.

4.3 Dissipative structures and bifurcations in reacting systems—theoretical considerations

The pioneers, whose work has just been outlined, provided pointers and directions for what has become a veritable explosion of interest in reaction systems that can become organized in space and/or time; and in sequences of biochemical reactions and higher level cellular processes and interactions that actually exhibit such organization. For a time after Turing's work interest in, for example, oscillating chemical systems continued to be somewhat muted, in spite of the much earlier discovery of a purely inorganic chemical reaction displaying a periodically changing reaction rate in homogeneous solution (Bray 1921). But the discovery in 1958 by Belousov (1958) of the spatial and periodic order to be observed in the oxidation, by bromate in the presence of the cerous–ceric ion couple, of citric, malonic, and other carboxylic acids containing a methylene group, and its detailed study by Zhabotinsky from 1964 onwards, stimulated a renewed interest as it came to be realized that such behaviour was not an isolated curiosity but had parallels elsewhere when certain kinetic conditions were fulfilled. (See Zhabotinsky (1964, 1980) and Tyson (1976) for a survey of the Belousov–Zhabotinsky reaction. For general reviews, see: Higgins 1967; Nicolis 1971; Degn 1972; Nicolis and Portnow 1973; Noyes and Field 1974; Noyes 1980; also two Russian books (ed. Frank (1967) and Sel'kov (1971), the Faraday Society Symposium No. 9 (1974) and the 1980 (Vol. 84, No. 4, pp. 295–418) issue of the *Berichte der Bunsengesellschaft für Physikalische Chemie* devoted to the kinetics of physico-chemical oscillations.)

During the 1960s interest in such phenomena was further stimulated by the recognition of biological periodicities—the so-called 'biological clocks' (Brown, Hastings, and Palmer 1970) and 'circadian rhythms' (Queiroz 1974; Aldridge 1976)—and of oscillatory phenomena in biochemistry, beginning with the observation of oscillations in photosynthesis (Wilson and Calvin 1955), in glycolysis (Duysens and Amesz 1957; Ghosh and Chance 1964; Chance, Estabrook, and Ghosh 1964; Chance, Hess, and Betz 1964), and in peroxidase-catalysed oxidations (Yamazaki, Yokota, and Nakajima 1965), the first clear example of an oscillatory reaction in an open soluble-enzyme system. These and other developments of the 1960s have been well reviewed by Hess and Boiteux (1971). These experimental discoveries generated intense theoretical interest, which took many authors back to at least some of the pioneering work already described (it is intriguing to note what was overlooked!), and the spate of inquiry and discovery continues unabated. In particular, the theoretical investigation has required renewed mathematical analysis of non-linear differential equations and of 'bifurcation' in their solutions (see, for example: Minorsky 1962; Hanusse 1973; Murray 1977; Cronin 1977; Tyson and Othmer 1978; a useful geometric interpretation by Stucki 1978, pp. 145–8); Haken (1978); the papers of a 1977 conference on bifurcation theory and its applications, Gurel and Rossler 1979; Fife 1979; Rapp 1979b; Iooss and Joseph 1980; Segel 1981; Eckmann 1981; and Hofstadter 1981, a helpful introduction for the non-mathematician). It has also required a more thorough investigation into the theory of fluctuations, especially those near to bifurcation points (for useful general accounts in the present context see: Nicolis and Prigogine 1977, especially Part III; Herschkowitz-Kaufman and Erneux 1979; Escher 1980; and Nicolis 1981). Such 'bifurcation' means 'a qualitative change in the topological structure of a dynamical system as a consequence of a shift in some critical parameter' (Decker 1979).

The systems that give rise to organization in time and/or space and to multiple steady states that are of pertinence to biological processes with which we shall be principally concerned here are those in which both chemical reactions and diffusion are occurring. (Oscillating membrane phenomena involving oscillating potentials and ion movements are, of course, exceedingly important in certain biological contexts but will not be considered further in this volume. For an analysis and review of the application of bifurcation theory to two-dimensonal reaction-diffusion systems, see Schiffmann (1980). This volume has its emphasis rather on those more general physico-chemical principles and models that might illuminate the origin and nature of biological organization.)

The conservation of a particular chemical species in a particular locality can be expressed in the following kinetic equation:

$$\frac{\partial X_i}{\partial t} = v_i(\{X_j\}) + D_i \nabla^2 X_i, \qquad (4.29)$$

where X_i and D_i represent, respectively, the concentration and diffusion coefficient of chemical species X_i and $\{X_j\}$ represents the concentrations of all the chemical species present. The first term v_i is that function of all these concentrations which is the rate of formation of X_i by chemical reaction and the second term, which involves the space coordinates, is the time change of X_i through diffusion and so involves the space-derivatives ($\nabla^2 X_i$) of its own concentration. The function v_i, the rate of formation of X_i by chemical reactions, is usually a non-linear function of the concentrations X_j, so (4.29) constitutes a set of non-linear partial differential equations and it is from this basic feature that the mathematical complexities originate. These complexities will not be elaborated here and the reader is referred to the sources quoted above (see p. 138 and other particular references as occasion requires their citation). In the following only a qualitative account will be given of the main types of solution to these equations that investigation has discovered to be possible. Steady-state solutions, often one such unique solution, in which all the participating chemical substances are at a steady concentration (i.e. all $dX_i/dt = 0$) are frequently possible, in addition to the mathematically trivial equilibrium solution, if the system is an open one in the thermodynamic sense, that is, if matter and energy can leave and enter it. Such steady-state solutions are homogeneous in composition and distribution throughout the system. When equations such as (4.29) are non-linear these steady-state solutions, or solution, can in appropriate circumstances become unstable for critical values of the controlling parameters—ranges of concentrations, as well as rate constants and diffusion coefficients. These last two may themselves depend on medium conditions of temperature, ionic strength, pH, viscosity, etc. for solutions in water. Beyond these critical values of the parameters new solution(s) of the equations become possible. The process of passing through such a critical value is called *bifurcation* and beyond this point the new alternative organization of the system, which is by now far from equilibrium, constitutes a dissipative structure in the sense of Chapter 2, where it has already been given a thermodynamic significance and interpretation. Bifurcation theory is, as already hinted, itself a developing branch of mathematics and will not be entered into here (see references above, p. 138; Rapp (1979b) on its relation to control theory and metabolic regulation; and, for the bifurcation analysis of non-biological systems, see Nicolis and Auchmuty (1974) and Sattinger (1973)). The general idea is illustrated in Fig. 4.8. The ordinate represents a characteristic property of the solution of equations such as (4.29), e.g. the space- or time-average of the concentration of particular substances, or the amplitude of their oscillations, if the solution is periodic even below the bifurcation point. The abscissa is a single parameter, λ, which has a critical value, λ_c such that at $\lambda < \lambda_c$ there is a single solution to the equation, on what is called the *thermodynamic*

140 THE KINETICS OF BIOLOGICAL SELF-ORGANIZATION

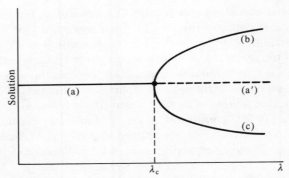

Fig. 4.8. The first primary bifurcation: (a) stable thermodynamic branch; (a') branch (a) becoming unstable beyond λ_c; (b) (c) new stable branches of solutions arising through bifurcation at λ_c. (From Nicolis 1981, Fig. 4.)

branch, (a), when there is a uniform distribution of the system's properties (including the X_i) through time and space. In this range, any perturbation of the system is damped down and it returns to its state before the perturbation (it is 'asymptotically stable'). But when $\lambda \geq \lambda_c$, new stable solutions, represented by the lines (b) and (c) become possible, *instead* of the continuation (a') of (a). So that the original solution (e.g. a steady state) represented by (a) and (a') loses its stability when λ exceeds λ_c and the two new states are stable. The system follows the paths (b) or (c), in most cases with equal probability, i.e. symmetry is possible, but not always (Nicolis 1981). Which path is followed beyond λ_c involves an element of chance inherent in the fluctuations that occur in the system near to and just beyond the bifurcation point at λ_c. Beyond this point, the new solutions (b) and (c) are asymptotically stable and which of the two paths is followed at the bifurcation point determines the future evolutionary possibilities of the system that in this regard stores, as it were, the memory of the crucial fluctuation which set it upon path (b) or (c) at that critical point (or just beyond it). Actually, the diagram in Fig. 4.8 is only the simplest of possible bifurcation diagrams (one that leads to symmetry-breaking at λ_c) (see Nicolis and Prigogine 1977, section 7.6; Nicolis 1981). The various other forms nevertheless have the broad feature referred to above—namely, that at least below a certain critical *range* of λ that is close to λ_c (rather than *point* in these other cases), there is one solution on the thermodynamic branch; beyond this range, two (or more) different solutions become possible and stable, instead of that which characterizes the thermodynamic branch. Various patterns of relationship between the stability of the state corresponding to the thermodynamic branch and the two other states can prevail in the vicinity of λ_c and can lead to hysteresis without symmetry-breaking and to sub-critical effects, that is why a 'range' around λ_c is referred to; but these variations do not vitiate the general picture of bifurcation just given. For the basic point is that beyond a critical instability of the thermodynamic branch, which may be a steady state (or, in some circumstances, an oscillating system), there arise new states with the appearance of ordered 'dissipative' structures of a new type relying on continuing exchange of matter and energy with their environment. These new structures can exhibit organization in space and/or in time, or can exhibit multiple, and new, steady states. We now turn to a qualitative discussion of the various possibilities that have been envisaged as a result of mathematical investigation.

Reaction in the absence of diffusion

The simplest case that might produce organization in time is that of two concentration variables, X_1 and X_2. The equation (4.29) then reduces to

$$\frac{dX_1}{dt} = v_1(X_1, X_2); \qquad \frac{dX_2}{dt} = v_2(X_1, X_2). \tag{4.30}$$

The steady state allowed by (4.30) corresponds to a 'singular point', i.e. particular values of X_1 and of X_2 (a point in the phase plane (X_1, X_2)). Its stability may be studied by the same methods as used for mechanically oscillating systems, namely by determining how the system responds to an infinitesimal displacement from the steady state ('normal mode analysis', see Nicolis 1971). If X_1^0 and X_2^0 represent the steady-state concentration of X_1 and X_2 (when $dX_1/dt = 0 = dX_2/dt$), then let

$$x_1 = X_1 - X_1^0 \quad \text{and} \quad x_2 = X_2 - X_2^0. \tag{4.31}$$

For an infinitesimal perturbation from the steady state

$$\frac{dx_1}{dt} = a_{11}x_1 + a_{12}x_2$$

$$\frac{dx_2}{dt} = a_{21}x_1 + a_{22}x_2. \tag{4.32}$$

The coefficients $\{a_{ij}\}$ depend only on rate constants and concentrations and equations (4.32) have normal mode solutions of the form $(x_1)_0 e^{\omega t}$ and $(x_2)_0 e^{\omega t}$, where ω is a frequency. Substitution of these values for x_1 and x_2 into (4.32) shows that the condition that they are not zero is

$$\begin{vmatrix} a_{11} - \omega & a_{12} \\ a_{21} & a_{22} - \omega \end{vmatrix} = 0. \tag{4.33}$$

This 'secular equation' (4.33) in general has two eigenvalue solutions ω' and ω'' and x_1 and x_2 will be linear combinations of terms in $e^{\omega' t}$ and $e^{\omega'' t}$.

Stability is indicated if ω' and ω'' both have negative real parts so that the perturbations x_1 and x_2 decay to zero with time; but if they have positive real parts then the perturbations will grow exponentially and the system will depart from the steady state. Moreover if ω' and ω'' contain imaginary parts then the presence of growing or decaying oscillations about the steady state may be deduced. Stability criteria (Nicolis 1971) allow the signs of the eigenvalues to be deduced from the a_{ij} values and so the ranges of stability and instability can be related to ranges of values of rate constants and concentrations. 'Instability', so deduced, does not guarantee that oscillations will necessarily occur but, with only two intermediates, there are theorems (see Minorsky 1962) application of which can determine whether or not stable limit cycles will be attained.

The application of these methods confirms that oscillations around steady states can occur only if the system is far from equilibrium (outside the 'linear' range already referred to in Chapter 2). Undamped oscillations around a *stable* steady state can occur with an infinity of frequencies, and amplitudes, that depend on the initial perturbation. The first Lotka (1910) model is an example of oscillation involving an infinite number of extrema of monotonically *decreasing* amplitude (Class II of Schmitz' (1973) classification of chemical oscillation); and the second Lotka model (1920a,b) is another such, in this instance exhibiting an infinite number of oscillations of *constant* amplitude (Class III of Schmitz (1973)) in a sustained open system. (N.B. Schmitz' 'Class I' oscillations involve a *finite* number of maxima and minima and finally decay back to the steady state as described by Denbigh et al. (1948), the relevance of whose work to biological systems has already been discussed (section 4.2)).

If the singular point in the (X_1, X_2) phase plane that is a solution of (4.30) represents an *unstable* steady state, then it may constitute the focus, or node, for sustained oscillations in X_1 and X_2 in time, with a unique amplitude and frequency, regardless of initial conditions. This then constitutes what is called a *limit-cycle* and it is these that occur, in contrast to the Lotka–Volterra type, when stable oscillations are observed in biochemical systems (see section 4.5.1). Such stable limit-cycles, which can arise only in non-linear systems, are depicted as closed trajectories in the concentration-space around the singular point, as for example in Fig. 4.9 which shows a theoretical plot corresponding to the changes of ATP and ADP concentrations in the phosphofructokinase-catalysed reaction of the glycolytic pathway (for further discussion of this model and system see section 4.5.1 and p. 179). It should be noted that the system moves round these limit-cycles in a well-defined direction for a given set of parameters: so, in this sense, in limit cycles there is a *breaking of temporal symmetry*— a direction in time has been defined.

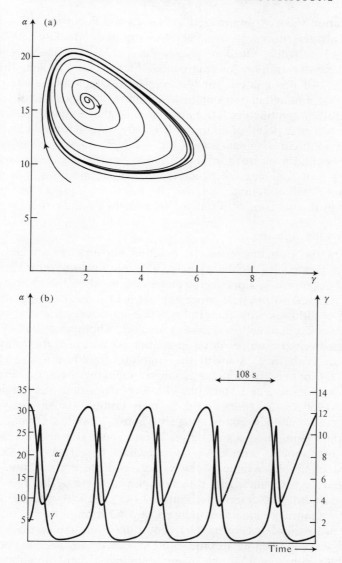

Fig. 4.9. (a) Evolution to limit-cycle solution in the α-γ phase plane in the homogeneous case when the stationary state is an unstable focus: the limit-cycle can be reached from outside. α, γ, are normalized concentrations of substrate and product of an allosteric enzyme (PFKase). See equation (4.52). (b) Time variation of α and γ. (For values of constants assumed and full equation see Goldbeter and Lefever 1972, from which this figure is taken (their Fig. 2).)

Bifurcation theory (summarized by Nicolis and Portnow (1973), section I.C.3, with references to Andronov, Vitt, and Khalkin (1966), Andronov, Leontovitch, Gordon, and Maier (1971) and Hanusse (1972)) illuminates the conditions of emergence of stable limit-cycles. The pertinent results of this analysis for biochemical oscillations are the demonstration that it is undamped small oscillations at a critical set of values of the controlling parameters (cf. λ_c in Fig. 4.8) that initiate the transition from the 'singular point' of an *unstable* steady state to a *stable limit-cycle*; and that multiple singular points, i.e. *multiple steady states*, are also possible beyond a bifurcation point. Stable limit-cycles clearly represent an example of a dissipative structure, in the thermodynamic sense, but now elaborated and made explicit by consideration of the chemical kinetics, in the absence of diffusion, that might produce them.

Reaction with diffusion

It is often the case that chemical reaction networks or sequences that would not give rise to oscillatory behaviour as a result of their kinetic and chemical organization alone, nevertheless do so when the chemical kinetics are coupled to physical processes, such as diffusion. Moreover, the coupling of diffusion with chemical reaction introduces a spatial vector so that homogeneous reaction–diffusion systems, whether oscillating or not, can in some circumstances develop spatial asymmetries that may themselves vary with time. So with the coupling of diffusion and chemical reaction the possibility of *space symmetry-breaking* arises (cf. Prigogine, Lefever, Goldbeter, and Herschkowitz-Kaufman 1969; see Gmitro and Scriven (1966) and Othmer and Scriven (1969) for an analysis that considers also chemical engineering situations). The basic equation is still (4.29) and, being non-linear, its solution gives rise to the problems already mentioned. With only *one variable*, the frequency ω in the exponent of the eigenvalue is always real and the perturbations evolve monotonically with time, with diffusion playing a stabilizing role, so that the first instability is determined only by the chemical kinetics. Thus no intrinsic spatial dependence is induced beyond instability.

With *two variables*, as in the absence of diffusion, the secular equation (4.33) is of second degree and the solutions (i.e. ω' and ω'', see above) can be complex conjugates. If the ω are *imaginary*, then the solutions are time-dependent and, on bifurcation beyond a point of instability (at which a controlling parameter $\lambda = \lambda_c$), they are also periodic with a definite intrinsic frequency. These periodic solutions can, beyond the bifurcation point ($\lambda \geqslant \lambda_c$), give rise to *stationary* or *propagating waves* in the concentration variables and thus, at any instant of time, spatial symmetry can be broken, for the space is subdivided with regions of different concentrations of X_1 and X_2.

When the ω are *real*, the steady state solutions prevailing below the bifurcation may give way, beyond this bifurcation point $(\lambda \geq \lambda_c)$, to solutions characterized by an intrinsic wavelength in a hitherto spatially homogeneous system. This *space symmetry-breaking* can take various forms (Nicolis 1981) such as: *standing waves*, which may even be *localized*, i.e. the concentration of a substance falling virtually to zero outside certain limits drawn within the total space of the system; *propagating waves* in the concentrations of X_1 and X_2, and even *solitary waves* in an infinite medium. *Time symmetry-breaking* can also occur for these reaction–diffusion systems beyond a bifurcation with the setting up of *limit-cycles*. Moreover, as in the absence of diffusion, new *stable steady states* can emerge and these may be *multiple*, in which case the possibility of hysteresis occurs in the transition between states as the controlling λ is first increased and then decreased.

With *three or more variables* new possibilities arise. One may have more than one region of instability in the parameter λ and it is possible for different modes, with different intrinsic wavelengths or with different frequencies, to interact near critical values of λ. Such interactions can lead to quasi-periodic or chaotic behaviour and the idea of 'cascading bifurcations', a whole succession of bifurcations, is now being explored (e.g. Nicolis 1981). These are associated particularly with the possibility of *varying more than one control parameter* and diverse and complex forms then, apparently, become possible (Erneux and Hiernaux 1980; Erneux, 1981)—e.g. *rotating waves* and sudden decreases in wavelength in systems of small dimensions, which may be of interest in relation to biological morphogenesis. There is even the possibility that the kind of ordering beyond bifurcations that we have been discussing, especially in relation to systems of two variables, might give way to the irregular oscillations of *chaotic behaviour* with characteristic scales varying from the molecular to the macroscopic. In that case, the order of dissipative structures would be sandwiched between two regimes of the controlling parameters, in one of which thermal chaos prevailed (with microscopic space and time scales); and, in the other, non-equilibrium, turbulent chaos with wider and larger characteristic scales.

The role of fluctuations

Up to this point, the theoretical considerations have all been 'deterministic' that is, they have been based on the macroscopic equations of chemical kinetics and diffusion. Such an approach is entirely justified when systems are close to equilibrium, for application of Einstein's formula for the probability distribution of fluctuations (see section 2.6), which is based on the Boltzmann definition of entropy (section 2.1), shows that the range of the fluctuations in any parameter (such as a concentration X_i) is negligibly small compared with its average, macroscopic value. The actual distribution of such fluctuating variables in ideal systems is in fact a Poisson distribution

$$\Pr\nolimits_X = e^{-\bar{X}}\frac{(\bar{X})^X}{X!}$$

where $\Pr_X$ is the probability of the variable having the value X when its mean value is $\bar{X}$. This distribution also prevails for non-equilibrium situations when the equations for a chemical process are linear (though, as we saw, this is, in fact, rarely the case).

However, such a description is bound to be inadequate for non-linear systems, especially as they approach a bifurcation point; and beyond such a point it clearly breaks down, for it provides no criteria for the system 'choosing' between the alternative available states, such as (b) and (c) in Fig. 4.8. For this reason considerable attention has, in recent years, been focused on the stochastic processes involved in fluctuations in non-linear systems, especially near to critical points—whether in phase transitions (Haken 1975, 1977a,b, 1978) or at bifurcations of the kind described in the preceding sections (see the survey of Nicolis and Prigogine 1977, Part III for extensive references; Walgraef, Dewel, and Borckmans 1980; Nicolis 1981). The mathematical details are beyond the scope of the present treatment but some indication of how it is that fluctuations may come to be seen as the triggers of instability, and thus of self-organization, is called for.

The phenomenon of phase transitions (Landau and Lifshitz 1957) is already a well-known example of a situation where small fluctuations are amplified and, after amplification, drive the system to a new phase, and a similar role is played by finite fluctuations in first-order transitions involving nucleation and metastability. There is a striking analogy between such transitions and non-equilibrium transitions at a bifurcation point (Haken 1975, 1977a,b, 1978), for in the systems beyond the 'critical' points, the fluctuations increase, departing from a Poisson distribution, and finally drive the system into a new macroscopic state, with new average values of its parameters. Thus fluctuations 'drive the averages' and, if the resulting system is more organized, Prigogine's concept of 'order through fluctuation' is applicable. In the critical region the departure from a Poisson distribution can be sufficiently marked that long-range spatial coherences appear and space symmetry-breaking in the new structure beyond the critical point then occurs.

The theoretical problem arises from the fact that in systems of more than one variable operating in the non-linear range far from equilibrium, there is no suitable 'potential' (such as an excess free energy) that can be adequately defined for use in the Einstein formula (see section 2.6) for the probability of a fluctuation. For in these circumstances, the variance of fluctuations changes qualitatively as their scale increases. Small-scale fluctuations can still be accommodated by the Einstein formula, provided the reference for the 'potential' in the probability distribution is now the non-equilibrium state itself. This is the condition of *local equilibrium*, already discussed (section 2.3). But because the reference state is not a fixed one (being the non-equilibrium state), the state variables can evolve in time and be inhomogeneous with respect to space as a result of the non-equilibrium constraints, even though the form of the probability distribution describing local equilibrium is Einsteinian. Moreover, large-scale fluctuations over the range of macroscopic volumes comparable to the size of the system itself can occur, not controlled by an Einstein-type formula, with the results described in the preceding paragraph.

Nicolis (1981) and Lefever and Prigogine (1981) have examined more fully the relation between the dynamics of fluctuations in systems far from equilibrium and the macroscopic behaviour—the former in relation to internal fluctuations generated spontaneously by the system itself and the latter in relation to external noise and its effect on instabilities (see also Horsthemke and Malek-Mansour 1976). The systems referred to are maintained in non-equilibrium by applied constraints which evoke a question presented thus by Nicolis (1981):

> reaction diffusion systems ... can exhibit coherent behaviour which manifests itself in the form of regular spatial patterns or of rhythmic phenomena emerging abruptly in a hitherto homogeneous and stationary system. Both phenomena imply sharp and reproducible correlations between distant parts of the system. How can such coherence arise in a dilute solution of molecules behaving almost like hard spheres, and hence incapable of recognizing each other over more than a few Angstroms?

He goes on to show that, in the expression for the rate of change with time of the first moment of the probability function that represents the concentration of a reactant, it is terms arising from deviations from a Poisson distribution that are entirely responsible for this rate of change differing from that predicted by the deterministic rate equation—and these non-Poissonian terms would be zero for an ideal solution at thermodynamic equilibrium. Moreover,

> One can show that a spatial *correlation function* between distant parts of the system will emerge whose amplitude is directly proportional to the non-Poissonian part of the fluctuations. In summary, deviations from the Poissonian are the triggers that enable the system to deviate from the macroscopic regime and hence to choose between many available solutions in the presence of bifurcation phenomena. Normally, in the limit of a large volume, one may neglect them. But in some cases, the deviation from the Poisson behaviour becomes comparable to the average value itself.... Note that diffusion, which is always present, tends to re-establish the Poissonian in a scale of the order of

the mean free path.... However, the competition with chemical kinetics may give rise to a deviation from the Poissonian in a macroscopically large subvolume of the system. Subsequently, this large fluctuation may propagate throughout the system and drive it to a new macroscopic state. Molecular dynamics studies, whereby the chemical reactions and the thermal motion of the molecules are simulated on a computer, confirms the existence of different laws governing fluctuations of different scales (Boissonade 1979). (Nicolis 1981)

(For a more quantitative approach reference should be made not only to this paper of Nicolis (1981) but also to: Nicolis and Turner 1979; Nicolis and Malek Mansour 1980; and Walgraef et al. 1980).

4.4 Conditions for self-organization and oscillations

Self-organization

In our earlier discussion of the thermodynamics of dissipative structures, the joint conditions that were proposed (end of section 2.7; see also secton 2.9) as prerequisites for self-organization to occur in a system were; (i) the system must be *open* to the flux of matter and energy; (ii) it must not be at equilibrium but preferably *far from equilibrium*; and (iii) it must be *non-linear* in its flux–force relationships, i.e. there must be strong coupling between its processes. And we quoted (p. 62) the 'theorem' of G. Nicolis (1974) to the effect that a system satisfying these conditions (and with a defined entropy) would be asymptotically stable near to equilibrium and, more significantly, would display the possibility of becoming unstable at a critical distance from it.

To what extent are these three criteria satisfied by living systems?

(i)′ It is a truism about living organisms that they can only live if there is a continuous inward flow of nutrient chemical substances (many of which are carriers of chemical free energy) and an outward flow of waste products. In addition to this interchange of matter and energy carried by matter in chemical bonds, many systems of cells in living organisms are also directly dependent on a supply of energy in the form of light and substances (such as oxygen) that will release energy from the chemical substances the cells imbibe. So living systems are clearly *open* systems in the sense (i).

(ii)′ All chemical reactions removed at all significantly from equilibrium display a non-linear relationship between their driving forces (affinities, i.e. free-energy differences) and actual rates. But the chemical processes of living organisms display non-linearity more readily (i.e. closer to formal equilibrium and at lower concentrations) than most inorganic processes because: (a) they involve allosteric enzymes that respond, positively or negatively, in a cooperative fashion to small changes in the concentrations of reactant metabolites and controlling ('effector') sub-

CONDITIONS FOR SELF-ORGANIZATION AND OSCILLATIONS

stances; (b) they are organized in networks and pathways in which there are many kinds of positive and negative feedback, i.e. cross-catalytic and cross-inhibitory influences exerted either by involvement of the reactants themselves or by chemical control over enzyme activities (this category includes the self-instructive 'hypercycles' (Fig. 5.8) of Eigen and Schuster (section 5.4) postulated as a plausible stage in the molecular origin of life); (c) the concentrations of the catalytic proteins, the enzymes, is usually very much less than that of the particular reactant metabolites themselves, so that the kinetic description of any such stage can be simplified by assuming a steady state for the enzymatic forms and complexes. This both reduces the number of variables and makes the resulting equations highly non-linear; and, finally (d) many of the chemical processes in living cells and organelles have a strongly vectorial character because of electrical fields generated *in situ* and because of diffusion processes either across membranes (with their markedly directional character controlled by the membrane proteins) or from one part of a cell, organelle, or organisms to another over a wider span than simply across membranes.

(iii)' The controlled biochemical reactions of living organisms usually operate a long way from equilibrium—indeed equilibrium for a living organism is synonymous with its death! Even reactions tending to be close to equilibrium are often displaced away from it by the way membranes serve to create marked concentration gradients.

Thus it is not surprising that self-organization as dissipative structures is much more frequently observed in systems of biochemical reactions than in other branches of chemistry, as we shall see below (section 4.6) from the wealth of biochemical and the relative paucity of purely chemical examples. The evidence that such systems are behaving as dissipative structures often takes the form of the observation, under particular conditions, of the occurrence of oscillations in the concentration of one or more reactants (metabolites) which is in itself an indication that there has been a change of regime, the substitution of one kind of molecular organization by another. So we must now examine more explicitly the prerequisites for oscillations and whether or not these are fulfilled in living organisms, since the criteria for self-organization just stated were derived from thermodynamic considerations which usually (as we saw in section 2.6) describe only the conditions that permit, but do not predict, the transitions to new dissipative structures.

Oscillations

Can one be more specific about the types of non-linear kinetics that might give rise not only to dissipative structures, in general, but to oscillations, in particular? Nicolis and Portnow (1973) have examined this question to which an answer may be given for homogeneous systems (no diffusion) of two concentration variables (X_1 and X_2) and the general conclusions of their mathematical analysis have already been described (section 4.3 'Reaction in the absence of diffusion'). One of the requirements that undamped oscillations should occur at the critical point is that, with reference to the notation of equations (4.30), at least one of the $(\partial v_i/\partial X_i)_0$ should remain positive around the critical point. They go on to show (Nicolis and Portnow 1973 p. 370 (section C3)) that this condition is physically fulfilled if at least one of X_1 and X_2 catalyses its own production, either directly through an *autocatalytic reaction step* or indirectly by *activation* of a substance which produces it, e.g. in the sequence (cf. (4.23) of Lotka)

$$\begin{aligned} A + X_1 &\rightarrow 2X_1 \\ X_1 + X_2 &\rightarrow 2X_2. \end{aligned} \quad (4.35)$$

A further condition for oscillations is that $(\partial v_1/\partial X_2)_0$, $(\partial v_2/\partial X_1)_0$ (where subscripts zero refer to a steady state) should remain negative in some neighbourhood and this condition is equivalent to: *either* one of X_1 and X_2 taking part in the autocatalytic process producing the other; *or* the reaction sequence involving at least one cross-catalytic step whereby X_1 acts as a catalyst to produce X_2 (or vice versa), followed by X_2 being converted to X_1 (directly or via another catalyst), e.g. in

$$\begin{aligned} A + X_1 &\rightarrow X_1 + X_2 \\ X_2 &\rightarrow X_1 \end{aligned} \quad (4.36)$$

—for which the detailed rate equations indicate the possibility of substrate (X_2) inhibition in the second reaction.

With more than two variables, it has been shown (see also section 4.5 below) that inhibition alone, without additional catalytic steps, can give rise to sustained oscillations. Nicolis and Portnow (1973, p. 371) conclude: '... positive and negative feedback, sometimes combined with cross-catalysis, is a necessary prerequisite for the existence of stable, sustained oscillations'. But these conditions, which are also included amongst those for self-organization, are fulfilled by the biochemical networks of living organisms (see (ii)' (a), (b), and (c) above), so that the existence of stable, sustained oscillations in living systems is to be expected. The properties of living systems favour the occurrence of non-equilibrium oscillating instabilities.

Nothing has so far been suggested about the *time scale* of such oscillations, and in fact, as we shall see, living systems manifest oscillations over a very wide range, with frequencies stretching continuously over ten orders of magnitude—from those at the molecular level (the glycolytic and mitochondrial oscillations), through the epigenetic and cellular to the physiological (e.g. in the 'clock' of large systems such as that of the heart muscle). The development of such oscillations may indeed have evolutionary significance, since they would display efficient response times and damping rates that would allow the system so controlled to react in an adaptive manner to continuously changing external conditions (Hess and Boiteux 1971, p. 253); and furthermore they might form the basis for spatial differentiation where gradients of chemical potentials are produced by oscillating chemical systems (Prigogine *et al.* 1969; Prigogine 1967).

4.5 Models of kinetic self-organization

The mathematical developments described above (section 4.3) and the realization that the conditions for self-organization, via temporal and spatial periodicities, are fulfilled by biochemical and biological systems (section 4.4) has led, since the early 1960s, to detailed application of these mathematical analytical tools to many plausible models of such systems—more indeed than have yet found their experimental application and confirmation. But this wide-ranging exploration of the theoretical possibilities has been very fruitful in making clear how biological self-organization might arise at the new level of complexity that emerges in such systems in a way consistent with the underlying physico-chemical principles and yet manifesting new principles of organization and their associated conceptual elaboration. Moreover, a number of these theoretical models have, as we shall see in the following section (4.6), been directly vindicated by their ability to interpret many features of, for example, oscillating phenomena in biochemical and biological systems. In this section an indication will be given of the kind of modelling that has been undertaken in the last decade or so, since the 'pioneering' phase already described in section 4.2.

In addition to the valuable volume of G. Nicolis and I. Prigogine (1977) on 'Self-organisation in non-equilibrium systems' which is an essential resource in this area, a number of important volumes of collected papers have appeared, often associated with conferences: *Oscillatory processes in biological and biochemical systems*, Vol. I (Frank 1967) and Vol. II (Sel'kov 1971) in Russian; *Biological and biochemical oscillators* (Chance, Pye, Ghosh, and Hess 1973); *Physical chemistry of oscillatory phenomena*, a Faraday Society Symposium (1974); *Membranes, dissipative structures and evolution* (Nicolis and Lefever 1975); *Oscillatory phenomena in biological systems*, a meeting report by Boiteux, Hess, Plesser, and Murray (1977); *Kinetic physikalisch-chemischer Oszillationen*, 1979 Discussion (in German and English), Bunsen Gesellschaft für Physikalische Chemie; and 'Cellular oscillators', a review volume of the *Journal of*

Experimental Biology (1979). Useful survey articles have been written by: Hesse and Boiteux (1971), 'Oscillatory phenomena in biochemistry'; Nicolis and Portnow (1973), 'Chemical oscillations'; Othmer (1977), 'Current problems in pattern formation'; Hess, Boiteux, Busse, and Gerisch (1975), 'Spatio-temporal organization in chemical and cellular systems'; Goldbeter and Caplan (1976), 'Oscillatory enzymes'; Aldridge (1976), 'Short range intercellular communication, biochemical oscillators and circadian rhythms'; Cronin (1977), 'Some mathematics of biological oscillation'; Tyson and Othmer (1978), 'The dynamics of feedback control circuits in biochemical pathways'; Hess, Goldbeter, and Lefever (1978), 'Temporal, spatial and functional order in regulated biochemical and cellular systems'; Hess (1978), 'Oscillations: a property of organized systems'; Berridge and Rapp (1979), 'A comparative survey of the function, mechanism and control of cellular oscillators'; Boiteux, Hesse, and Sel'kov (1980), 'Creative functions of instability and oscillations in metabolic systems'; Hess and Boiteux (1980), 'Oscillations in biochemical systems'; and Rapp (1979a) has compiled an atlas of cellular oscillators.

The theoretical studies of kinetic models may be conveniently subdivided into the modelling of purely temporal self-organizations (section 4.5.1), in which transport processes play no explicit part; and that of spatial self-organization (section 4.5.2), which can accompany and be superimposed on the temporal when transport processes, especially diffuion, are also significantly involved. This latter possibility has, as we shall see (section 4.6.2), only quite recently been observed in the biochemical reaction system of glycolysis (Boiteux and Hess 1971, 1978, 1980) and these observations have further encouraged theoretical modelling of biological self-organization.

4.5.1 *Temporal self-organization*

The biological role of sustained oscillations, i.e. periodicity in time, in biochemical systems, for example in glycolysis (see section 4.6.1), is not at all clear, although it seems improbable that it is entirely unconnected with the slow 'circadian' rhythms of whole biological organisms. But, whatever the functional role of such oscillations, they are of direct pertinence to the ability of molecular systems to become self-organizing, whether at the molecular or the supra-molecular level. For 'organization' is the property of being able to retain and exhibit a pattern, of molecules or of structures made up of molecules, through time. Now *sustained* oscillations are the basis of the continuity of all observable patterns and structures in time, for the amplitudes of spatial oscillations are usually below the predetermined level of discriminating observation at which structures and patterns are discerned—and some spatial oscillations are always occurring at any finite temperature (and are not absent even at absolute zero). So the possibility of molecular and higher-order systems having the kinetic properties that allow self-organization reduces largely, in the first instance, to the possibility of their displaying in time sustained oscillations of the concentrations of their constituents. The important and additional possibility of *spatial* pattern-formation is, of course, highly significant and will be considered in the next major sub-section (4.5.2). However the possibility of sustained temporal oscillations is a primary prerequisite, kinetically speaking, of all self-organization, and its mathematical feasibility has already been expounded (section 4.3).

Such undamped, sustained oscillations appeared to be generally ruled out for linear (first-order) chemical reaction systems in the pioneering studies of Denbigh et al. (1948), already mentioned (section 4.2), and this conclusion was also drawn by Hearon (1953) for an arbitrarily organized linear reaction system containing no irreversible steps and by Meixner (1949), on thermodynamic grounds. Subsequently it has been shown (Seelig 1970) that a system, consisting of (i) a chemical reaction, whose rate is linearly dependent on the concentrations of a catalyst, and (ii) a chain of successive first-order reactions, whose input is the product of (i) and whose output is its catalyst, can display undamped purely sinusoidal oscillations in reaction rates and concentrations of intermediates. Seelig describes such systems as characterized by a *closed loop* in the flux of *information* but an *open chain* in the flux of *matter*, the catalyst. So non-linearity is not itself an indispensable condition for undamped oscillations, even though the sequential autocatalytic systems which a number of the pioneering investigators (see section 4.2) demonstrated to be capable of sustained oscillations were, in fact, non-linear (e.g. Lotka (1920a,b) and Moore (1949), as well as Bak (1963)). At this stage of development of ideas on oscillating chemical systems it seemed that it was the features of linkage between *autocatalytic* reactions that was the all-important necessary, though not sufficient, condition for sustained oscillations. It was Spangler and Snell (1961, 1967) who demonstrated that a hypothetical but realistic system of two catalytic reactions, cross-coupled by means of classical competitive inhibition reactions (Fig. 4.10), could indeed display sustained periodic behaviour. The parameters η (Fig. 4.10) are the stoichiometric numbers for the association of products (P or P') of one reaction sequence with the catalysts (B' or B) of the other to form inactive complexes (I' and I). When $\eta = 2$, and non-linearity is thereby introduced, the partial derivative of the steady-state reaction flux, J_{ss} (at constant A'), with respect to the concentration of the source substance A, i.e. $(\partial J_{ss}/\partial A)_{A'}$, can become negative over a limited range of A values (between the vertical lines 1–2 and 3–4 in Fig. 4.11). If A is varied cyclically over a range that includes this negative slope region, with A' constant, then the system exhibits a hysteresis loop (points 1 to 2 to 3

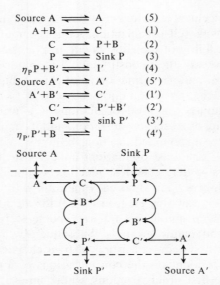

Fig. 4.10. Model cross-coupled catalytic kinetic system of Spangler and Snell. (From Spangler and Snell 1967, Fig. 1, p. 383.)

to 4 to 1) under the steady-state conditions. If the influx of A˙ is fixed, making the concentration A a dependent variable, then this negative-slope region (1 to 3) of A is characterized by instability. For any positive fluctuation in A, causes a decrease in J_{ss}, so that A builds up even more, the fluctuation being thereby amplified, until A reaches point 1 when the rate J_{ss} increases suddenly to point 2. This rise of J_{ss} from point 1 to 2

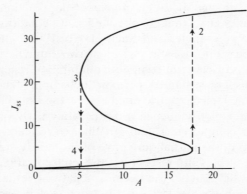

Fig. 4.11. Steady-state reaction flux, J_{SS}, plotted against reactant concentration A, with constant A' for the reaction system of Fig. 4.10. (Fig. 1 of Spangler and Snell 1961—who give details of rate constants, etc., employed.)

MODELS OF KINETIC SELF-ORGANIZATION

then steadily reduces A to its value at point 3 and then suddenly drops to point 4. Again A builds up to point 1 and so the cycle repeats, i.e. oscillates, as a result of the instability of the region of A between 1–2 and 3–4. (All of this assumes steady-state conditions and A' constant.) Spangler and Snell were also able to show that the frequency of this oscillation is less dependent on temperature than are the individual reaction steps and could even become independent, as seems to be the case for a number of observed biological periodic phenomena. Again, as observed with biological periodicities, that of the chemical system could also be made subject to external synchronization and modulation, e.g. by altering the reactant influx or the total catalyst concentration with a different phase or frequency. This model, very early in the development of theoretical studies, demonstrated that 'biological periodicity may be based entirely on chemical kinetic systems, without resort to diffusive or phase boundary effects' (Spangler and Snell 1961, p. 458).

A very comprehensive and general chemical mechanism for producing oscillatory kinetics, without invoking transport effects, was developed by Higgins, first briefly in connection with the oscillation of the glycolytic intermediates in yeast (Higgins 1964) and then in a wide-ranging and comprehensive theoretical treatment (Higgins 1967). In order for a chemical mechanism to exhibit oscillatory behaviour, he deduces that it is necessary that certain general types of reaction pathways exist. Let X be some chemical whose *net* rate of production (v_X) is determined by some arbitrary set of pathways I and let Y be another chemical with the corresponding v_Y and II. Let the collected sets of pathways I and II be represented by large arrows and let small arrows from X and Y directed toward the large arrows indicate the influence of X and Y on the net fluxes, v_X ($= dX/dt$) and v_Y ($= dY/dt$), where X and Y denote respective concentrations. The terms 'act' and 'inh' written on such small arrows represent activating and inhibiting effects, respectively (and are equivalent to the *signs* of the partial derivatives $\partial v_X/\partial X$, $\partial v_X/\partial Y$, $\partial v_Y/\partial Y$, $\partial v_Y/\partial X$). A generalized 'two-body' (i.e. two-independent-variables) oscillatory mechanism would then be represented as in Fig. 4.12. The conditions, illustrated in Fig. 4.12, that oscillations *can* (not must) exist are then (Higgins 1964, 1967):

(1) One of the substances (say X) must activate its own production (assuming Y remains fixed), i.e. increasing concentration of X will *tend* to increase the net rate of production of X.

(2) The other substance (Y) must tend to inhibit its own net production, which is normally the case for most reactions in which the rate of removal of a reactant increases with its own concentration. So the self-couplings (of X and Y, respectively) must be of opposite character (sign).

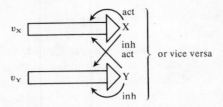

Fig. 4.12. General oscillatory reaction system. (From Higgins 1967, p. 32.)

(3) There must be cross-coupling of *opposite* character. If increasing Y tends to activate net production of X, then increasing X must tend to inhibit the net production of Y, or vice versa. Conditions (1), (2) and (3) apply to the *net* rates of production of X or Y, so many different specific mechanisms can satisfy them. Thus, since one can write

$$\xrightarrow{v_1} X \xrightarrow{v_2}$$
$$\xrightarrow{v_3} Y \xrightarrow{v_4},$$

where v_1 and v_3 are specific rates of production, and v_2 and v_4 of removal, then

$$v_X = v_1 - v_2 \quad \text{and} \quad v_Y = v_3 - v_4.$$

So the *net* rate v_X can be increased either by increasing v_1 or by decreasing v_2 and similarly with v_Y.

The conditions (1), (2), and (3) are necessary but not sufficient conditions for oscillation to occur. Another condition is:

(4) The cross-coupling terms must be greater than the magnitude of the product of self-coupling terms, i.e.

$$\left|\left(\frac{\partial v_X}{\partial Y}\right)\left(\frac{\partial v_Y}{\partial X}\right)\right| > \left|\left(\frac{\partial v_X}{\partial X}\right)\left(\frac{\partial v_Y}{\partial Y}\right)\right|. \tag{4.37}$$

Additional conditions concern the application of (4.37) to systems possessing regions of different character (in the sense of Fig. 4.12)—it has to be applied 'in an average sense' (Higgins 1967, pp. 32, 33)—and the stability of the points, in concentration space, about which the limit-cycle oscillates. Higgins (1967) then applied his criteria to a number of different types of reaction: sequential reactions, a back-activation oscillator, forward-inhibition oscillators, a back-activation/back-inhibition oscillator, and to a forward-activation/forward-inhibition oscillator. Examples of some of these are discussed below in the light of treatments subsequent to that of Higgins, which still remains, however, one of the most comprehensive, primary sources for such studies. He has summarized his 'Oscillatory Rule of Thumb' thus

> When the following conditions are satisfied for two variable chemical systems, then there exist values of the SCV (structural control variables] for which sustained oscillations can be obtained. The conditions are (1) that singularities exist for finite values of the variables (X, Y); (2) that the variables can satisfy the oscillatory net flux diagram [Fig. 4.12] in the region of the singularity as determined by the dominance relations; and (3) that Δ [the r.h.s. minus the l.h.s. of (4.37) with actual signs taken into account] *can* be made positive at the singularity (Higgins 1967, p. 44, (124).)

Higgins' general treatment provided, as he put it, 'a plethora of specific mechanisms easily constructed in terms of known reactions' (Higgins 1967, p. 52) that could give rise, in homogeneous systems, to oscillating reactions, the essential requirement for which is, he showed, the existence of feedback (positive and/or negative) in the reaction pathways. Oscillations in homogeneous systems were not dependent on reaction processes coupling with transport or absorption processes, as such, as had been previously thought (Bak 1963). This analysis led naturally to the concept of a 'source of oscillations', namely, the subset of reaction pathways and chemical variables which possess this necessary feedback—then oscillations generated at this 'source' could be projected into connected reactions with a variety of consequences, parallel in many ways to the response of an electrical circuit to a generator of electrical oscillations. Conversely, according to Higgins, if a perturbation of a single chemical substance in any chemical system leads to any form of oscillatory response, then the system must contain closed reaction pathways which provide feedback. This may occur through bimolecular reactions in which a substance at one stage in a sequence is also involved in an earlier step ('stoichiometric feedback') or through inhibition or activation of specific catalysts by small amounts of reactants ('non-stoichiometric feedback'). Such processes are now known to be widespread in biochemical systems, for they form one of the underlying mechanisms of control in such reaction networks (see the Symposium of the Society of Experimental Biology, XXVII, 1973).

The methods available for analysis of the stability of chemical reaction networks have now reached a high level of mathematical sophistication and applicability with the help of computers (well surveyed by Stucki 1978). Clarke (1974, 1976, 1980), in particular, has developed a form of graph theory applicable to such problems and Stucki (1978, pp. 116–29) has given an account of the algebraic part of Clarke's method.

In the following sub-sections (of 4.5.1) we will be describing models of various types of kinetic systems that are capable of sustained oscillations and that fall broadly within the categories of Higgins (models of spatially

Autocatalysis

Even the simplest autocatalytic process can lead to the possibility of multiple steady states in a reaction system, and so to the possibility of abrupt transitions and hysteresis phenomena. Consider (following Decker 1979) the following autocatalytic/second-order-in-products (ASOP) reaction

$$A + 2B \underset{k_2}{\overset{k_1}{\rightleftharpoons}} 3B, \qquad (4.38a)$$

followed by

$$B \underset{k_4}{\overset{k_3}{\rightleftharpoons}} P, \qquad (4.38b)$$

where the overall reaction is

$$A \rightleftharpoons P, \qquad (4.38c)$$

and so is the conversion of A into P through the intermediate B which also catalyses its own formation in (4.38a). If the concentrations of A and P (also represented by the same italic letters) are kept constant by the system being open to infinite reservoirs of A and P, then both equilibrium and steady-state ($dB/dt = 0$) solutions can be found, and these latter can be triple, since in certain ranges of $k_4 P/k_2$ there are three real solutions for B. These multiple steady states can only arise far from equilibrium (Nicolis and Prigogine 1977, pp. 170–3). A steady-state diagram representing the concentration of the autocatalyst-intermediate B on the ordinate against $k_4 P/k_2$ on the abscissa looks very much like Fig. 4.11, with an intermediate region displaying multiple steady states and hysteresis (1–2 to 3–4 on Fig. 4.11). Because of the possibility of hysteresis, the system has an intrinsic excitability since, as in Fig. 4.11, fluctuations of the concentration of P in the 'unstable' region of negative slope can lead to abrupt transitions from the lower to higher branches, and vice versa, even if the *average* value of P lies well within this region (corresponding to that between 1–2 and 3–4 on Fig. 4.11).

The ASOP reaction of (4.38a) in an open system can be combined in various ways with conventional reactions, and with inflow and overflow involving external sources and sinks. Again following Decker (1979), these can be represented with their respective rate constants thus:

$$\xrightarrow{a_1} A \qquad \underset{\text{shunt}}{\xrightarrow{d}} B \qquad A \xrightarrow{a_2}$$

$$\xrightarrow{b_1} B \qquad B \underset{\text{pumping}}{\xrightarrow{p}} A \qquad B \underset{\text{overflow}}{\xrightarrow{b_2}} \qquad (4.39)$$

and
$$A + 2B \xrightarrow{c} 3B.$$

Some of the possibilities are depicted in Fig. 4.13, as examples of a 'complete system' (Fig. 4.13(a)). In addition to the case mentioned above (4.38), there are three kinds of arrangements (I–III below) that can lead to oscillations.

I. With $p > 0$ and $d = 0$ (Fig. 4.13(b)), the system is that of the well-known trimolecular reaction studied by the Brussels School, and so

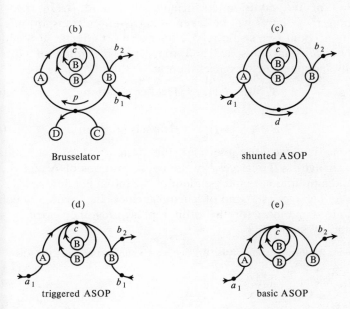

Fig. 4.13. Different types of open systems involving autocatalysis/second-order-in-products (ASOP) reactions. Small dots (.) represent reactions; A, B, C, D are reactants; a_i, b_i, c, d, p are reaction constants. (From Decker 1979, Fig. 1.)

called the 'Brusselator'. In the notation used†, so far, this set of reactions is:

(1) Source, $S \to B$.   Irreversible   (4.40a)
(2) $B + C \to A + D$ Irreversible   (4.40b)
(3) $A + 2B \to 3B$ Irreversible   (4.40c)[as (4.38a)]
(4) $B \to P$, final product Irreversible   (4.40d)[as (4.38d)]

There is inflow and overflow of B, and an auxiliary pumping reaction (2) whereby B is pumped into an energetic state A. The process may be regarded as A periodically discharging, at a critical value of A, through the cooperative ASOP reactions (3) back to B. If the boundary conditions are time-independent, the reactions are proceeding irreversibly (i.e. are very far from equilibrium) and $D \to 0$ and $P \to 0$, then it is still possible to define steady states which may be achieved by conceptually continuous transitions from the equilibrium state, i.e. they are 'on the thermodynamic branch'. But states on this branch prove not to be necessarily stable, because of the cubic non-linearity introduced with the rate equations via (3), which is ASOP. The kind of instability that occurs, at fixed kinetic constants taken as unity, depends on the type of perturbations that are considered, the values of the relevant concentrations of S and C and the values of the diffusion coefficients of A and B. In the present sub-section, these latter will be taken as zero—so no transport effects are taken into account (see section 4.5.2 for transport effects). If S and C are maintained constant in space and time, and the initial fluctuation is practically independent of space, then if

$$C > (S^2 + 1), \quad \text{Decker notation} \quad (4.41a)$$

i.e.

$$B > (A^2 + 1), \quad \text{Brussels notation} \quad (4.41b)$$

fluctuations become unstable and the system undergoes undamped periodic changes, a 'limit-cycle' in the concentrations of A and B, whose period and amplitude are independent of the initial conditions (Fig. 4.14). The system behaves as a kind of 'chemical clock' in a markedly coherent, self-organized manner, for the orbit representing the 'clock' is always asymptotically stable.

† The more familiar notation of the Brussels group depicts these reactions as:

(1) $A \to X$ (4.40a*)
(2) $B + X \to Y + D$ (4.40b*)
(3) $Y + 2X \to 3X$ (4.40c*)
(4) $X \to E$ (4.40d*)

with D and E both going to zero. (The notation in the main text, and that of the Brussels group are related thus (Brussels notation second): $A \equiv Y$; $B \equiv X$; $C \equiv B$; $D \equiv D$; $S \equiv A$; $P \equiv E$.)

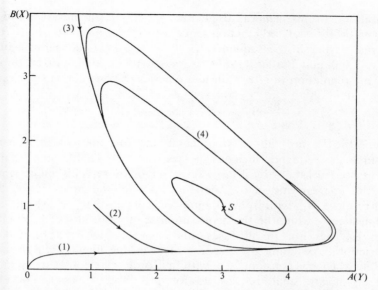

Fig. 4.14. Approach to a limit-cycle in the 'Brusselator' reaction system (4.40) for $S = 1$, $C = 3$ for different initial conditions (Decker notation): (1) $A = B = 0$; (2) $A = B = 1$; (3) $A = 0$, $B = 10$; (4) $A = 3$, $B = 1$. The axes represent concentrations of A and B during the limit-cycle. Notation of the Brussels school in brackets. (From Glansdorff and Prigogine 1971, Fig. 14.4.)

II. With $p = 0$ and positive d or b_1 (4.39), two systems (Fig. 4.13(c) or (d)) are obtained that reflect a more chemically natural situation: a steady supply of substrate A, which gives the product B through an ASOP reaction. At low concentrations of the autocatalyst B, A tends to accumulate followed by a sudden periodical discharge at a critical value, initiated by B from the first-order shunt d (Fig. 4.13(c)) or the inflow b_1 (Fig. 4.13(d)). Either positive d or b_1 secures a finite concentration of B for a restart of the discharge. Asymptotically stable limit-cycles in the concentrations of A and B are the result. In view of these results II, the general system (Fig. 4.13(a)) should also allow oscillatory solutions when there is also a small overflow a.

III. If $p = d = b_1 = 0$, we have a limiting case ($\alpha = 1$) of the oscillator of Sel'kov (1968), developed to account for oscillations in glycolysis (see below under 'Substrate inhibition').

End-product inhibition

The oscillatory system of Spangler and Snell (1961, 1967) has already been instanced as a case of cross-coupled competitive inhibition by one enzyme of another, and Higgins (1964, 1967), as we have seen, includes inhibitions in his general oscillatory reaction schemes. Inhibition in enzymatically catalysed reaction sequences by end-products of a particular enzyme earlier in the sequence has been of direct biochemical interest since Yates and Pardee (1956) showed it to be an important mode of enzyme regulation *in vivo*. Consider a series of consecutive reactions

$$S_0 \xrightarrow{E_0} S_1 \xrightarrow{E_1} S_2 \ldots S_{i-1} \xrightarrow{E_{i-1}} S_i \ldots \rightarrow S_n \qquad (4.42)$$

where the E_i are the enzymes catalysing formation of S_{i+1} from S_i. Assume the sequence includes at least one irreversible step, that S_0 is maintained constant (by a large reservoir or by replenishment) and that E_0, the first enzyme in the sequence, is inhibited in its action by the end-product S_n, with some degree of cooperativity p which, in the simplest case, would be the actual number of molecules of S_n combining with E_0 (but not in all cases, and there is no need to make this restriction). Assuming the enzymes are working linearly where the $S_i \ll K_i$, the Michaelis constants, i.e. with first-order rate constants k_i, and that the dS_i/dt are small relative to forward- and back-rates of enzyme operation, then (Morales and McKay 1967)

$$dS_1/dt = \frac{k_0 S_0}{1 + \alpha S_n^p} - k_1 S_1$$

and

$$dS_i/dt = k_{i-1} S_{i-1} - k_i S_i. \qquad (4.43)$$

Morales and McKay simulated these equations on an analogue computer and found that for certain ranges of values of the parameters of the system, an initial transient phase could be followed by sustained oscillations for an indefinitely long time, with unsymmetrical wave forms of the S_i and with amplitudes dependent on the k_i, but increasing with p if these were fixed.

This kind of system was of interest in connection with other-than-metabolic oscillations. For Goodwin (1963, 1965) had already proposed that, at the epigenetic level, oscillations in the rate of enzyme *synthesis* could be controlled by repression of a genetic locus G. This locus (G) was regarded as producing a messenger RNA (M) which, by combination with ribosomes (to make polysomes), then gives rise to an active enzyme (E), that catalyses the formation of a product (P), m molecules of which combine with each locus G, via a metabolite pool, to repress its genetic activity:

$$G \rightarrow M \rightarrow E \rightarrow P \qquad (4.44)$$
$$\uparrow_{m} \quad \text{Repression} \quad |$$

Goodwin's equations can be cast into the general form (Griffith 1968)

$$dM/dt = \frac{1}{1+P^m} - \alpha M$$
$$dE/dt = M - \beta E \qquad (4.45)$$
$$dP/dt = E - \gamma P,$$

where α, β, and γ are positive. Goodwin (1965) showed by an analogue-computer analysis of these equations (with $m = 1$) that, under determined restricted ranges of the parameters, non-linear oscillations could occur and he examined the synchronous interlocking and subharmonic resonance that could arise from the interaction of such oscillators.

These initial studies of Goodwin and of Morales and McKay continued to be of prime significance even when Griffiths (1968) later showed that some of the supposed oscillations deduced by Goodwin for the system (4.44), with $m = 1$, had arisen out of errors in analogue simulation on the computer, and when Walter (1969a) showed that the sustained oscillations deduced by Morales and McKay for the system (4.42) were the result of computer artefacts. However, oscillations could be obtained in the system (4.44), under certain conditions, when $m > 8$ (Griffiths 1968). This was, of course, not at all likely when m represented the number of molecules of inhibitor combining with one genetic locus G, but does not exclude the possibility of the conditions for oscillation being fulfilled if more than one gene were repressively coupled together. Be that as it may, the work both of Goodwin and of Morales and McKay stimulated a succession of fuller mathematical analyses of sequential systems with feedback inhibition, such as (4.42) and (4.44), and their associated kinetics (4.43) and (4.45). In particular, Walter (1969a,b, 1970) has worked out the conditions for sustained oscillation to occur in the form of a table of values p (in (4.43)), that allow this, as a function of the number (n) of substances in the sequence (4.42). It transpires both from his computer simulations (Walter 1969a,b, 1970), and from explicit analytical solutions by others (Viniegra-Gonzalez 1973; Higgins, Frenkel, Hulme, Lucas, and Rangazas 1973) of the equations, that if macromolecular control circuits involve a sufficient number of kinetically important steps then oscillations can occur at biochemically feasible, low values of the inhibitor 'index' p in (4.42) and (4.43). Thus when $n = 8$ to 10, then, for oscillations to occur, $p \geq 2$ but when $n = 3$, then p has to be equal to or greater than an unlikely value of 9 (Viniegra-Gonzalez 1973, Table 1). Thus, as Walter (1969a, p. 872) says, 'the idea suggested by Goodwin (1963, 1965) and Morales and McKay (1967) (i.e. sustained oscillations can arise naturally in "controlled" biochemical systems [describable by (4.42)]) is fundamentally sound'—even if the joint values of p and n

allowing such oscillations are different from those originally computed. If there are time delays, e.g. because of transport of S_0 from one location to another where the reaction sequence (4.42) occurs, then the value of p required for oscillation is reduced, in the absence of enzymatic destruction or removal (Landahl 1969). More general conditions for sustained oscillations in these systems have been developed (Hunding 1975), the calculations refined and extended by harmonic-balancing analysis to related biochemical networks containing convergent pathways, multiple loops, time delays, etc. (Rapp 1975, 1976; Mees and Rapp 1978), and topological methods have been applied (Tyson 1975). However, in spite of the prevalence of this kind of regulation through *negative* (inhibitory) feedback in such enzymatically controlled biochemical reaction sequences, none, in contrast to those with *positive* (activating) feedback, have been shown, in practice, to exhibit oscillations and so the interest in this kind of system has not been sustained.

Substrate inhibition

One of the simplest inhibitory enzymatic reaction schemes that Higgins (1967) considered was (his Table I, equation 8; Table II, rate law (d); and Figure 9D):

$$S + E \xrightarrow{k_1} ES$$

$$ES \xrightarrow{k_2} E + P \qquad (4.46)$$

$$S + ES \underset{k_4}{\overset{k_3}{\rightleftharpoons}} ES_2$$

where S = substrate, E = enzyme (initial concentration E_0), P = product, ES = active, enzyme-substrate complex, and ES_2 = inactive complex, so substrate is inhibiting formation of P by removing active ES. Assuming a steady state, with constant *total* enzyme concentration, Higgins deduced

$$-\frac{dS}{dt} = \frac{k_1 E_0 S}{1 + \left(\dfrac{k_1}{k_2}\right)S + \left(\dfrac{k_1 k_3}{k_2 k_4}\right)S^2}, \qquad (4.47)$$

and this shows that dS/dt plotted against S rises fairly steeply to a maximum, from which it declines more slowly in a curve convex to the S axis. If the rate of transport of the substrate into the well-stirred reaction system is given (Degn 1968) by:

$$\frac{dS}{dt} = K_S(S_x - S), \qquad (4.48)$$

where K_S is a constant and S_x is the concentration of S in the reservoir, then dS/dt falls linearly with S. This line usually cuts the curve of (4.47) at three points (Degn 1968, Fig. 1), one of which corresponds to an unstable stationary state and the other two to stable ones. Degn (1968) showed that such stationary states existed when the substrates were oxygen (from a gaseous reservoir) and NADH and the enzyme was horse-radish peroxidase.

More generally, if one also allows for first-order decay of the product, and also for product inhibition one obtains (Sel'kov 1967, in Frank 1967; Sel'kov 1968):

$$\frac{dS}{dt} = K_S(S_x - S) - f(S, P) \qquad (4.49a)$$

$$\frac{dP}{dt} = \mu[f(S, P - K_P P], \qquad (4.49b)$$

where $f(S, P)$ is a decreasing function of both S and P, such as (4.47), and μ is related to the stoichiometry and enzymatic constants. If the degree of inhibition is equal to or greater than 2, Sel'kov (1968) showed that limit-cycles, i.e. oscillations, are possible, if certain other parametric conditions are satisfied.

Another particularly fully-analysed model of a substrate-inhibition system is that of Seelig (1976a,b). His reaction scheme is

$$X + Y \rightarrow P + Q, \qquad (4.50a)$$

catalysed by M (an enzyme or a transition metal, say) via

$$M + X \rightleftharpoons MX \qquad (4.50b)$$

$$MX + Y \rightarrow M + P + Q \qquad (4.50c)$$

or

$$MX + Y \rightleftharpoons MXY \rightarrow M + P + Q. \qquad (4.50d)$$

If this is an open system, far from equilibrium, and one in which excess of X can render MX inactive (or at least a less reactive complex) by

$$X + MX \rightleftharpoons MX_2, \qquad (4.50e)$$

i.e. by substrate inhibition, then the system displays: hysteresis in the concentration of X as a function of the rate of its own inflow (when Y is constant); and in certain ranges of the parameters, spontaneously undamped oscillations (which can, in the case of sufficiently large spatial dimensions, also become standing periodical dissipative structures or running chemical waves, see section 4.5.2 below). Seelig (1976a) discusses how such a system might be realized chemically in practice, since it includes no improbable bimolecular or autocatalytic stages.

Other kinetic models that include substrate inhibition have been proposed in combination with various kinetic features, for example: with product activation (to explain glycolytic oscillations, Sel'kov (1968)); with both product activation and allosteric enzymatic properties, also to explain glycolytic oscillations (Goldbeter and Lefever 1972); and with allosteric and isoteric enzyme mechanisms to explain hysteresis of the input characteristic (reaction rate vs. concentration of another noninhibitory substrate) and sustained oscillations (Kaimachnikov and Sel'kov 1977).

Activation by product

In a reaction sequence, one of the reaction products may activate an enzyme earlier in the chain thereby stimulating its own synthesis and that of its precursors, back to the enzyme in question. The equations of Sel'kov given in (4.49) apply again, with the first term on the r.h.s. of (4.49a) replaced by a more general term for the rate of entry of substrate, and the function f now an *increasing* function of the product P, at least within a prescribed range of concentration. Sel'kov (1968) found that sustained oscillations of the limit-cycle type can arise if the degree of product activation is greater than unity (i.e. P occurs in $f(S, P)$ to a power greater than one).

Cooperativity in allosteric enzyme action

Although substrate inhibition by itself can be 'the cause of oscillations, independently of the specific molecular mechanism of action of an enzyme [i.e. whether it is allosteric or isoteric]', according to Kaimachnikov and Sel'kov (1977), yet in the combination of substrate activation or inhibition *plus* product activation *plus* allosteric enzyme action the specific contribution of the enzyme mechanism cannot be so readily dissociated from the oscillatory behaviour of the whole system, when the necessary conditions are fulfilled.

The earliest attempt to examine the role of cooperative, 'allosteric' behaviour of a 'dimeric' enzyme on its ability to undergo oscillations in its activity, when this was enhanced by the product, was that of Goldbeter and Lefever (1972), who had in mind specifically the oscillating activity of phosphofructokinase (PFK) in the glycolytic pathway and its activation amongst other things by its product, adenosine diphosphate (ADP); its inhibition by excess of one substrate, adenosine triphosphate (ATP); and its activation by the other, fructose-6-phosphate (F-6-P). Their treatment (further elaborated by Goldbeter and Nicolis 1976) goes beyond those already mentioned (e.g. of Higgins (1964) and Sel'kov (1968)) in explicitly involving the parameters characterizing the cooperativity of an allosteric enzyme (i.e. PFK). They assumed that structurally the allosteric enzyme consisted of two 'protomers', using the terminology of Monod *et al.* (1965), and that, as a dimer, it can exist in two conformations, R and T, which may differ in their catalytic activity and affinity toward the substrate. The transition between these two forms is reversible and concerted,

$$R_0 \underset{k_2}{\overset{k_1}{\rightleftharpoons}} T_0 \tag{4.51}$$

and the allosteric constant

$$L = k_1/k_2 = (T_0/R_0)_{\text{equilibrium}} \tag{4.52}$$

where the zero subscripts denote no binding of any ligand.

Initially the assumption was made that R was active and T was inactive as enzyme, that the substrate binds to both forms while the product (a positive effector) only to the active form T, but in their 1976 paper Goldbeter and Nicolis treated the more general case when both have activity. The forms of R carrying the substrate were taken to decompose irreversibly to yield products, so the enzyme was regarded as operating far from equilibrium.

Environmentally, the system was regarded as open, with substrate supplied at a constant rate and product removed at a rate proportional to its concentration (factor, k_S). The system and the various kinetic constants for binding and dissociation of the substrate (S) and product (P) for the R and T forms are given in Fig. 4.15, in which R_{ij} denotes the R-state binding i molecules of S and j molecules of P. The substrate exhibits differential affinity towards the two enzyme conformations (K-effect) and states R and T differ in their catalytic properties (V-effect—so it is a mixed K–V system in the terminology of Monod *et al.* 1965).

Application of conservation equations, kinetic equations based on fast equilibration of the enzyme with respect to metabolites, and the equations of Monod *et al.* (1965), finally yields (Goldbeter and Lefever 1972; Goldbeter and Nicolis 1976) the following equations for the time-dependence of the reduced concentration of S (viz. $\alpha = S/K_R$) and of P, (viz., $\gamma = P/K_R$):

$$\frac{d\alpha}{dt} = \sigma_1 - \sigma_M \Phi \tag{4.53a}$$

$$\frac{d\gamma}{dt} = \sigma_M \Phi - k_S \gamma, \tag{4.53b}$$

where

$$K_R = d/a \quad \text{(see Fig. 4.15)}, \tag{4.53c}$$

$$\sigma_1 = v_1 K_R^{-1}, \tag{4.53d}$$

$$\sigma_M = 2kD_0 K_R^{-1}, \tag{4.53e}$$

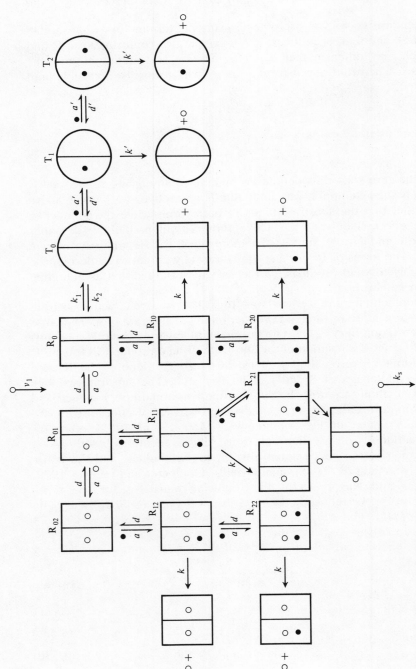

Fig. 4.15. Model of an allosteric dimer enzyme activated by the reaction product. The substrate (●) is supplied at a constant rate; the product (○) binds exclusively to the active R state of the enzyme and is removed proportionally (constant k_S) to its concentration. a, d are association and dissociation constants of the enzymatic complexes; k, k' are irreversible decomposition-rate constants of enzyme forms R and T. (From Goldbeter and Lefever 1972, Fig. 1.)

with k as in Fig. 4.15 and $D_0 = \Sigma R + \Sigma T$, and

$$\Phi = f(\alpha, \gamma, L, k/d, k'/d', k'/k) \qquad (4.53f)$$

i.e. Φ is a function of the substrate and product concentrations and of the various rate constants depicted in Fig. 4.15.

Stability analysis of these equations (Goldbeter and Lefever 1972; Goldbeter and Nicolis 1976; see also Babloyantz and Nicolis 1972) shows that, neglecting all diffusion effects, there are conditions when the steady state ($\dot{\alpha} = 0 = \dot{\gamma}$) can become unstable and that near and beyond this critical situation (that involves specific values of functions of the terms and functions appearing in equation (4.53) above) the steady state is enclosed in α–γ space by a limit-cycle (Fig. 4.9(a)), corresponding to sustained oscillations in time (Fig. 4.9(b)) of the concentrations α ($= SK_R^{-1}$) and γ ($= PK_R^{-1}$). This limit-cycle is stable since it can be reached from inside, starting from the unstable steady state, or from outside.

Goldbeter and Lefever (1972) were able to construct stability diagrams showing the ranges of the allosteric constant L which would allow limit-cycles to occur, as functions of c ($= K_R K_T^{-1}$ for substrate $= (d/a)/(d'/a')$), in Fig. 4.15, i.e. the ratio of dissociation constants of the ligand from the R and T forms); of ε ($= k/d$); and of σ_1 (4.52d). These diagrams each contain three domains: one in which the system remains stable; a second in which limit-cycle oscillations can occur; and a third containing no physically acceptable steady state, usually corresponding to values of the allosteric constant, L, that are large but still often actually observed with allosteric enzymes. The diagrams also show that substrate inhibition of the enzyme is not a prerequisite for limit-cycle behaviour (the system can oscillate if $c = 0$ and $k'/k = 0$) but other relationships between values of parameters allowing oscillations are too complex to allow simple generalizations and are best expressed in the form of the stability diagrams. Broadly, it is clear that the positive feedback of the product and the cooperativity of the ligand interactions with the enzyme are the two main features of the system responsible for instability of the stationary state and so for the occurrence of oscillations.

Cooperativity is conveniently, if not precisely, expressed in accounts of allosteric enzyme behaviour in terms of the 'Hill coefficient' n_H which is defined as

$$n_H = d \log(\theta/(1-\theta))/d \log \alpha \qquad (4.54)$$

where θ is the fraction of its maximum value of either the rate of enzyme reaction, or of its binding of substrate. When $n_H = 1$, there is no cooperativity, no oscillations and Michaelis–Menten kinetics are obeyed; if $n_H \gtrless 1$, there is positive or negative cooperativity, respectively. Goldbeter (1977) investigated the relation between values of n_H for the allosteric dimer model of Fig. 4.15, with positive feedback from the reaction product. He was able to show that sustained oscillations are associated for this model in the unstable steady state with large values of n_H, close to 2, the number of protomers in the enzyme molecule, and he concluded that positive enzyme cooperativity plays a primary role in the mechanism of such periodicities controlled by an allosteric enzyme activated by its product. Increase in the number of promoters, and so in the cooperativity, still allows oscillations (Venieratos and Goldbeter 1979; Goldbeter and Venieratos 1980), when the substrate and product concentrations evolve on comparable time-scales, but with a decrease in amplitude and with a decrease in the permitted values and range of substrate injection rates (σ_1). These results contrast with that obtained for models of enzyme systems inhibited by end-products, when the amplitude of any allowed periodicities increases with the degree of cooperativity of the regulated enzyme (Walter 1970).

Covalent enzyme modification

One of the most important modes of metabolic control, other than that by allosteric enzymes, is the reversible covalent modification of an enzyme. Martiel and Goldbeter (1981) have analysed the conditions under which sustained oscillations could develop in a biochemical system regulated autocatalytically by a reversible, covalent enzyme modification, where the product of the modified enzyme activates the enzyme that catalyses the covalent modification. This analysis was undertaken because such a situation could reasonably be postulated as occurring in the slime mould *D. discoideum*, in which adenylate cyclase (or guanylate cyclase) would then putatively be regarded as activated through phosphorylation by a protein kinase which would itself be subjected to positive feedback through activation by cyclic AMP (or cyclic GMP). Their results indicate that sustained oscillations around a non-equilibrium unstable steady state can occur as an effect of positive feedback exerted through covalent modification of a key enzyme. Furthermore, for certain values of the controlling parameters close to those that produce oscillations, the system was excitable since it could amplify in a pulsatory manner chemical perturbations exceeding a threshold. Although there is no evidence at present that adenylate cyclase is in fact controlled by covalent modification, these results are clearly relevant to the interesting slime-mould system with its extraordinary pattern-forming abilities (discussed further below in section 4.6.2) and are of general significance in that they add a new mode of enzyme regulation to those that have previously been shown capable of producing periodic or excitable behaviour (reviewed by Goldbeter and Caplan 1976).

The foregoing summarizes some of the kinetic models of systems that can give rise to patterns of concentration-variation in time and, in particular, sustained oscillations, since these latter, as we shall see, occur in biological systems and even irregular temporal variation may also be of significance in morphogenesis. Models not discussed include those that involve branchings (see Nicolis and Portnow 1973, Section III A6); those that take account of imposed fluctuations (e.g. Hahn et al. 1974; see also Stucki 1978, pp. 171–3); and oscillations exhibited by reactions in a membrane that involve both chemical and diffusion coupling and auto-catalysis (Katchalsky and Spangler 1968; Caplan, Naparstek, and Zabusky 1973), e.g. the oscillations in the pH recorded by a glass electrode coated with a protein–papain membrane and immersed in a high pH solution of a synthetic substrate, benzoyl-L-arginine ethylester (Naparstek, Thomas, and Caplan 1973). It is apparent that the *stability* of any biochemical system of interlocking reactions is of prime importance in understanding how it functions *in vivo*. A valuable guide to the stability analysis of biochemical systems has been developed by Stucki (1978), who provides not only the mathematical background but practical guidance on the application of computers to the analyses.

4.5.2 *Spatial self-organization*

The fundamental mathematical possibility of spatial pattern-formation ('space-symmetry breaking') in reaction–diffusion systems, broadly governed by a set of equations such as (4.29), has already been discussed (section 4.3). It should be stressed that much of the mathematical interpretation (see Murray 1977; Fife 1979) of spatial phenomena in an ecological context, concerning the interactions of biological populations with diffusion (Levin 1977) is formally within the same class. Here we now consider some of the kinetic models which have explicitly taken transport processes, especially diffusion, into account in analysing the evolution of particular chemical and biochemical reaction systems—following the fundamental clues provided by Turing (1952). These were further expounded by Gmitro and Scriven (1966) and Othmer and Scriven (1969), with interesting comparisons to chemical-engineering situations in which transport and transformation processes inevitably mutually interfere in ways dependent on the geometry and scale of open-systems boundaries (e.g. whether one-dimensional rods, fibrils, and tubules, or two-dimensional sheets, membranes, or surface layers). One of Turing's reaction schemes (his second, see Fig. 4.6) was also analysed thermodynamically by Prigogine and Nicolis (1967) by the methods of the Brussels school, and, as already described (Section 4.2) they too concluded that this system (Fig. 4.6) had well-defined space symmetry-breaking instabilities beyond certain critical affinities, which then led to new stable steady states that would be inhomogeneous. An important development which generalized further the genuine validity and biological significance of space symmetry-breaking behaviour, was made by the Brussels school (Prigogine *et al.* 1969) when they showed that three particular catalytic biochemical systems that had previously been shown to be capable of *temporal* chemical oscillations, could also become unstable with respect to diffusion and so exhibit spatial patterns and inhomogeneities. The three systems in question were: the dark reaction in photosynthesis (Chernavskaia and Chernavskii 1961); the substrate-and-product-inhibited system of Sel'kov (1967) as given in equation (4.49) above but with diffusion terms, $D_S(\partial^2 S/\partial r^2)$ and $D_P(\partial^2 P/\partial r^2)$ added to (4.49a) and (4.49b), respectively, to allow for diffusion of substrate S and product P ($r =$ the single space coordinate); and a product-activated enzyme reaction, again that of Sel'kov (1968, see preceding section, under 'Activation by product'). The characteristic lengths of the spatial inhomogeneities so arising in these systems are of the order of 10^{-2} to 10^{-4} cm, which are large with respect to molecular dimensions, and so manifest a *macroscopic* level of spatial order.

The 'Brusselator'

The possibility of formation of spatial order in an autocatalytic reaction system has been firmly established by allowing for the effects of diffusion in the trimolecular reaction scheme (4.40) and (4.40*), the so-called 'Brusselator' (Tyson 1973):

(1) $A \xrightarrow{k_1} X$ (4.55a)

(2) $B + X \xrightarrow{k_2} Y + D$ (4.55b)

(3) $Y + 2X \xrightarrow{k_3} 3X$ (4.55c)

(4) $X \xrightarrow{k_4} E$, (4.55d)

using here the notation of the Brussels school for this reaction scheme (previously given under (4.40*), rather than that of Decker in (4.40) and Fig. 4.13 (see footnote, p. 160)). From now on, we shall employ this notation of the Brussels school. (For their treatment in detail, in addition to the references cited in the text and diagrams see also; Lefever 1968; Prigogine and Lefever 1968; Lefever and Nicolis 1971; Herschkowitz-Kaufman and Nicolis 1972; Nicolis and Auchmuty 1974; Auchmuty and Nicolis 1975, 1976; Herschkowitz-Kaufman 1975.)

As before, A, B, D, and E are all open to the outside, and for simplicity, set all kinetic constants to unity, and so all equilibrium constants too (i.e. $(A/E)_{eq} = (B/D)_{eq} = 1$). The system is regarded as evolving only in a single space dimension, r, so

$$dX/dt = D_X \, \partial^2 X/\partial r^2 + \text{reaction terms,}$$

(where D_X = diffusion coefficient of X) and similarly for dY/dt. In the homogeneous steady state, these two rates of change are zero and $X_0 = A$ and $Y_0 = B/A$. For such steady states of the system, fluctuations are damped to zero, even if initially oscillating, so the steady state is stable with respect to arbitrary perturbations. If $D \to 0$ and $E \to 0$, and the system is far from equilibrium and so reactions are proceeding irreversibly, and the boundary conditions continue to be time-independent, then it is still possible to define steady states which may be achieved by continuous (conceptual) transitions from the equilibrium state: that is what 'being on the thermodynamic branch' means. But states on this branch prove not necessarily to be stable, depending on the type of perturbations which occur and the values of the concentrations and diffusion coefficients (we are using fixed kinetic constants of unity). Some of the possibilities are as follows:

(a) If diffusion terms, as already discussed in 4.5.1, are suppressed, so the initial fluctuation is practically independent of space; and A, B, D ($\to 0$) and E ($\to 0$) are constant in space and time, and in (4.41b) $B > (A^2 + 1)$, fluctuations become unstable and the system undergoes undamped periodic changes, a 'limit-cycle' whose period and amplitude are independent of the initial conditions (Fig. 4.14). The system is therefore behaving as a chemical clock in a markedly coherent manner.

(b) If, with A, B, D, and E still constant in space and time as before, the initial fluctuations vary with space, the thermodynamic steady state again becomes unstable at a critical value of B which depends on A, D_X and D_Y. If B exceeds this value a stable steady state finally emerges in which the steady stable concentrations of X and Y vary periodically in space, r, with a constant maximum and minimum amplitude and with a definite 'wavelength' Fig. 4.16; (cf. the final, long phase of the Zhabotinsky reaction.) Thus a spatial structure has arisen spontaneously in a previously homogeneous system, under conditions which are far from equilibrium and in which a minimum level of dissipation is required for it to be established. It has a lower entropy than the uniform steady-state and is stabilized by the flow of energy and matter from the outside environment. This is a good example of 'order through fluctuations' (cf. p. 42).

(c) If A is initially distributed *non*-uniformly, in space, while B is still uniform in its distribution (a more realistic model), then non-uniform distributions of A, X, and Y are set up which are not, in this case,

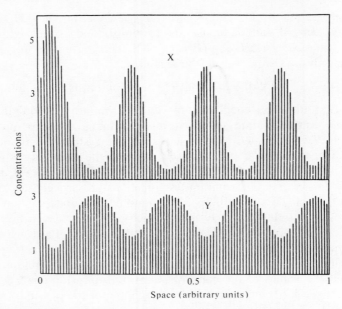

Fig. 4.16. Space-dependent steady state distribution of X and Y for the trimolecular autocatalytic reaction system (4.55) ($\equiv$ 4.40, 4.40*), arising beyond a symmetry-breaking instability. The concentrations of X and Y in the boundaries are maintained equal to the homogeneous steady-state values $X = 2$, $Y = 2.62$. Values of other parameters are: $A = 2.00$, $B = 5.24$, $D_X = 1.6 \times 10^{-3}$, $D_Y = 8 \times 10^{-3}$. (From Prigogine and Nicolis 1971, Fig. 4, after Lefever 1968).

periodic in form and are symmetrical about the spatial centre. The stability of these spatial distributions is different for different sets of initial parameters, as illustrated in the non-equilibrium 'phase diagram' of Fig. 4.17 (see legend for details) which has been drawn to show the situation for different combinations of values of B and D_Y, at constant A, D_A, and D_X. Interesting features emerge for this system in domains II and III of this 'phase diagram', in which the steady state is no longer stable.

In II, a dissipative structure may be established for which the distribution is stable in time, but which is spatially periodic and *localized* within the total available space (Fig. 4.18) and sometimes also *duplicated* i.e. with a plane of symmetry. It determines, as it were, its own boundaries, within the space of the whole system. This is only one amongst several forms of dissipative structures which are solutions to the equations under these conditions, the one being established depending on the type and position of the initial fluctuation. This now represents the introduction of a 'historical' element in determining the evolution of the system, along

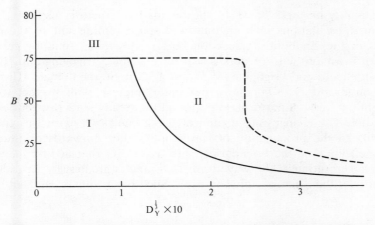

Fig. 4.17. Stability properties of the trimolecular autocatalytic reaction system (4.55)—see text. Non-equilibrium steady-state solutions are shown as a function of the amount of B present and of the diffusion constant (D_Y) of Y. Domain I: steady-state stable with respect to fluctuations in composition. Domain II: fluctuations increase monotonically. Domain III: fluctuations amplified and develop into oscillations. Steady-state unstable in II and III. (Conditions: average $A = 14.0$, $D_A = 1.97 \times 10^{-3}$, $D_X \equiv 1.05 \times 10^{-3}$ and see text.) (From Prigogine and Nicolis 1971, Fig. 7.)

with causal laws developed from the initial differential equation of the rate processes. The importance for biogenesis is apparent.

If B is large and D_Y is close to D_X then (with values of D_A, D_X and the mean values of A, X, and Y similar to those which generated the state depicted in Fig. 4.18), the system moves into domain III and a new kind

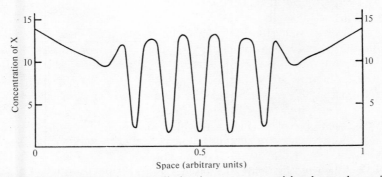

Fig. 4.18. Localized steady-state dissipative structure arising beyond a point of instability in the autocatalytic reaction system (4.55) (see text). Conditions: $D_X = 1.05 \times 10^{-3}$; $D_Y = 5.25 \times 10^{-3}$; $B = 26.0$; mean $X = 14.0$; mean $Y = 1.86$. (From Prigogine and Nicolis 1971, Fig. 8.)

176 THE KINETICS OF BIOLOGICAL SELF-ORGANIZATION

of instability emerges in which the dissipative structure displays simultaneously oscillations in both time and space reminiscent of the early stages of the Zhabotinsky reaction before the stable, spatially-periodic pattern is established (for details see, for example, Prigogine and Nicolis (1971); Nicolis and Prigogine (1977, pp. 93 ff.)). So the domain III, when B is large and $D_X \sim D_Y$, is of particular interest, with respect to the evolution in time of perturbations around the steady-state distributions of X and Y. The changes are complicated but can be described as corresponding to the propagation of wavefronts either outwards or inwards, after reflection at the boundaries, with a velocity that at first increases with concentration at the wavefront itself, and is greater, by an order of magnitude, than that of diffusion itself. Such behaviour is depicted more

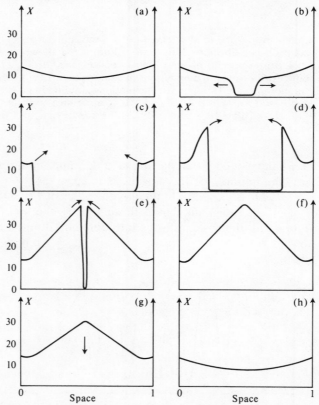

Fig. 4.19. Characteristic stages of evolution of the spatial distribution of X, for the system (4.55), during one period of the wave. $D_A = 1.95 \times 10^{-2}$, $D_X = 1.05 \times 10^{-3}$, $D_Y = 66 \times 10^{-5}$, $B = 77$, $\bar{X} = \bar{A} = 14$, $\bar{Y} = B/\bar{A} = 5.5$. (From Herschowitz-Kaufman (1973) reproduced by Goldbeter and Nicolis 1976, Fig. 1.6.)

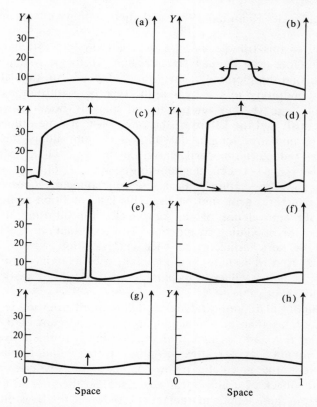

Fig. 4.20. Evolution of Y in the system (4.55) under the same conditions as in Fig. 4.19. (Fig. 1.6 of Goldbeter and Nicolis 1976, from Herschowitz-Kaufman 1973.)

precisely in Figs. 4.19 and 4.20 (from Herschkowitz-Kaufman (1973) reproduced by Goldbeter and Nicolis (1976)) which show, respectively, how X and Y vary with time at particular spatial locations along the one dimension (r) of diffusion. Notice that in this case, too, there can be some degree of localization and also of duplication (e.g. Fig. 4.19(c), (d); Fig. 4.20(c), (d)). In the fuller and more general treatments (e.g. those of Nicolis and Prigogine 1977, or of Erneux and Herschkowitz–Kaufman 1975) the requirement that kinetic constants be unity is relaxed; the quantities t, X, Y, A, B, and the D_i then refer to scaled variables.† The possibility of spatially rotating waves has also been demonstrated (Erneux

† The scale variables (t, X, Y, A, B, D_i) are related to the ordinary scales of time, concentrations, etc. represented by a prime, thus: $t = K_4 t'$; $X = (k_3/k_4)^{\frac{1}{2}} X'$; $Y = (k_3/k_4)^{\frac{1}{2}} Y'$; $A = (k_1^2 k_3 / k_4^3)A'$; $B = (k_2/k_4)B'$; $D_i = D_i'/k_4$.

and Herschkowitz-Kaufman 1977) and their stability examined (Erneux 1981).

Analysis of this trimolecular, autocatalytic system (4.55) has clearly shown, therefore, that the new stable regime which the system ultimately reaches beyond the thermodynamic branch depends on: the values of the boundary conditions; in some cases, the type of perturbations acting on the system (e.g. whether temporally or spatially homogeneous or inhomogeneous); and the values of the rate constants, the diffusion coefficients, D_i, and the length, L, of the system along the diffusion dimension—indeed, more particularly, on the ratio $D_i//L^2$ which measures the coupling between neighbouring spatial regions, i.e. only if the system is above a certain critical size can coupled, coherent, spatio-temporal structures arise and persist. This last condition amounts to the approximate equivalence of the time-scales for diffusion through the system and for oscillation in reaction. The combination of these conditions can be very stringent, so each real reaction system, even if it corresponds to a well-analysed model, has to be examined carefully with respect to the actual values of its parameters to see if any spatio-temporal organization is possible.

This trimolecular, autocatalytic model is more relevant to biological reaction systems than at first might appear on account of the rarity of *tri*molecular reactions, as such. For the non-linearity this introduces into the rate equation is a feature of other entirely bimolecular reaction schemes (e.g. the glycolytic pathway) when this involves positive or negative feedback of a non-linear character. So the general picture that emerges from these kinetic mathematical studies of the Brussels school on the trimolecular reaction system (4.55) are valuable in indicating that such non-linear reaction systems, open and far from equilibrium, can, when transport processes are also involved, give rise to: (i) a standing concentration-wave exhibiting limit-cycle behaviour—and so a possible basis of biological oscillations; (ii) a steady-state regime exhibiting a spatial organization, which may be localized and even duplicated, and which is dependent on the type of the initial fluctuation that brought it about—so providing a possible basis for the system being able to store information (i.e. have a 'memory') or, at least, introducing a historical element into the determination of the course of its future evolution; and (iii) a propagating concentration wave with an abrupt wave front moving in space at speeds greater than that of diffusion, and sharply localized in space and time—and so able to perform regulatory functions or to act as transmitters of information in the form of chemical signals. The role of reaction–diffusion kinetic processes in biological self-organization has therefore been shown by the analyses of this hypothetical model to be a distinct possibility for non-linear biochemical reaction schemes that involve not only inhibitions and activation, but also transport processes.

Substrate inhibition

Mimura and Murray (1978a) have introduced the possibility of diffusion into the very feasible, and potentially applicable, substrate-inhibition model of Seelig (1976a, b), already described (section 4.5.1, p. 165) by allowing the two species it involves (X and Y) to diffuse along one particular coordinate. They show that, by introducing such diffusion terms, the steady state can become unstable through a diffusion-driven instability and so can result in finite-amplitude spatial structures, in finite domains, with zero flux-boundary conditions. In a spatial context in which the parameters give limit-cycle behaviour, the existence of travelling wave trains is possible, and so are threshold phenomena resulting in propagation of highly stable 'solitary' waves ('solitons') of finite amplitudes in one of the substrates (Britton and Murray 1979). Such solitary waves travel without change of shape, at speeds orders-of-magnitude faster than a pure diffusion process (cf. Murray 1977) and can pass through cell walls and so, because of the threshold behaviour of the reaction mechanism, give rise to avalanche effects. Such a phenomenon clearly provides a basis in reaction–diffusion systems for the transmission of chemical signals ('information') from one part of an organism to another in a single pulse. Another reaction–diffusion model giving rise to similar effects (oscillations and steady-state spatial structures) has been devised in which an enzyme is immobilized within an artificial membrane structure through which substrates can diffuse (Kernevez, Joly, Duban, Bunow, and Thomas 1979; Hervagault, Friboulet, Kernevez, and Thomas 1980): this, too, can give rise to threshold waves (Britton and Murray 1979).

Allosteric enzymes

The system of Goldbeter, Lefever, and others (Fig. 4.15), combining a cooperative allosteric enzyme with substrate activation or inhibition and product activation, has already been discussed (section 4.5.1, p. 166) and its ability to give sustained oscillations, in the absence of diffusion, has been related to the various controlling parameters which, in this case, include the cooperativity of the enzyme. Their original analysis (Goldbeter and Lefever 1972) assumed periodic boundary conditions, i.e. that the value of a dependent variable x at a point r satisfies $x(r) = x(r+L)$, where L is a characteristic dimension of the system—and this ensures that the system undergoes spatially uniform limit-cycles, independently of diffusion. However, when such periodic boundary conditions are replaced by the assumption that the concentrations of the participating molecules are maintained constant at the boundaries (and all other conditions are as before), dynamic space-dependent dissipative structures are shown to arise (Goldbeter 1973). These space-dependent structures take various forms according to the dimensions of the system, the diffusion coefficients and the maintained boundary concentrations of substrate and product, α

180 THE KINETICS OF BIOLOGICAL SELF-ORGANIZATION

and γ respectively, in their 'reduced' form (eqn (4.53)). The space-dependent structures are: time-independent regimes (which actually are still on the 'thermodynamic branch'); and, beyond a critical point of instability of the thermodynamic branch, standing waves and propagating waves. In the domain of parameters that gives rise to sustained oscillations in the homogeneous case, different kinds of regime may be realized successively by increasing the size of the system, for given values of the diffusion coefficients and boundary conditions (concentrations, injection and outflow rates). For still larger dimensions, the solution becomes quasi-homogeneous in space and tends again to the limit-cycle observed in the absence of diffusion effects.

The regimes dominated by diffusion and by the chemical reactions are determined by the relation of the actual length, L, of the system to a

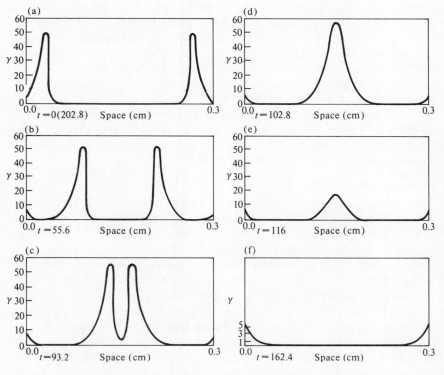

Fig. 4.21. Propagating concentration wave ($L \simeq 10\,L_c$). Distribution of the product concentration (γ) as a function of position at successive time intervals for the allosteric enzyme system of Fig. 4.15 when diffusion is also taken into account. $L = 0.3$ cm; $L_c \sim 2.5 \times 10^{-2}$ cm (eqn. (4.56)); $L/L_c \simeq 12$. (From Goldbeter 1973, Fig. 2, where numerical values of the relevant parameters are given.)

critical length (see Glansdorff and Prigogine 1971)

$$L_c = \left(\frac{8\alpha_0 D_\alpha}{\sigma_1}\right)^{\frac{1}{2}}, \qquad (4.56)$$

where α_0 is the reduced substrate concentration in the steady state and $\alpha = \alpha_0$ at $r = 0, L$; D_α is the diffusion coefficient of the substrate; and σ_1 is the 'reduced' injection rate of substrate $= v_1/K_R$ (4.53d). Taking values of the relevant parameters that pertain in glycolytic pathways, the interpretation of which is in mind in all these model studies, leads to the following schematization (Goldbeter and Nicolis 1976):

$L \ll L_c$: propagation of the boundary conditions inside the system: time-independent regime homogeneous in space.
$0.1 < L/L_c < 1$: time-independent regime on the thermodynamic branch, not necessarily homogeneous in space.
$1 < L/L_c < 10$: standing wave.
$L \simeq 10 L_c$: propagating wave (an example is given in Fig. 4.21).
$L > 10^2 L_c$: oscillations quasi-homogeneous in space.

For a given dimension, the transition between a time-independent solution and a standing wave can be attained by increasing the diffusion coefficient, D_γ, of the product or by lowering the substrate concentration at the boundaries. Other relationships, when the substrate injection rate σ_1, is such as to satisfy the homogeneous stability condition, are given in Table 4.3.

Coupled oscillations

Living systems display innate rhythms at many different levels both in adult organisms (Goodwin 1963, 1965; Bunning 1960, 1973; Aschoff

Table 4.3. Behaviour of the allosteric enzyme system (Fig. 4.15) as a function of the system dimension L and the diffusion coefficients of substrate (α) and product (γ).

L (cm)	$D_\alpha = 3D_\gamma$ $= 3 \times 10^{-6}$ cm^2 s^{-1}	$D_\alpha = D_\gamma$ $= 10^{-6}$ cm^2 s^{-1}	$D_\gamma = 3D_\alpha$ $= 3 \times 10^{-6}$ cm^2 s^{-1}
10^{-2}	Time-independent regime	Time-independent regime	Quasi-stationary wave
10^{-1}	Spatio-temporal regime	Propagating wave	Propagating wave
1	Quasi-homogeneous oscillations	Quasi-homogeneous oscillations	Quasi-homogeneous oscillations

From Goldbeter and Nicolis 1976, Table 5.1, where numerical values of the relevant parameters are given.

1965; Sweeney 1969; Pavlidis 1973; Winfree 1973a) and in developmental processes (Davidson 1968; Harris and Viza 1972; Bonner 1974). The more detailed analyses of models of biochemical oscillators, an outline of which has just been given, has encouraged a number of authors to examine closely the effect of coupling such biochemical oscillators, in particular those that depend on kinetically non-linear processes, have limit-cycles, and are weakly coupled (e.g. amongst others, Winfree 1967; Pavlidis 1971). The question being asked has been formulated by Winfree (1967) as 'What modes of temporal organization—if any—could result from weak interactions in a population of innately oscillatory devices?'. These analyses have been broadly successful in demonstrating that systems of such coupled oscillators can display many of the characteristics evidenced by biological oscillators such as phase resetting (cf. Winfree 1975), 'entrainment' (the ability to adjust frequency in response to a periodic influence from outside (cf. Andronov, Vitt, and Khaikin 1966)), and splitting of rhythms into new frequencies. The possibility of the development of *spatial* patterning in such systems of coupled oscillators soon became apparent (as it did in Turing's 1952 seminal study) and has been developed as a plausible basis for the spatio-temporal control of developmental processes and pattern formation (e.g. Goodwin and Cohen 1969; Robertson, Drage, and Cohen 1972) and of mitosis (Tyson and Kauffman, 1975, though they also thought a spatially homogeneous limit-cycle could explain the mitotic cycle in *Physarum*); and for a spatial patterning in two diffusion-coupled cells in each of which an oscillatory glycolytic process is occurring (Ashkenazi and Othmer 1978). Perhaps the most detailed study of such possibilities has been made in connection with the developmental transitions in the slime mould, *D. discoideum* (see below section 4.6.2).

Pattern formation

It is widely recognized that one of the basic problems of biology is that of differentiation, namely, how a single type of fertilized egg cell can differentiate into the many different types present in the adult organism. This problem, of how gene expression can be inhibited and this inhibition released in a temporally ordered manner, is only rivalled in its complexity and ability to baffle by the problem of how cells of a given (now presumed differentiated) type can become organized into different patterns in space. As Wolpert (1978) puts it:

> Among the vertebrates as a whole—fishes, amphibia, reptiles, and mammals—there is some variation in cell type, but the key to the different organization of all these forms does not lie in the cells as such: it lies in how these basic building units are arranged in space during development.

Gierer (1974) has also formulated the problem thus:

In the past 20 years much has been learned about how the sequences of nucleotides in DNA we call genes are transcribed into equivalent sequences of "messenger" RNA and how these latter sequences are then translated into specific enzymes and other proteins. Biologists have also discovered mechanisms by which the transcription of genes can be turned on and off. This understanding, however, has shed little light on morphogenesis: the origin and development of form. By what sequence of steps do the enzymes spun in individual cells modify the shape and function of the cells themselves and ultimately dictate the complex architecture in which thousands, millions and even billions of cells are marshalled? What are the biochemical and biophysical processes that generate macroscopic patterns and specify proportions?

One response to these questions is to conceive that the cells come to 'know' their position and are programmed to differentiate appropriately on the basis of such positional knowledge—i.e. first the cells are assigned 'positional information' and then they interpret that information according to their genetic programme. This idea of Wolpert (1969, 1971, 1981) allows both for different distributions in space of the primary information and for variety in its consequences as a result of the variegated nature of the cells receiving the information with respect to their particular, genetically controlled, developmental programmes. There is, according to Wolpert (e.g. 1978), good evidence for this kind of development in biological systems and his articles (1969, 1971, 1978, 1981) and those of other authors should be consulted for fuller substantiation (e.g. Gierer 1974; Goodwin 1963; Goodwin and Cohen 1969; Ciba Symposium 1975; Bryant, Bryant, and French 1977; Othmer 1977; Wolpert 1977; Cowan 1980. See particularly the proceedings of the Royal Society Discussion of 1981). The relevant point here, is the connection between this basis of biological pattern formation in 'positional information' and the foregoing account of spatio-temporal patterning in kinetic systems involving reactions and transport (mainly diffusion) processes. For we have reported the considerable evidence from model studies and their mathematical analysis (often involving bifurcation theory) that such initially homogeneous systems—in appropriate ranges of the controlling parameters and when they are open, non-linear, and far from equilibrium—can develop patterns of concentration of reactants in space that can be stable or propagating or oscillating over longer time periods than those characteristic of the reactions themselves.

Such spatial distribution of substances, that can be 'morphogens' (cf. Turing 1952), that can interact with the reaction systems of cells to induce chains of consequential reactions, could clearly act as the positional informers in the way Wolpert postulates as necessary for pattern formation, i.e. they could act as 'pre-patterns'. Since they can arise spontaneously, this kind of dissipative structure, already fully discussed in Chapter

2, provides perhaps an even more convincing basis for pattern formation than some of those subsequently proposed in relation to Wolpert's hypothesis, e.g. active transport (Babloyantz and Hiernaux 1974), phase shifts between two periodic signals (Goodwin and Cohen 1969), or a diffusion gradient of the conventional kind (Crick 1970; Gierer & Meinhardt, 1972).

The last-mentioned (Gierer and Meinhardt 1972; also Meinhardt and Gierer 1974, Gierer and Meinhardt 1974, and Gierer 1981) constructed a model for cell differentiation in which it is proposed that an activator and inhibitor, both capable of diffusing through tissues, are required. The activator stimulates production of both itself and the inhibitor, which interferes with activation and diffuses more rapidly than the activator, and so has a larger range (hence this is called the 'lateral-inhibition' type of pattern formation). Their mathematical analysis takes account of activation and inhibition reactions together with diffusion and the system was shown to be capable of generating patterns of areas (in the two-dimensional case) of different levels of activation because of the pattern of activator concentration (e.g. Fig. 4.22(a)). Gierer has described these results in qualitative terms as follows.

> In the model the properties of the receptor protein are chosen so that activator molecules, once they are released, can enhance further release. Thus a local head start of activator concentration above the average value can trigger further activation. The simultaneous increase in the release of inhibitor provides a supply of inhibitor molecules that diffuse rapidly and limit total activator production in a wider area. An increase of activator in one region can thus occur only at the expense of a decrease in other regions, provided that both are within range of the inhibitor. The process confines activation to a part of the total area. Furthermore, the formation of secondary centers of activation is inhibited in the vicinity of existing or developing centers of activation. Eventually pattern formation comes to an end and the pattern becomes stable, either because the diffusion of the activator prevents further confinement or because in the activated area nearly all the receptor-protein molecules have occupied membrane sites so that the activator concentration cannot increase further. (Gierer 1974, p. 53).

In this way a 'bristle' type of two-dimensional pattern can be generated when the total area exceeds the range of both activator and inhibitor (Fig. 4.22(a)). In a cylindrical surface, regular spacing of activated areas, analogous to the spacing of leaves in plants and buds in hydras, can be obtained (Fig. 4.22(b), (c), (d)).

This model assumes the existence of a shallow gradient in the distribution of one, at least, of the components of the system (activator or inhibitor sources or particulate structures) and Gierer and Meinhardt offer no explanation how this could come about. This particular assumption has been evaded by Babloyantz and Hiernaux (1975) who have

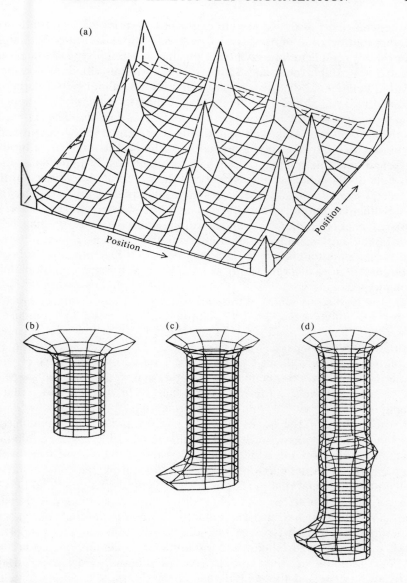

Fig. 4.22. Computer drawings showing how the 'lateral-inhibition' model of Gierer and Meinhardt (see text) can generate different patterns. On a nearly uniform sheet, the model can produce activated regions with a nearly equal spacing that might result, for example, in the formation of bristles (a). Starting from a cylindrical surface, the model is capable of producing a pattern (b), (c), (d) resembling the pattern of hydra buds, appearing alternately on opposite sides of the cylinder. (From Gierer 1974, p. 54, after Meinhardt and Gierer 1974.)

worked out the possibility of pattern formation more fully in relation to the distribution of a 'morphogen' in a three-component chemical system that can induce bacterial enzymes of the Jacob–Monod type (Monod and Jacob 1961; Babloyantz and Sanglier 1972) without assuming any preexisting homogeneity of distribution. They showed that smooth variation in the gradient of the morphogen can generate a very discontinuous response curve for the induced enzyme and so sharp differences between successive cells placed in such a gradient—and so incipient differentiation of boundaries between previously alike cells. Moreover in a diffusion-governed field, the number of cells in which a change is induced depends critically on the length of the system and the boundary conditions (cf. the characteristics of reaction–diffusion systems described above, p. 181). But the distinctive feature of their model is that it allows the generation of 'polarity', i.e. of gradients in concentration (and so of 'sources' and 'sinks' of morphogens), purely by the interaction of reaction and diffusion processes; and that if there are two or more reacting and diffusing substances (cf. Sattinger 1973), a stable pattern—and so, ultimately, structure—can emerge.

It is only wise to refer, at this juncture, to the suggestion of Bunow, Kernevez, Joly, and Thomas (1980) that an instability model of Kauffman, Shymko, and Trabert (1978), for sequential compartment formation in *Drosophila* wing discs based upon reaction–diffusion instabilities, may be difficult to validate because of the sensitivity of wing patterns to the physico-chemical parameters and to the assumed shape of the domain. However, Murray (1981) has been able to show that a reaction–diffusion system that can be driven unstable by diffusion could be responsible for the laying down of the pre-pattern for animal coat markings. He considered in detail a specific, practical, substrate-inhibition reaction mechanism and demonstrated that the geometry and scale of the domain, the relevant part of the integument, during the time of laying down plays a crucial role in the resulting structural patterns—and he was able to relate these to the coat colour distribution in a number of animals. Bard (1977, 1981) has also explored the range of patterns that can be generated by a Turing-type reaction–diffusion system and has particularly studied the role of thresholds and initiating and boundary conditions and he was also able to show great similarities between the theoretically derived patterns and those observed on the flanks of certain mammals, mainly cats and zebras.

This kind of approach based on the reaction–diffusion interaction (see also Erneux and Hiernaux 1980) is typical, as was mentioned at the beginning of this sub-section 4.5.2, of a whole class of mathematical analyses of reaction–diffusion systems conceived in wider terms than purely chemical reaction and linear diffusion of substances controlled by a gradient of chemical potential. Not only can the transport processes of substances be extended to include other modes of transport (including gradients of electrical potential), but also the reference of the mathematics can be widened to purely ecological systems. Thus, for example, Mimura and Murray (1978b) have shown how stable, spatial heterogeneity (patchiness) can arise in certain predator–prey situations (bounded domain with zero boundary flux) when the 'diffusion' of the prey is small compared with that of the predator. The concept of 'symmetry-breaking', the generation of stable, spatial patterns, through the interplay of reaction rates and diffusion, has also had application not only to the problem of pattern formation in embryology and development but also to neurobiology (see Cowan 1980 for a review of some fascinating new developments, as well as of the embryological work). For it turns out that the properties of nets of neurons are very similar to those of the diffusion-coupled reaction systems we have been discussing. A sheet of coupled excitatory and inhibitory neurons can, if excitation exceeds a critical point, undergo symmetry-breaking bifurcations to new states of temporal oscillations, travelling waves, or standing spatial patterns (Wilson and Cowan 1973; Cowan 1980), as in dissipative structures—and in the case of the visual cortex these can be exactly correlated with drug-induced visual hallucination patterns (Ermentrout and Cowan 1979; Cowan 1980).

4.6 Kinetic self-organization in biochemical and biological systems

The previous sections have demonstrated, albeit in a somewhat theoretical and generalized manner, how systems of biochemically feasible reactions can, if they have certain general features, and if particular parameters fall in defined ranges, display regular cycles of concentration changes in time (time-symmetry breaking). Furthermore, if transport, mainly diffusion, processes can also occur, spatial patterns may appear (space-symmetry breaking) which can themselves sometimes change with time. Observation of any such phenomena implies that a bifurcation point has been passed in a system moving away from its equilibrium state to and beyond a steady-state solution of the differential equations that control variation in reactant concentrations. Such a symmetry-breaking system is a dissipative structure, in the thermodynamic sense, and can only be maintained under suitable non-equilibrium conditions (Chapter 2, and section 4.4).

Symmetry-breaking phenomena occur in practice at various levels in the hierarchy of biological organization. So one can designate as 'dissipative' systems of many different kinds: initially homogeneous systems of soluble enzymes (including cell-free extracts of yeast etc. undergoing glycolysis); systems of mitochondria (and of whole cells in glycolysis); systems of cells, whether in suspension, cultures, or *in vivo*, such as the variety of situations listed in Table 4.4, below, of Rapp (1979a); physiological systems (Mackey and Glass 1977); systems consisting of functioning organs; embryonic and differentiating systems; ecological systems, i.e. systems of populations of living organisms, as we have already mentioned (section 4.5.2); and so, perhaps, even evolutionary

Table 4.4. Summary of 'An atlas of cellular oscillators' compiled by Rapp (1979a).

Category	Range of period of oscillations observed	Number of experimental papers reported in this category
I Oscillations in enzyme-catalysed reactions		
(A) Miscellaneous systems	1–10 m	14
(B) The glycolytic oscillator	2 s–3 h	47
(C) Oscillatory ion movements in mitochondria	1–30 m	20
(D) Oscillations in photosynthesis	4 s–80 m	2
II Oscillations in protein synthesis (epigenetic oscillations)	40 m–10 h	17
III Oscillations in cell-membrane potential (see also subsequent sections)	0.25 s–2 m	12
IV Oscillations in secretory cells	0.5 s–5 m	23
V Neural oscillators		
(A) Oscillations in neurotransmitter content and release	2 s–5 m	9
(B) Membrane-potential oscillations in single neurons	10 ms–20 s	75
(C) Central nervous system oscillators (EEG)	20 ms–48 s	18
VI Muscle oscillations		
(A) Skeletal	0.05 s–70 s	8
(B) Cardiac muscle	0.1 s–2 s	59
(C) Smooth muscle	2 s–hours	65
VII Oscillations in cell movement, growth, and development		
(A) Periodic cell movement in *Physarum polycephalum*	1–3 m	29
(B) Periodic movement during aggregation in *Dictyostelium and Polyspondylium*	1.5–10 m	20
(C) Periodic mitosis in *Physarum polycephalum*	8–12 h	4
(D) Periodic spore release and growth in Ascomycetes	6–26 h	4
(E) Periodic events in development (*Hydra*, etc.)	6 s–20 h	11
VIII Miscellaneous	0.5 s–30 m	6

systems of organisms. Thus there is very extensive evidence for the occurrence of kinetic self-organization in living systems which are therefore both symmetry-breaking and dissipative. It is convenient to look first at the much more extensive evidence for the occurrence in biochemical and biological systems of oscillations, i.e. time-symmetry breaking, and then at the more restricted, but highly significant, evidence for space-symmetry breaking.

4.6.1 *Temporal self-organization*

The range of possible periods of purely temporal oscillations in biological systems varies from the long-period oscillations, of circadian rhythms (Bunning 1960, 1973; Hastings and Schweiger 1975; Aldridge, 1976) to the cellular oscillations whose periods are a few minutes or less, seconds even, such as those depicted in Fig. 1.4. Within a single organism, say a mammalian vertebrate, every major organ system can exhibit sustained oscillations under the right conditions, covering periods (e.g. in man) from a thousandth of a second to a month or year. Even when the longer duration oscillations (e.g. circadian rhythms) were excluded, Rapp (1979a), in a valuable 'Atlas of cellular oscillators', had to report over 450 *experimental* papers on biochemical and biological oscillators with periods of an hour or less. His listing of cellular oscillations consisted of eight principal sections and is summarized in Table 4.4, together with the range of periods of oscillation observed within each category—and the number of papers published up to his 1979 compilation. This atlas is a striking indication of both the vast range of experimental work in this field and of quickening interest during the 1970s. He did not include non-biological chemical oscillators (Tyson 1976; Winfree 1975), ecological rhythms (May 1974, 1979) and oscillations of mainly clinical interest (Glass and Mackey 1979) in his atlas.

In another valuable survey, Berridge and Rapp (1979) have reviewed the high-frequency oscillations produced by specific cell types and for which both physiological and biochemical explanations are beginning to appear. They summarized the nature and proposed function of some of these cellular oscillators in the form shown in Table 4.5, which includes both time and space symmetry-breaking processes. Often, however, the cellular oscillatory activity, when spontaneous, has no obvious function and may simply reflect, they suggest, the dynamic nature of cellular control mechanisms.

Within any biological oscillatory activity there are usually only one, or a few, mechanisms whose basic, oscillatory, instability generates the observed rhythm and sometimes, even within this basic mechanism, there is a further concentration of control: thus glycolytic oscillations may well drive a number of cellular oscillations observed at a 'higher' level of

Table 4.5. A summary of the nature and the proposed function of some cellular oscillators.

Tissue or cell type	Nature of oscillation	Proposed function
Dictyostelium discoideum	Periodic release of cAMP and contractility	Aggregation and differentiation
Physarum polycephalum	Shuttle-streaming	Distribution of materials and chemotaxis
Acetabularia	Periodic action potentials	May provide positional information during regeneration
Macrophages	Membrane potential hyperpolarizations	Contractility and possibly chemotaxis
L-cells	Membrane potential hyperpolarizations	Contractility
Aplysia burster cells	Bursts of action potentials	Release of neurohormone
β-cells	Bursts of action potentials	Release of insulin
Anterior pituitary	Action potentials	Release of hormone
Smooth muscle	Slow wave potential changes	Pacemaker activity for myogenic rhythm
Sino-atrial node	Action potentials	Cardiac contraction

From Berridge and Rapp 1979, Table 1).

complexity, yet within itself it is controlled by the oscillatory behaviour of the 'master' enzyme phosphofructokinase, PFK. Sel'kov (1971, p. 5) has even argued that circadian rhythms are generated by a self-oscillating biochemical system (a 'cell clock', he call its) and explains that the longer period of the circadian rhythms could be generated if the period of the controlling cell-clock system were lengthened by a large 'buffer' capacitance in that system. (For example, a large reservoir of polysaccharides can lengthen the temporal changes in fructose 6-phosphate in glycolytic oscillations to periods of the order of several hours to a day (Sel'kov 1972, 1980).)

Because a wide range of cellular processes is carried along, 'entrained', by the basic oscillator(s), it is often difficult to isolate these processes and one has to rely on identifying the input and output properties of the oscillatory system (Fig. 4.23). There seem to be two distinct kinds of cellular oscillators: those located in the surface membrane and those in the cytoplasm. As illustrated in Fig. 4.23, these two types of oscillator may well interact, especially if they have components in common, such as calcium, and they can certainly co-exist—indeed they are unlikely ever to operate entirely independently of each other.

Because of its key role in metabolism and because it is the most thoroughly understood biochemical oscillator, it is worth describing a little more fully the oscillations of the glycolytic pathway.

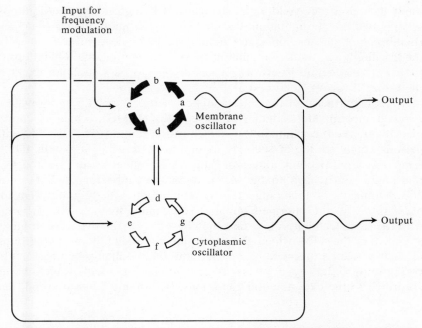

Fig. 4.23. The location and relationships of cellular oscillators. A membrane oscillator composed of a variable number of components (a–d) is responsible for generating a rhythmical output usually in the form of fluctuations in membrane potential. A chemical output may also be generated by various cytoplasmic oscillators (d–g). The two oscillators might be linked to each other by sharing a common component (d). The frequency of these oscillators can be adjusted by a variety of input signals which interact with specific components of the oscillators. (From Berridge and Rapp 1979, Fig. 2.)

Glycolytic oscillations

The first definite observations of oscillations in the concentration of NADH in glycolysis were reported by Duysens and Amesz (1957) and later in intact yeast cells and extracts from them (Ghosh and Chance 1964; Chance, Estabrook, and Ghosh 1964; Chance, Hess, and Betz 1964). Detailed accounts of these observations and the subsequent intensive study of this system, particularly by B. Hess and co-workers at Dortmund, are available in a number of survey articles (e.g. Garfinkel and Hess 1964; Hess and Boiteux 1971; Boiteux and Hess 1975; Goldbeter and Caplan 1976; Hess *et al.* 1978; Berridge and Rapp 1979; Boiteux *et al.* 1980; Hess and Boiteux 1980; Boiteux and Hess, 1981). Glycolysis is the best studied and understood of all instances of periodic behaviour at the metabolic level, for it clearly has importance: it is one of

192   THE KINETICS OF BIOLOGICAL SELF-ORGANIZATION

the principal energy-yielding mechanisms of living cells; it is ubiquitous, so may well be the metabolic 'clock' underlying many longer biological rhythms; and, because phosphofructokinase has indeed turned out to be the 'oscillophore', the key regulator enzyme controlling the oscillations of the whole biochemical network, it has proved to be a suitable object of theoretical analysis (section 4.5).

In yeast extracts, all the intermediates of the glycolytic pathway (e.g. reduced nicotinamide adenine dinucleotide (NADH), which is conveniently observed) oscillate in the concentration range 10^{-2} to 10^{-3} M with periods of the order of seconds to minutes, that increase with fall in temperature. Injection rates of substrates (glucose or fructose 6-phosphate, F6P) have to be within a critical range for oscillations to occur—outside of this range the system adopts steady states. During oscillations, all the intermediates undergo change in their concentration with the same frequency, but differing phase angles relative to each other, as shown in Fig. 4.24, which also illustrates that they fall into two groups, in each of which the maxima and minima of the changes coincide. The two groups differ by a phase angle ($\Delta\alpha$ in Fig. 4.24) which varies according to the experimental conditions. Ghosh and Chance (1964) and

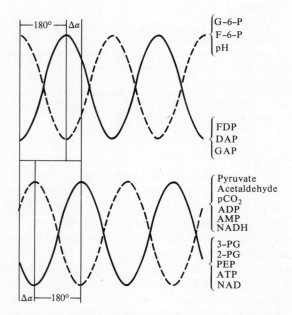

Fig. 4.24. Oscillations of the concentrations of glycolytic pathway intermediates. The amplitudes are normalized. (From Boiteux and Hess in Chance *et al.* 1973, Fig. 3.)

Higgins (1964) suggested that the 'oscillophore' was phosphofructokinase (PFKase, E.C.2.7.1.11) and this has since been adequately substantiated. This enzyme is an allosteric protein, with non-Michaelian kinetics at pH 6.9. It catalyses the quasi-irreversible reaction

ATP + F6P → ADP + FDP (fructose 1,6-diphosphate).

The enzyme is inhibited by its substrate, ATP, and is activated by both its product ADP and also by AMP: it is the concentrations of these adenine nucleotides, rather than FDP and F6P (which also activates it), that control its activity in yeast. A summary of these processes has been given in a particularly accessible way in a diagram of Berridge and Rapp (1979) reproduced here as Fig. 4.25.

Intact yeast cells also display oscillations in the concentrations of glycolytic intermediates both as single cells (Chance, Williamson, Lee, Mela, De Vault, Ghosh and Pye in Chance *et al.* 1973) and in populations of cells (Ghosh and Chance 1964; Hess and Boiteux 1971). These *in vivo* oscillations usually have longer periods than in the cell extracts. It is not yet clear whether such glycolytic oscillations occur in cells other than yeast. In active aerobic respiration, a continuous supply of ATP would, in fact, tend to damp them out. So the wider occurrence of such oscillations is still an open matter though, if they occur, they would undoubtedly be very influential in driving other systems, through fluctuations in the concentration of ATP and hydrogen ions, for example. Another possible biological function of glycolytic oscillations has been suggested by Tornheim (1979) from his observations that such oscillatory behaviour in skeletal muscle extracts involves a different combination of factors and is generated by an AMP-dependent, autocatalytic activation of PFKase by the product FDP. This type of glycolytic oscillation has the advantage of maintaining a higher average ATP/ADP ratio than it would in the steady state that occurs in the absence of FDP-activation of PFKase—and, in the context of conditions of skeletal muscle *in vivo*, there are also other distinct advantages in maintaining a high ATP/ADP ratio (Tornheim 1979).

No other oscillating system of biochemical reactions has been studied as thoroughly and is as well understood as the glycolytic pathway, but it is perhaps not without significance that oscillations have been observed in the three principal pathways of energy supply—glycolysis, photosynthesis in the dark (Table 4.4, I(D)), and in mitochondrial respiration (Table 4.4, I(C)). The glycolytic system, viewed in conjunction with the successful theoretical analyses of it that are described above (section 4.5.1), moreover provides clear evidence of the propensity for a system of biochemical reactions that involves cross-inhibition and cross-activation

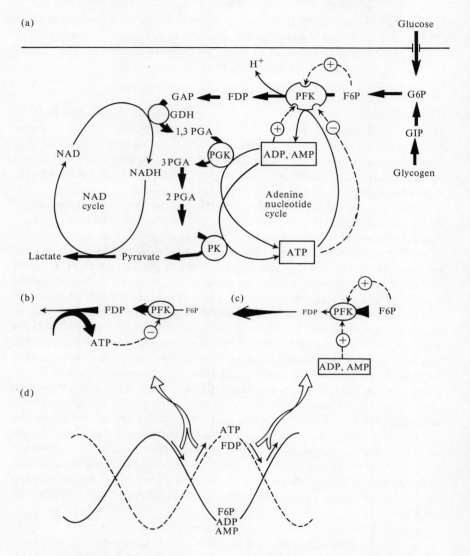

Fig. 4.25. The glycolytic oscillator. (a) A summary of the glycolytic pathway (thick arrows) together with the NAD and adenine nucleotide cycles (thin arrows). The dotted lines represent the allosteric control of phosphofructokinase (PFK) by ATP, ADP, AMP, and fructose 6-phosphate (F6P). The circles indicate those enzymes which seem to be important for oscillatory activity: GDH—glyceraldehyde dehydrogenase; PGK—phosphoglycerate kinase; PK—pyruvate kinase. (b) and (c) Changes in the concentration of key intermediates at two points during the oscillatory cycle. Large-face lettering of metabolites is used to indicate high concentrations while smaller lettering indicates a comparatively low

to self-organize. Such a system is clearly able to display temporal patterns of cooperative, indeed 'corporate', behaviour of the kind that are distinctive of living organisms and that have their basis in the joint character of the kinetic features of the individual processes *and* their mutual systematic cross-connections in the system.

4.6.2 *Spatial self-organization*

We have already seen (section 4.5.2) that systems that can be temporally self-organized—that is display 'time-symmetry breaking', such as oscillations—often also display the possibility of developing spatial patterns. Such 'space-symmetry breaking', we saw, could take various forms: standing, propagating, and solitary concentration waves, steady-state regimes with a spatial organization that may be localized and even duplicated, and other types of pattern formation. It seems that the possibility of space-symmetry breaking, with its potential consequences for spatial pattern formation, results from an underlying temporal organization.

In sub-section 4.5.2, we also saw how the incorporation of transport (mainly diffusion) processes into certain kinds of non-linear differential equations for reacting systems could lead to such space-symmetry breaking in a number of biochemically and biologically pertinent models, as well as in more hypothetical systems such as the 'Brusselator'. Such developments are clearly relevant to the organization of biochemical reactions *in vivo*, the spatial aspects of which have been usefully classified by Hess (1980) into the following two categories.

(1) The *microspace* is classified as the basic reaction space of biochemical reactions in an enzyme or a membrane-bound enzyme (see Table 4.6 which summarizes a number of properties of microcompartments rather similar in size in enzymes or membrane-bound enzymes). Within this microspace, also described as the 'active site compartment', elementary cyclic and concerted chemical reactions occur, such as acid–base catalysis or redox reactions coupled to diffusion processes through its boundaries. The dynamic nature of the involvement of the spatial dimension at this micro-level is apparent both in the increasingly detailed stereochemical accounts of the functioning of the active sites (now available through X-ray crystallography and nuclear magnetic resonance studies) and in the

---

concentration. In one part of the cycle (b) PFK is being switched from an active to an inactive state by ATP whereas later in the cycle enzyme activity is switched back to the active state as ADP, AMP, and F6P begin to accumulate (c). (d) As PFK is switched back and forth between its two activity states, the glycolytic intermediates oscillate with some components 180° out of phase. (From Berridge and Rapp 1979, Fig. 3.)

THE KINETICS OF BIOLOGICAL SELF-ORGANIZATION

Table 4.6. Microscopic order in enzymes and membrane-enzymes.

Systems	Reaction space		Time	Mechanism		
	Size (cm^3)	Boundary (cm)	Rate constant ($M^{-1} s^{-1}$)	Diffusion	Transformation ($10^3 s^{-1}$)	Forces
Enzymes	10^{-21}	$4-20 \times 10^{-8}$	$10^7 - 10^5$	3d and 2d*	Proton and electron cycles	Electrostatic
Membrane enzymes			$10^9 - 10^{11}$	2d and 1d		Electrostatic photoinduced charge separation

* d = dimensions
After Hess 1980.

investigation and theoretical analysis of reaction and diffusion in membrane-bound systems. It is not proposed here to expand on these essentially molecular dynamics (for references and an account of his own highly pertinent work on the proton-pumping of the bacteriorhodopsin molecule of purple membrane in *Halobacterium halobium*, see Hess 1980).

(2) The *macrospace* is classified as the basic reaction space of cellular processes defined by cellular compartments surrounded by stable membranes (see Table 4.7). Global transformations are localized in compart-

Table 4.7. Macroscopic order

Systems	Space		Time	Mechanism		
	Size (cm^3)	Boundary	Rate constant† (s)	Transport	Transformation	Control
Mitochondria lyosomes, peroxisomes, vacuoles	10^{-12}	Membrane–protein	10^{-3}	Diffusion, chemical waves, 1:1 translocation	Metabolic cycles, biosynthesis, transcription and translation, information transfer	Allostericity, covalent modification
Bacteria	10^{-10}	Membrane–protein and	10^{-1}			
Cells	10^{-9}	Cellular wall	10^1			

† Approximate turnover times.
After Hess 1980.

ments and interactions between these compartments occur through controlled transport across membranes. The general properties of such cellular macrospace are summarized in Table 4.7 (which excludes the cells with large cellular bodies, such as moulds, fungi, neural cells, or muscles). The size of the cellular bodies and the rate constants of metabolic processes vary over three to four orders of magnitude. How in each cellular compartment, complex transformations are organized in cycles, controlled by feedback networks and coupled to other compartments by transport processes is increasingly well understood. But differentiation *within* any one 'compartment', as statically defined, can also occur both on a temporal and a spatial scale. The former we have already encountered in the oscillations of glycolysis and its associated alternation with glyconeogenesis (for references, see Hess 1980). Observations of spatial order have now been reported for the first time in a biochemical system, again the glycolytic one.

Dynamic compartmentation in glycolysis

The spontaneous transition of a chemical reaction system from homogeneity to spatio-temporal order, that is, generation of periodic variations in space, has been observed with a number of non-biological systems, in particular that of Belousov and Zhabotinsky (Busse 1969; Winfree 1972, 1973b (spiral, rotating and concentric ring waves); see also Tyson 1976, 1977). But it is only relatively recently that a *biochemical* system has been demonstrated experimentally to undergo space-symmetry breaking and the formation of *spatio*-temporal order. These observations, which go back to reports in 1971 of spatial gradients of pyridine nucleotides in initially homogeneous, oscillating yeast extracts (Boiteux and Hess 1971), were made directly visible by high-resolution ultraviolet photographic techniques (Hess *et al.* 1975; Boiteux and Hess 1978; Hess and Plesser 1979; Boiteux and Hess 1980; Hess, Boiteux, and Chance 1980; Hess 1980). These techniques were applied to monitoring the transmission of ultraviolet light ($\lambda = 340$–390 nm) by a glycolysing yeast extract which was 1.2 mm deep contained in a circular quartz dish (of 3.2 cm diameter) under conditions that excluded stirring and convectional mixing. This allowed the two-dimensional pattern of the NAD/NADH ratio to be observed (NAD = oxidised nicotinamide adenine dinucleotide). It was found that a single local trigger signal (an injection of 0.1 micromole pulse of ATP into the centre of the dish) at first (up to 7 min) led to the expected enhancement of the oxidized form, NAD, in a diffusive expanding circle about the central injection point, on account of the rise in 1,3-diphosphoglycerate which increases the NAD/NADH ratio, through the action of glyceraldehyde diphosphate dehydrogenase (E.C.1.2.1.12). This would be expected as the ATP

198 THE KINETICS OF BIOLOGICAL SELF-ORGANIZATION

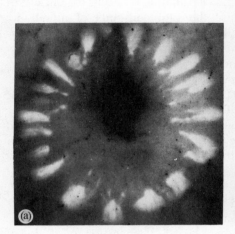
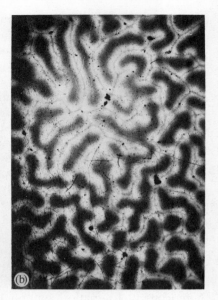

Fig. 4.26. Spatial dissipative structures in yeast extracts (from Boiteux and Hess 1980, Figs. 5, 8). UV-photographs of the pyridine nucleotide pattern. Reduced pyridine nucleotides yield dark structures. (a) Signal transmission. A pulse of 0.1 micromoles ATP was injected via a capillary glass tube into a glycogen-degrading extract at $t = 0$ m. The photograph was taken at $t = 14$ m. (Depth of layer, $d = 1.2$ mm, the dimensions corresponding to the photograph are approximately 11 mm square.) (b) Transformation of the pyridine nucleotide pattern. The oscillating yeast extract was stirred to obtain optical homogeneity at $t = 0$ m. Photograph at $t = 14$ m. (Depth of layer, $d = 1.2$ mm; the dimensions corresponding to the photograph are approximately 13 mm square.)

diffused. However, such bulk diffusion could not explain a subsequently observed pulse (from 8.5 m) of increased ATP concentration, pushing the system into oxidized state and propagating at a constant speed more than two orders of magnitude faster than that attributable to diffusion alone (Fig. 4.26(a)). This indicated an interaction of reaction and diffusion of the kind suggested by the model calculations (section 4.5.2), and was confirmed when continuous injection of AMP gave rise to concentric travelling waves of NADH/NAD patterns. This led Hess and co-workers to look for more stable, possibly periodic, spatial patterns under conditions when temporal glycolytic oscillations could occur in the stirred bulk solution. Now, of course, the extracts were unstirred, and, as before, substrate saturation and depletion were both avoided. Optical homogeneity was ensured by an initial stirring and this marked zero time. An initial sequence of structural development was distinguished by

Boiteux and Hess (1980; see also Boiteux *et al.* 1980) from the subsequent consecutive periods of cyclical transformation of a fairly stable pattern, corresponding to that appearing in the final stage of the initial cycle. The initial cycle included a stage, which did not arise again, with a roughly polygonal pattern strongly reminiscent of the Bénard phenomenon that is manifest when thin layers of fluid are heated. This Bénard phenomenon results from hydrodynamic non-linearity and instability (Chandrasekhar 1961), and constitutes an example of a purely physical dissipative system exhibiting 'order through fluctuation' (Glansdorff and Prigogine 1971, Chapters XI and XII). Perhaps in the glycolytic case, too, some purely hydrodynamic instability may have prevailed at this early stage but, in any case, it was only temporary for a less regular pattern soon took over more and more of the total area; this itself finally disappeared when the bulk solution completed its first NADH cycle, and approached the maximally reduced stage of its temporal oscillation. From this point on, a pattern reappeared that looked very much like the last stage (the 'less regular pattern' above) of the initial cycle and changed its basic structure fairly slowly, as depicted in Fig. 4.26(b), which shows the second and third cycle (see legend for details). The sequence, of which Fig. 4.26(b) is part, is an example of stable temporal and spatial oscillation in the glycolysing yeast extracts. The temporal oscillations of the bulk solution continued while its spatial structure also appeared and vanished periodically.

Boiteux and Hess (1980) state that the following features of their observations are in agreement with the models of temporal oscillations and spatiotemporal patterns in glycolytic systems: the structures are flux dependent; no excitability, signal transmission, wave propagation or pattern formation is observed in the yeast extract without substrates that can enter glycolysis; wave propagation and pattern formation have been observed so far only under conditions when the glycolytic system is also exhibiting temporal oscillations (and so only when there is a controlled input of glycolysable substrates and when supply is such that feedback control is operating); both temporal oscillations and spatial structure formation are influenced by any measure affecting the control characteristics of the primary 'oscillophore' (PFK) or the balance of the feedback transmitter system (adenosine phosphates). All of this agrees with the general character and predictions of the models of glycolytic spatio-temporal organization already discussed (section 4.5.2). However there remains 'a considerable gap between the rather general theoretical predictions and the multiplicity of the experimentally observed phenomena' (Boiteux and Hess 1980, p. 198). The previous models of glycolytic oscillations that incorporated diffusion did so only in one-dimensional terms (see section 4.5.2, e.g. Goldbeter 1973) and suggestively led to

200 THE KINETICS OF BIOLOGICAL SELF-ORGANIZATION

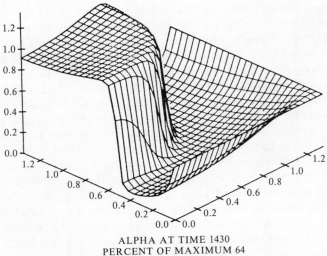

ALPHA AT TIME 1430
PERCENT OF MAXIMUM 64

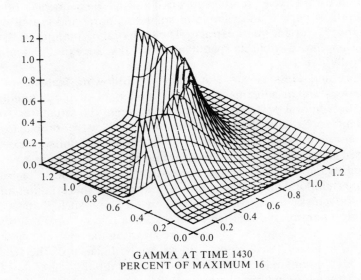

GAMMA AT TIME 1430
PERCENT OF MAXIMUM 16

Fig. 4.27. Isometric perspective of an intermediary state of oscillatory glycolytic diffusion in two dimensions. The *coordinates* of the *x,y-plane* are given in mm. The *z-axis* represents the normalized concentrations of α and γ, scaled to obtain isometric perspective. Time is computer time in seconds. The isometric drawings show a quarter of the total solution by symmetry from corner-forward position and corner-right position (90° rotation) to give a comprehensible presentation of the wave form in these dimensional perspectives. The waves of ATP (*top*) and

distance-dependence, as well as temporal oscillations (Fig. 4.21). So a first step to cope with the complexity of the experimental system was to extend the model calculations to a two-dimensional system, that is, coupling the oscillophore, PFK, to two-dimensional diffusion processes of its substrate, α, and activating product, γ (Hess et al. 1980). Partial differentials of α, γ governing diffusion with respect to two-dimensional coordinates (x, y) now appear in the equations for the production and removal of α and γ. These equations, together with the usual rate equation for an allosteric enzyme (cf. Goldbeter 1973), were reduced to first order and then solved by means of a computer program for each of the discrete 1352 points on a 26×26 square mesh area representing the system, with concentrations of α (ATP), γ (ADP), injection rates fixed within the experimental range and with appropriate enzymatic parameters. The system was thereby shown to produce spatiotemporal oscillations of the reactant concentration with a three-minute period at steady-state boundary conditions. These latter determine the waveform of the oscillations (Table 4.8) which Boiteux et al. (1980) classified as: purely temporal (bulk oscillation, no variation with space); purely spatial (concentrations of reactants vary in space, not in time, i.e. standing waves); and spatiotemporal (concentrations of reactants varying in space and time, i.e. travelling waves). This classification is very much along the lines already deduced from earlier model calculations (section 4.5.2, e.g. Table 4.3 and associated discussion in the text) but has now quite explicitly dealt with the two-dimensional variation of reactant concentration and has been able to represent this variation in an isometric perspective (see Fig. 4.27, which gives a typical example of the wave forms of a travelling wave obtained from the model after reaching the limit-cycle). Although only a tentative step in understanding the complexities of the biochemical system of glycolysis with respect to its dynamic spatial organization, these calculations do demonstrate at least that a region of instability in an enzymatic reaction can indeed break the homogeneity of a solution, leading to complex spatio-temporal patterns, and that the ratio between boundary volumes and reactivity space in biological organization is a

---

ADP (*bottom*) proceed from right to left in each period and are represented at 5-s intervals. The figure shows a snapshot of the wave progress roughly halfway through the given territory, with ADP forming a steep ridge with gradients in front and rear, whereas ATP travels like a reversed flood with a steep front gradient followed by a trough such as are seen in weather maps. The different wave forms of the two components result from the fact that the removal of α from the system is proportional to γ^2 so that when γ rises sufficiently, α decreases rapidly, promoting the wave propagation. (From Hess et al. 1980, Fig. 3a,b and text on p. 162.)

Table 4.8. Effect of boundary conditions on waveform in two-dimensional analysis of spatiotemporal patterns generated during glycolysis.

Boundary concentration	Area of mesh domain/area of boundary domain	Glycolysis in boundary domain	Diffusion in boundary domain	Wave-form
Steady state	1	Yes	No	Travelling[a]
Variable	1	Yes	No	Travelling[b]
Variable	1	No	Yes	Standing[c]
No transport	Unlimited	Yes	Yes	Temporal[d]

[a] Travelling waves from boundary to centre.
[b] Travelling waves from centre to boundary.
[c] Standing waves.
[d] Temporal oscillations of bulk solution.
From Boiteux *et al.* 1980, Table 1; see also Hess 1980, Table 5, and Hess *et al.* 1980, Table 2.

highly significant factor in determining the possible patterns that can form (cf. Table 4.8 and the critical length of eqn (4.56)). For the organization of biochemical processes in time and in space is based on a 'hierarchical principle of reaction space and boundary conditions' (Hess 1980, p. 89). Each of the isolated levels of Table 4.9 retains its own specific and autonomous functions and structure, and it is the coupling of these together through reaction and diffusion that manifest the cellular, tissue, physiological and other properties at higher organizational levels. The next section describes a particularly well-examined example of an oscillator operating in its effects at the cellular level.

Table 4.9. *Organization in space: compartments.*

Hierarchy	Location	Dynamic states
microspace boundary	active sites and channels Debye-Hückel layer	multistability (bi and tri) electrical potential
macrospace boundary	cytosol, mitochondria, lyosomes, peroxisomes membranes	chemical fluxes, gradients and waves electrical membrane potential
territorium boundary	multicellular organization destruction, vesicularization	chemical fluxes, gradients and waves dynamic transport

From Hess 1980.

A cellular oscillator: aggregation and communication in a cellular slime mould

Most living cells exist either as independent entities, such as bacteria, or as organized cell communities with a relatively stable morphology, within which the dynamic processes occur. However, there is a particular group of amoebae—the cellular slime moulds—that have the ability to exist in both forms. For example, *Dictyostelium discoideum* (with which we shall alone be concerned here) operates as individual cells while it is ingesting bacteria as its food in its natural environment (forest soil), growing by phagocytosis and dividing by binary fission (growth phase in Fig. 4.28). However, when its food supply of bacteria is depleted and starvation sets in a remarkable transition occurs whereby the population of hitherto individual cells begins collectively to move towards a number of focal points ("aggregation phase", Fig. 4.28) either in radiating streams and whorls of cells or, in the dense populations on an agar surface, in concentric or spiral rings (Figs. 4.29 and 4.30). In a few hours it has transformed itself into a coordinated slug-shaped mass of cells. This 'slug', of up to 100 000 cells, can move, under stimulus of unidirectional heat and light signals, from the original bacteria-depleted location to another more open one during a period of several days and over distances of several centimetres, if need be. After this 'slug' migration (morphogenetic development phase, Fig. 4.28) it undergoes a series of changes of shape ranging from that of an upright finger to that of a Mexican hat. The 'slug' then settles down on to its rear and transforms itself (culmination, Fig. 4.28) into a delicate stalk holding up a spore that is resistant to heat and drought and that can remain dormant until a more nutritionally auspicious moment arrives. Thus the cells survive by cooperation in order to begin again their individual growth and multiplication, until starvation again intervenes and the process repeats itself. The whole process, apart from the migration phase, takes about twenty-four hours (Bonner 1967). The complexity of the mechanisms that are operating in this fascinating and remarkable example of self-organization and communication at the cellular level may be judged from the fact that at least fifty genes are directly responsible for the unimpeded operation of the aggregation phase alone—and perhaps another hundred are peripherally involved in this phase (Williams and Newell 1976)—and that from 200 to 100 000 amoebae can aggregate together from distances of over 20 mm. It is this ability to communicate over such a long range, orders of magnitude greater than that of cells adjacent to each other in most developing systems, that has excited so much interest. The life-cycle and other general biological characteristics of this particular slime mould, *Dictyostelium discoideum*, have been reviewed by Bonner (1967) and Newell (1971). A useful survey for the non-specialist of the extensive

204 THE KINETICS OF BIOLOGICAL SELF-ORGANIZATION

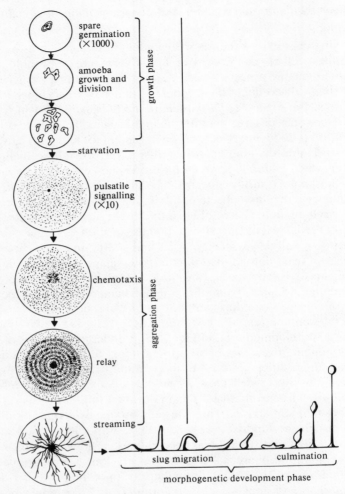

Fig. 4.28. The life-cycle of *Dictyostelium discoideum* showing the growth phase, aggregation phase and morphogenetic development phase. (From Newell 1977a, Fig. 1.)

work on this organism has been written by Newell (1977a), who has also reviewed studies of the biochemistry of its aggregation (1977b) and of the molecular mechanisms underlying its attraction and adhesion (1982). (The account given here is much indebted to these reviews of Dr P. C. Newell who also kindly made some of his photographs available.) Other reviews include those of: Bonner (1971), on 'Aggregation and differentiation in the cellular slime moulds'; Konijn (1972), on 'Cyclic AMP as a first messenger'; Gerisch, Malchow, and Hess (1974) on 'Cell communication and cyclic-AMP regulation during aggregation of the slime mould *D. discoideum*'; Robertson and Grutsch (1974) on 'The role of cAMP in slime mould development'; Loomis (1975) on '*Dictyostelium discoideum*: a developmental system'; Frazier (1976) on 'The role of cell surface components in the morphogenesis of the cellular slime moulds'; and Kay (1981) who reported on a 1981 conference on 'Gene expression and membrane changes in cellular slime moulds.

Current observations on the process of aggregation in *D. discoideum* have been briefly summarized as follows by Newell:

> The initial stimulus needed to start aggregation is starvation. This induces in some cells the ability to produce slow rhythmic pulses of the attractant cyclic adenosine monophosphate (cAMP) with an initial frequency of roughly one pulse every ten minutes. Meanwhile, the rest of the starving population produce cAMP receptors on their cell surface that enable them to perceive the pulsed signal. The cAMP signal does not diffuse very far from the centers of its production but is destroyed within 57 µm by phosphodiesterase enzymes (both free and membrane-bound) that are also produced by the starving amoebae. Two responses to this signal are noticeable. Firstly, the amoebae soon begin to move in the general direction of the signal source (and continue to move for 100 seconds covering a distance of about 20 µm) and secondly, about 12 seconds after receiving the signal, the amoebae themselves emit a pulse of cAMP. This pulse then diffuses away and reaches amoebae further out from the center. The amoebae closer to the center are prevented from relaying the signal inwards by their being refractory to further relay stimulation for several minutes after producing a pulse. (The relay refractory period.) By this system of relay, a series of waves of cAMP production, destruction and response move outward from the center as the amoebae move inwards [Figs. 4.29, 4.30, 4.31]. Such waves can be seen with the naked eye probably because of a difference in the shape of moving amoebae which tend to be elongated compared with still amoebae which tend to be more rounded and the waves are particularly clear when viewed as time-lapse movie films. Due to the relay of the pulse by the responding amoebae, their radial motion is unstable and amoebae tend to gather into streams which increasingly act as strong local sources of attraction [Fig. 4.30(e),(f)]. Amoebae can, at times, even be seen to move outwards from the center for a while in order to join a stream that happens to curve around behind them. Eventually, the amoebae in these moving streams all reach the aggregation center and a compact aggregate of cells is formed that then secretes a slime sheath over itself. The aggregation phase is now over and the morphogenetic phase of development begins. By rising vertically off the substratum, the aggregate converts itself into a finger-like body which flops over on to the substratum and migrates away as a 'slug' which eventually develops into the mature fruiting structure. (Newell 1977b, pp. 3 and 4).

Not all of the genera of the slime moulds use cAMP as the chemotactic agent (generically called 'acrasins') and, in those other genera (and also in *D. discoideum*), cAMP possibly has other functions connected with the morphogenetic phase. So the role of cAMP is complex, not least in the

Fig. 4.29. Concentric and spiral waves (seen using darkfield optics) of aggregating fields of *Dictyostelium discoideum* on an agar surface. The signalling centres which appear as bright dots are emitting pulses of cAMP at variable frequency. The bands of amoebae moving towards the centres appear bright and the stationary amoebae dark. In a number of cases the concentric banding pattern has given way to a spiral pattern. Such spirals are produced from a broken concentric band. A double spiral wave in the process of being formed from a concentric wave may be seen in the lower middle portion of the photograph. (From Newell 1982, Fig. 1.)

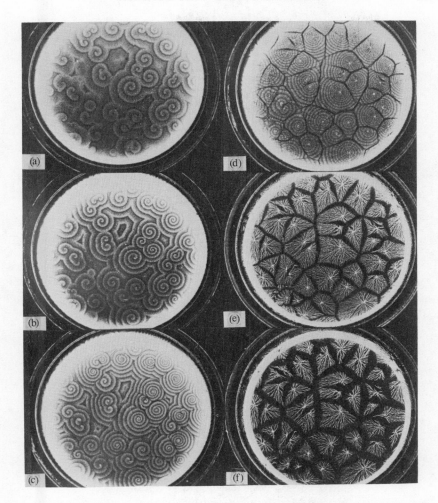

Fig. 4.30. Signalling patterns of *D. discoideum* showing the progressive increase in frequency of signalling, leading to the formation of territories and finally to aggregating streams. The photographs, taken at approximately 10-minute intervals show a field of 5×10^7 amoebae of the streamer mutant NP377 in a 50 mm diameter petri dish. The pattern seen with the wild type is similar but with narrower bright bands. (From Newell 1982, Fig. 3.)

aggregation mechanism itself which is conveniently divisible into the seven steps shown in Fig. 4.32.

Among all the complexities it is at least clearly demonstrable that periodic signalling occurs during the aggregation of *D. discoideum* and that cAMP can be identified as the naturally occurring chemotactic agent.

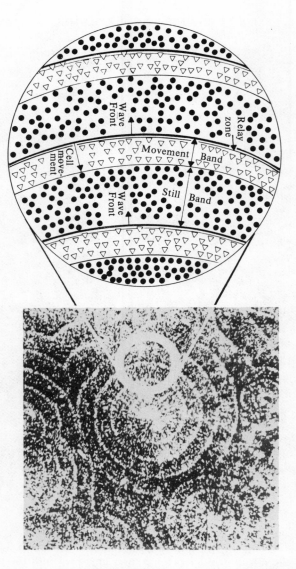

Fig. 4.31. Diagrammatic representation of the light and dark bands seen in relaying fields of amoebae using darkfield optics. The field of amoebae is divided into bands of centripetally moving cells (▷) and still cells (●) around the centre of attraction on the right of the diagram. Three wave-fronts are shown moving to the left in the diagram and represent the leading edges of the signals emitted by the centre. The concentric lines drawn within the moving and still bands represent the successive steps through which the wave front has passed in successive intervals starting from the release of the pulse from the centre. The width of the movement band is determined by the duration of movement of the amoebae after receiving a cAMP signal. The width of the still band is determined by the frequency of signal emission by the centre: the higher the frequency of pulsation, the narrower the band. (From Newell 1977a, Fig. 5.)

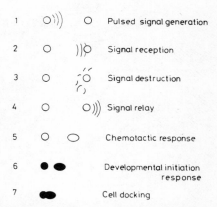

Fig. 4.32. The seven acts of aggregation *of D. discoideum*. The left-hand circle schematically represents a signalling amoeba in the aggregation centre (the three curved lines representing a single signal pulse being emitted) and the right-hand circle a responding amoeba. Steps 1–5 may be repeated several times before steps 6 and 7 are taken by the amoebae. (From Newell 1977*b*, Fig. 1.3.)

These autonomous oscillations in intra- and extra-cellular cAMP were shown to appear spontaneously in starving colonies prior to aggregation (Konijn, van de Meene, Bonner, and Barkley 1967; Gerisch and Wick 1975). Apparently there is a periodic increase in the net synthesis of cAMP which is the proximate cause of its periodic release. The oscillations (of period about 7–8 m) in intracellular cAMP are closely paralleled by oscillations in the activity of adenylate cyclase (that catalyses cAMP formation), of cytochrome *b*, intracellular cyclic guanosine monophosphate, pH, light scattering (i.e. cell shape changes, probably associated with oscillation in the intracellular level of calcium ions) and, as already mentioned, periodic release of cAMP. The relationships are complex (see Fig. 4.33, which is an attempt by Newell to depict some of the possible relationships), but there is no doubt that cAMP is a key intermediate since it can alter the oscillator as an input signal and is also one of its major output signals.

Various models have been proposed to explain how a periodic activation of adenylate cyclase could produce oscillations of cAMP (that of Martiel and Goldbeter has already been mentioned in section 4.5.1). These have been carefully and critically compared (Fig. 4.34) by Berridge and Rapp (1979). Some, such as that of Goldbeter and Segel (Fig. 4.34(b); Goldbeter and Segel 1977; Goldbeter, Erneux, and Segel 1978), can, by a careful choice of parameters that lead to periodic solutions, be notably concordant with the complex experimental observations. This particular model assumes that adenylate cyclase is an allosteric

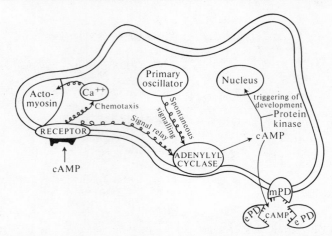

Fig. 4.33. Schematic representation of the mechanism of pulsatile signal generation, relay, chemotaxis, and the triggering of development. The primary oscillator periodically stimulates the adenylyl cyclase to produce cAMP pulses which are emitted and rapidly destroyed by extracellular (ePD) and membrane-bound (mPD) phosphodiesterases. Eventually, these pulses may activate protein kinases that trigger developmental events in the nucleus. Before all the cAMP is destroyed some is bound to receptors on nearby amoebae. There it induces chemotaxis (possibly by Ca^{2+} ion release) and production of a new pulse of cAMP (signal relay) by briefly activating adenylyl cyclase. (From Newell 1977a, Fig. 10.)

enzyme functioning in a way dynamically equivalent to the Goldbeter model of the glycolytic oscillator already described (sections 4.5.1 and 4.5.2). The model's principal inadequacy is its underestimation by an order of magnitude of the delay between intra- and extra-cellular cAMP peaks, but it could be adjusted to cope with this. It also has the weakness, like many of the models, of not taking into account the oscillation in cyclic GMP, for oscillation in guanylate cyclase activity could conceivably be driving those in adenylate cyclase. The model of Gerisch, Maeda, Malchow, Roos, Wick, and Wurster, (1977, see Fig. 4.34(c)) attempts to relate the activation of both of these enzymes. Another omission of all these models is, according to Berridge and Rapp (1979), their failure to consider calcium ions which seem to be essential to motility (Fig. 4.33). Rapp and Berridge (1977) tried to allow for this, too—but perhaps enough has been said to indicate that, although a number of plausible models illuminate various major features of the oscillatory system, and its control of aggregation as a pulsatile signalling system, much remains obscure and waits upon further experimental studies of, for example, the supposed allosteric behaviour of adenylate cyclase.

Events subsequent to the aggregation phase are no less complex (see for example, Frazier (1976), Kay (1981), Newell (1982)). Thus, for

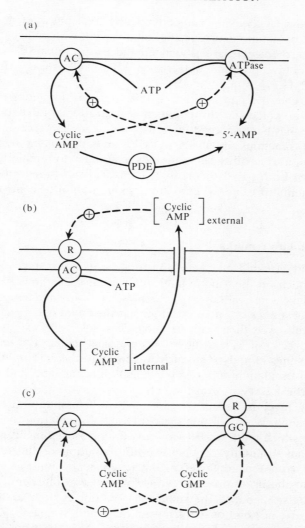

Fig. 4.34. A summary of some models which have been proposed to account for the periodic activation of adenylate cyclase (AC). All diagrams have been redrawn to facilitate comparison. (a) A cross-activation model (adapted from Goldbeter 1975). (b) Goldbeter–Segel model (redrawn from Goldbeter and Segel 1977). (c) The possible involvement of cyclic GMP in the activation of adenylate cyclase (modified from Fig. 14 in Gerisch *et al.* 1977). ATPase, ATP pyrophosphohydrolase; PDE, phosphodiesterase; R, receptor; GC, guanylate cyclase. (From Berridge and Rapp 1979, Fig. 18.)

example, the final 'docking manoeuvre', the final phase of aggregation, requires cell recognition and specific cell adhesion with which at least two types of macromolecule are associated ('lectin proteins' (discoidins and pallidins) and glycoprotein 'contact sites A') as well as Ca^{2+}/Mn^{2+} receptors (Newell 1977b, 1982).

However, complex and intricate as this system is, it is nevertheless one of the first in which the formation of a spatio-temporal pattern at the cellular level has been successfully attributed to the key role of oscillating biochemical reactions involving cAMP—and the phenomenon itself, which occurs also in other slime moulds, continues to remain a remarkable challenge both to all theorists of biological self-organization and to the experimentalist to unravel the interlocking web of its control mechanisms.

4.7 Concluding remarks

The foregoing account may have appeared, in its latter stages, perhaps to focus too much on the kinetic interpretation of oscillations *per se* at the metabolic level. For, although there now exist an immense number of experimental studies on such oscillations at the metabolic, epigenetic, and cellular levels, yet their precise biological role, and so evolutionary significance, has not by and large become at all plain. The exception to this is the slime moulds for which increasingly the interrelation of the biochemical oscillations (of cAMP, cGMP, calcium and hydrogen ions, adenylate cyclase, cytochrome *b*, etc.) and events at the cellular and intercellular level is becoming clear. Nevertheless, this vast range of experimental work establishing the existence of oscillations has a general significance which must not be overlooked. For oscillations can only arise if there is an instability of a steady-state solution of the fundamental differential equations that refer to the concentrations of metabolite, according to the usual procedures of chemical and biochemical kinetics. Such instability is indeed the sufficient condition for the existence of at least one periodic solution (Hastings, Tyson, and Webster 1977), whose stability is not thereby necessarily guaranteed. Now what has been amply demonstrated, both theoretically through consideration of appropriate models (section 4.5), and experimentally (section 4.6), is that such periodic solutions (i.e. oscillations) may be the basis and trigger for the appearance of new spatio-temporal patterns which did not exist in the steady states of systems of the particular kinds that actually occur *in vivo*. In other words, the experimental and theoretical studies of biological self-organization that focus on the kinetic aspects of living organisms reveal that instability in their kinetic systems can be the starting point of the formation of new forms of dynamic organization in space, in time, and in both jointly. This implies a new stress on 'the creative force of instability' (Boiteux *et al.* 1980). Such a viewpoint, it should be noted, has been reached from that of a purely kinetic interpretation of living systems and it is entirely congruent and consistent with the thermodynamic concept of 'order through fluctuation' in dissipative systems, which we saw in part I allowed a new understanding in thermodynamic terms of living organisms.

It would seem that the hopes of the pioneers of the kinetic approach, outlined in section 4.2, are, in principle at least, beginning to be fulfilled. Thus Boiteux, Hess, and Sel'kov in their review (1980) of the creative functions of instability and oscillations in metabolic systems involved in cellular regulation, stress

> the creative force of instability with respect to homeostasis and homogeneity as a new approach to the understanding of cellular regulation. Both homeostasis in time and homogeneity in space constitute the classic frame work for a textbook presentation of function and structure. However, we should be aware that this view of nature is reduced, and the observed states are always the product of random interactions between a system and its environment. Although exogenous fluctuations are largely filtered out by living systems, we observe in cells and multicellular bodies autonomous oscillations as well as dynamic inhomogeneity of matter. This means that the organisms create their own timing and spacing which might couple to external signals. The theoretical and experimental data reviewed allow one to conclude that the instability of metabolic systems may be of crucial importance for the temporal and spatial organization of living matter. (Boiteux *et al.* 1980, pp. 196 and 198.)

One of the somewhat unexpected fruits of the recent thermodynamic interpretations of living systems described in Part I was seen to be a new understanding of how they could have come into existence at all in a way not only consistent with thermodynamic principles but almost entailed by them. The kinetic interpretations have too, in their own way, provided new insights, and have suggested new paths of exploration, both conceptual and experimental, into the way existing living organisms are self-organized. But can kinetics contribute to our understanding of how self-organizing, self-reproducing living organisms might have arisen from apparently inchoate non-living matter? This is the question to which we turn in the next chapter.

5 Selection and evolution of biological macromolecules

5.1 The origin of life

THE question of the origin of life itself, of how life originated on the surface of our planet, has long exercized natural philosophers. A. I. Oparin in his influential work on *The origin of life on the Earth* (1936), and subsequently (1961, 1965) gave an account which, even if somewhat biased in the direction of a dialectical materialist interpretation of the history of ideas, is a fascinating survey of the ebb and flow of belief in the spontaneous generation of life from the inorganic and inanimate. The work of Pasteur in the early 1860s clearly established by experiment the impossibility of micro-organisms developing except from other micro-organisms. This clear-cut demonstration of the absence of spontaneous generation of life at the present occurred within a few years of the launching by Darwin and Wallace of the idea of natural selection as the general means whereby all living organisms evolved. Thus an increasing belief in the continuity and connectedness of all forms of life coincided with the establishment of the conviction that life was not *now* generated from the non-living. However that did, and does, not preclude the possibility that living forms developed from a non-living milieu as a prelude to evolution, for the absence of such observed occurrences today in the natural world is fully explicable by the vulnerability of any 'new', emergent prototypes of life to predation by those already established. So problems concerning the origin of life, in relation to the processes whereby non-living matter might become living—so evidencing metabolism, self-reproductivity and mutability (Oparin 1961)—and what the resulting living forms might be that could become the first unicellular organisms remain key issues today in the biological sciences. Lying behind the question of 'How did life originate from the inorganic in the past?' is, of course, the uniformitarian assumption expressed by Lyell in his *Principles of geology* that so influenced Darwin in his interpretation of biological history; namely, the assumption that it is only causes we can now see to be operative, or only causes extrapolated backward from known regularities and established properties of matter, that must be called upon to explain what happened in the past, whether in transformations of the Earth's crust or of the forms of living organisms. By this principle, that has proved so fruitful, our modern knowledge of kinetics and thermodynamics in relation to biological organization can properly be brought to bear on the problem of the 'origin of life', i.e. of how living matter could have developed from non-living.

Probable chemical developments in the prebiotic phase of 'chemical evolution' have been much studied in the last two decades and it now seems clear that the essential building blocks of macromolecules we know to be essential to evolved life could be present on the surface of the earth, or in its waters, in the eras when the transition from the non-living to the living would have taken place. These building blocks, including, in particular, the amino acids (to become proto-proteins) and energy-rich nucleoside phosphates (to become proto-nucleic acids and energy carriers), could form by the random collisions of molecules that underlie all chemical processes. Polymerizations are now known to have been able to occur under prebiotic conditions of the Earth's reducing atmosphere, with the use of various energy sources (not least the ultraviolet radiation of the Sun). Such polymerizations would have been uncontrolled by previously existing macromolecules and so would have led to either random monomer sequences or sequences only slightly different from random—and could not themselves constitute a self-reproducing system since they would not be carriers of information in such a random context. However, some at least of the macromolecules characterized by particular sequences must be assumed to have had autocatalytic properties—by virtue of the propinquity of certain chemical groupings or by themselves acting as templates. By 'autocatalytic', in this context, is meant the ability to increase the chances of formation of copies of a macromolecule of the same sequence as itself. The randomness of the molecular processes involved in such chemically feasible polymerizations and the randomness of the sequences of units in the polynucleotides and polypeptides so formed seem to militate against any possibility of these macromolecules carrying the kind of 'information' that is needed in the technical sense of information theory (see section 6.3). The 'instructions' (concerning, say, the sequences in macromolecules still being formed) need to be communicable from one 'generation' of molecules to the next in a situation where the 'value' of a given sequence to convey 'instructions' is still in the process of becoming both definable and defined. By 'instruction' here is meant simply the ability to reproduce autocatalytically its own sequence so that the macromolecules thus formed will themselves again act autocatalytically to continue the chain of copying. Information theory is chiefly concerned with the communication of information in a form where the mode of translating the information into further events (i.e. of 'reading the instructions') is clear-cut and fixed—as it is now established, but was not then, in the transfer of information in DNA to protein sequences via the well-understood and elaborated 'genetic code' of biological organisms. So information theory is not really a great help for interpreting the transition from disorganized, non-living matter to matter organized in a living form when 'meaning', i.e. the mode of transfer of information, is itself in the process of emerging from randomness and where there are still many choices available for ways of generating new information. The transition from non-living conglomerate to organized, 'living' matter—having at least the character of self-reproduction, and so inevitably of mutation and metabolism—could better be characterized as a *self*-organization with generation of information-carrying and reproduction with transmission of instructions. Realization of the jeopardy of decomposition into randomness in which the first, provisional, tentative, putative self-organized structures were placed points to the need for a *selection* process that implicitly 'evaluates' what might be called the 'candidate' biotic structures. This is the parallel, at the molecular level, of Darwin's principle of natural selection and it will need to be understood in terms of molecular parameters to be meaningful in this context of *molecular* self-organization. How selection rules can be based on chemical properties can be illustrated by appropriate games (Eigen 1971a, pp. 469 f).

The ability to *self*-organize to which we have been referring is but another description of the process of formation of the 'dissipative structures', the thermodynamic prerequisites of whose formation we discussed in Chapter 2. These prerequisites were: an open system, far from equilibrium, with non-linear relationships prevailing between fluxes and forces. Now one of the implications of the genesis of dissipative structures formed by autocatalytic chemical systems (e.g. the trimolecular, autocatalytic process of the 'Brusselator', and other systems described in section 4.5) is that particular features of an initiating fluctuation may impress themselves permanently in some aspect of the dissipative structure in which it results (cf. the *localized*, spatially periodic result of certain kinds of fluctuation in domain II of Fig. 4.17). The evolving network then stores and transmits information, but to arrive at this broad generalization only renders more acute the question of by what actual mechanisms a chemical network might initiate, transmit and store information. We know now that in biological organisms today, genetic information is stored in DNA and transmitted, via RNA, to proteins and by self-duplication, to progeny. But what kind of chemical network makes it possible for such an information-transmitting macromolecular system to emerge?. The thermodynamic theory of dissipative structures, although it gives meaning to (and even quantitative criteria for) 'order through fluctuation' does not, as is the wont of thermodynamics, provide any actual possible mechanisms for the emergence of ordered, biological hierarchies, in general, and of information-processing chemical networks, in particular. So we are bound to ask what *kinds* of changes are represented by the initiating fluctuations which the thermodynamics of dissipative structures now shows to be so potentially fruitful of new levels of stability, order, and complexity. In what respects do the new steady states differ from the old?

The Brussels school (Chapter 2) has in its examples been concerned to emphasize the real possibility of new spatial arrangements, new *structures* emerging when the conditions in dissipative systems are favourable (cf. the Zhabotinsky reaction). But a number of other authors (see section 6.2) have also been concerned to stress how the distribution and relations between functions, and the interlocking of structure and function, articulate a given hierarchy and discriminate between hierarchies. There is a *functional* as well as structural order to be found in biological hierarchies whose origin in the generation of primary information needs to be accounted for.

It is problems such as these to which Eigen and his colleagues have addressed themselves in their studies of selection and evolution among populations of informational macromolecules. Their analyses go beyond the thermodynamic considerations in applying kinetic and stochastic theory to such populations and greatly illuminate the thermodynamic ideas giving them a more 'local habitation and a name'.

The broad question to which Eigen and his colleagues at Göttingen have addressed themselves—in the first major article of Eigen himself (1971*a*) and in a more condensed and simpler account (Eigen 1971*b*), in a helpful survey and exposition by one of his colleagues (Küppers 1975), and in a more popular and highly illuminating book (Eigen and Winkler 1981*a*)—may be expressed thus. How was it possible for the order and complexity which underlie the self-reproducing biosynthesis of living cells to originate in that product of the long phase of chemical evolution, namely, the 'primordial soup' of essential monomers (amino acids, nucleoside phosphates, etc.) and the uncontrolled random polymers (polyamino acids and polynucleotides) into which they condensed? More particularly, how could information-carrying and information-transmitting macromolecules (the nucleic acids) and functional macromolecules (the proteins) have emerged together to constitute the first self-reproducing unicellular system, and so represent the 'living' coming from the 'non-living'? In this we have a twentieth-century version of the chicken-and-egg puzzle, for 'function' cannot build up any reproducible order without 'information', i.e. a system of transmission of instruction and of their execution; and 'information' can only acquire meaning through the 'function' for which it is coding.

5.2 Phenomenological deterministic treatment

The model system

The first stage of Eigen's treatment is purely phenomenological and gives a deterministic account, by means of ordinary chemical kinetic considerations, of the selection and evolution of informational macromolecules. It deals with a large population of macromolecules and so this stage of the enquiry yields mean values of new parameters of interest; e.g. the 'selective value', a measure of the ability of a macromolecule to be selected with a given information content, and the 'excess production' of macromolecules (these will both be defined later). Both these parameters are more precise and relevant to selection and reproduction at the molecular level than the thermodynamic ones and so supplement and go beyond the various functions of entropy discussed in Chapter 2.

The model systems envisaged in this stage of the enquiry is a formalization of the 'primordial soup' as a box containing an aqueous medium in which are dissolved both monomers and macromolecules (Fig. 5.1). The system is open to receipt of energy, in the form of energy-rich monomers,

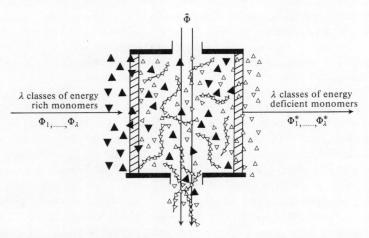

Fig. 5.1. Model system for a molecular self-organization. The box represents to a certain degree the primordial soup of biological macromolecules. The system is assumed to be open and far from internal equilibrium. In order to prevent the system from decaying to equilibrium one has to feed in free energy, e.g. in the form of energy rich monomers. The polymeric sequences ⎯◁◁◁⎯ inside the box which carry, as a consequence of the primary structure, 'biological information' are steadily assembled and decomposed. The energy deficient decomposition products Δ are eliminated continuously from the box. There exist two possibilities to maintain a steady state for the above system, namely (a) control of fluxes $\Phi_1, \ldots, \Phi_2, \Phi_1^*, \ldots, \Phi_2^*$ or (b) control of concentrations via the dilution flux $\hat{\Phi}$. (From Kuppers 1975, Fig. 3.)

such as the appropriate nucleoside phosphates, and is far from internal equilibrium. Polymeric sequences, which are not necessarily specified either as nucleic acids or proteins and which carry 'information', are assembled and decomposed in the box from which energy-deficient decomposition products are eliminated continuously. So processes of formation and decomposition are driven by positive affinities. To maintain a steady state the inward and outward fluxes of monomer reactants and decompositon products and/or the volume of the box may be controlled. It is supposed that the total formation rate of informational macromolecules exceeds the decomposition rate and that there is virtually no non-instructed production of any individual information carrier. The total number (n_ν) of macromolecules per unit volume containing ν monomers is then $\sum_i^N x_{\nu_i}$, where x_{ν_i} is the number of copies per unit volume of the ith class of such macromolecules and is distinguished by its primary sequence and so by the information it carries, and N is the number of possible classes (= number of possible sequences = λ^ν, where λ = the number of kinds of monomers, 4 for nucleic acids, 20 for proteins). Considering only macromolecules of a uniform length ν (and so dropping the subscript ν, for simplicity), the phenomenological rate equation for the concentration of the ith class of macromolecule is:

$$\frac{dx_i}{dt} = (A_i Q_i - D_i - \phi)x_i + \sum_{\rho \neq i}^{N} \psi_{il} \cdot x_i \qquad (5.1)$$

$$\quad\;\;(1)\;\;\;(2)\;\;(3)\qquad\quad(4)$$

The autocatalysis in this process is that of *self*-copying and the term (1) is the production rate of the ith class of macromolecules and is a product of: the 'amplification factor', A_i, namely, how many such macromolecules, regardless of their internal sequence, can be made on each of those of the ith class present at time t (as represented in the concentration x_i outside the bracket); and of a 'quality factor', Q_i, which is the fraction of the new macromolecules of length ν which are *exact* copies, that is, have the correct internal sequence characteristic of the ith class. So A_i represents a 'birth' and $(1 - Q_i)$ represents 'mutability'. 'Death' is represented by the term (2), the decomposition rate constant D_i times the concentration x_i; and term (3) represents what may be called 'enforced emigration', the diminution in the x_i (the same for all i) which results from dilution. Term (4) represents the contribution to the synthesis of the ith class of informational macromolecules which is made by wrong copying of other (l) informational macromolecules of the same length. Between them, terms (1), (2), and (3) constitute the essential characteristics of life: metabolism (A_i, D_i); self-reproduction (when $A_i \cdot Q_i > D_i$); and mutability ($Q_i < 1$). Here term (4) will be taken as negligible in order to simplify the presentation (Eigen and co-workers used this simplification in the initial stages of their study). The definition of A_i and Q_i is purely phenomenological for one could count the number of copyings per unit time that are instructed by a given macromolecular template and determine (by sequence analysis, say) the number of correctly made species i.

Selection under constraint

For there to be real competition for survival there must be constraints, so the treatment is primarily developed for constant overall organization, that is, under the constraint of *constant forces*, with $\sum_i^N x_i = $ a constant, n (i.e. the total number of information macromolecules is constant, as also is the total concentration of monomers). This constraint, of constant overall population (and so of overall affinity), can be introduced by adjusting the dilution flux to compensate for the *excess productivity*, $E_i = A_i - D_i$. Hence:

$$\sum_i^N (A_i - D_i)x_i = \text{total } excess\ productivity = \text{dilution flux} = \Phi = \phi \sum_i^N x_i.$$
(5.2)

The *mean productivity* is defined as

$$\bar{E} = \sum_i^N E_i X_i \Big/ \sum_i^N x_i (\equiv \phi),$$
(5.3)

and the 'selective value', W_i of the ith class of conformational macromolecules is defined as

$$W_i = A_i Q_i - D_i.$$

From these definitions, one obtains the following coupled set of differential equations:

$$\frac{dx_i}{dt} = (W_i - \bar{E})x_i,$$
(5.4)

ignoring term (4), which represents a 'repair of errors'. Equation (5.4) is inherently non-linear because each population variable, x_i, occurs in $\bar{E}$. So all equations are coupled via this term which provides a sliding and self-adjusting threshold value indicating the self-organization of the system.

Those species will grow for which $W_i > \bar{E}$, so that $\bar{E}$ represents a threshold value. The distribution of the population of macromolecules, as between the different kinds, i, of primary informational sequences, will shift (see Fig. 5.2) in such a way that $\bar{E}$ moves to steadily higher values, because those species whose $E_i < \bar{E}$ at any time will be eliminated and

PHENOMENOLOGICAL DETERMINISTIC TREATMENT

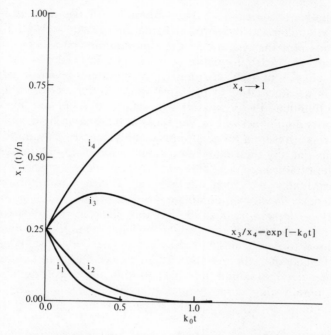

Fig. 5.2. Selection in a system of four competing species according to the solution of eqn (5.4): $W_1 = 1$, $W_2 = 4$, $W_3 = 9$, $W_4 = 10$. The four functions $x_i(t)$ have been normalized to the constant overall concentration n. They describe the time-dependence of the single concentrations. The time-scale has been reduced to a general first-order constant k_0. It is interesting to notice that $x_3(t)$ initially increases and passes through a maximum before it finally decays. (From Eigen 1971b, Fig. 1.)

this raises $\bar{E}$ until it reaches the maximum selective value to be found in all the species present, i.e.

$$\bar{E} \to W_{\max}. \qquad (5.5)$$

It should be noted that $W_{\max}$ may belong to a single information carrier if $A_{\max}$, $Q_{\max}$, and $D_{\max}$ are constants; or to a catalytic cycle, and so be a function of the concentrations of the members of the cycle; or it may include a whole hierarchy of reaction cycles, with concentration terms of higher order.

'Selection equilibrium'

In the preceding paragraph, selection has been characterized by an extremum principle, somewhat analogous to thermodynamic equilibrium, but here there are involved not time maxima of entropy, but 'optima', i.e.

maxima relative to alternative compositions, under the given constraints. The state characterized by (5.5) might be called 'selection equilibrium', provided one remains aware that it is a metastable state. For it stabilizes information for the reproduction of the 'fittest' in the population of macromolecules, but only so long as there is no selective advantage occurring in the fluctuating error distribution. So the species, whose W is W_{max}, will dominate the population in the final state of 'selection equilibrium'. But its dominance will never be total because it is not unique and there is always present a range of reproducibly occurring mutants, for any one of which at any time fluctuations may entail giving rise to a W that exceeds the established W_{max} (see section 5.3 below, especially pp. 227, 228). Thus a new W'_{max} will be established with a new informational macromolecular species predominant.

The process of self-organization, on this phenomenological treatment consists of: a process of *selection*, in which the molecular distribution narrows down to the species with the highest selective value, W_{max}; and a process of *evolution* wherein the system passes through a sequence of metastable selection equilibria, through the successive dominance of 'mutants', new 'wrong' copies with higher Ws. So

$$W_{max} \to W'_{max} \ldots \to W_{optimum}. \tag{5.6}$$

The criterion for optimal *and* stable evolution can then (Eigen 1971a, p. 479) be written approximately as

$$W_{max} > \bar{E}_{i \neq m}, \tag{5.7}$$

where $\bar{E}_{i \neq m}$ is a mean residual productivity, averaged over all species except that (m) dominant when W reaches its maximum (writing m for 'max' or maximum). It is broadly defined thus

$$\bar{E}_{i \neq m} = \frac{\sum_{i \neq m}^{N} E_i x_i}{\sum_{i \neq m}^{N} x_i}. \tag{5.8}$$

The approximate selection criterion (5.7) amounts to a physical definition of the Darwinian term 'fittest' as applied to these self-copying macromolecules. From (5.7), (5.8) and the definitions of n and E, it follows that

$$\text{stationary error fraction} = \left(1 - \frac{\bar{x}_m}{n}\right) = \frac{A_m}{E_m - E_{i \neq m}} (1 - Q_m). \tag{5.9}$$

If $\bar{x}_m/n$ is unity, then $W_m = E_m$ and $\bar{x}_m \to n$, representing the extreme of a selection process but with no opportunity for further evolution. If, however, $\bar{x}_m/n$ is to become zero, then the approximate value of the quality factor Q_{min} that would not ensure survival can be deduced from (5.9) with its l.h.s. as unity, namely:

$$Q_{min} = \frac{A_{i \neq m} + D_m - D_{i \neq m}}{A_m}. \quad (5.10)$$

Clearly Q_m must exceed this to ensure survival but not approach unity which precludes further evolution, so

$$Q_m > Q_{min}, \quad (5.11)$$

where Q_{min} is given by (5.10). So (5.11) constitutes an important *selection condition* for preserving the 'information' of the system. A low Q_m ($>Q_{min}$) produces both a large *variety* of mutants from which selection can occur and thus eventually allows a higher optimum value of W_m; and it speeds up the rate of evolution. So, briefly, the final optimum value $W_{opt}(=A_{opt}, Q_{opt} - D_{opt})$ is characterized by a compromise between the need for as small a value of Q as possible, to increase the rate of evolution, and for Q to be large enough to ensure survival, by maximizing W_m for that species.

The foregoing treatment has been developed for the constraint of constant overall organization, applied through keeping the total number of macromolecules constant. This is parallel to keeping constant the forces X_α, as they are denoted in the thermodynamic analysis of steady states (section 2.5). Alternatively the influx of monomeric material and the reaction flows could have been considered as maintained constant with the content varying. This would have corresponded, in the thermodynamic treatment, to keeping the J_α constant and varying the X_α—which is the situation assumed to prevail when $\delta_X \sigma$ (and $\delta_X P$) are defined and shown to obey the stability criterion for fluctuations, namely of being positive (see section 2.6 and eqn (2.70)). The system of phenomenological equations for this latter constraint has been worked out by Eigen (1971a) and, although the equations are more complex than for the other constraint, it was shown that the system again selects for a maximum 'selective value' among the population present at any one time, via a sliding threshold. Furthermore, in relation to the thermodynamic stability criteria (section 2.6), it can also be shown that each mutation leading to an increase in 'selective value' corresponds to a negative fluctuation in entropy production ($\delta_X \sigma < 0$), indicating the possibility of instability. Evolution at constant flows corresponds to a sequence of such instabilities.

5.3 Stochastic approach

Limitations of the deterministic treatment

As already stressed, this first stage of the treatment was deterministic and applied to the 'mean values' of E, etc. But two of the processes involved in molecular competition are non-deterministic in the sense of arising from an independent causal chain: the occurrence of specific mutants, which can be quantum-mechanically uncertain; and the amplification of an individual macromolecular species, which is subject to statistical fluctuation. If only one copy of a species existed at a given moment and it happened to decompose before it was copied then that species would be eliminated, whatever the *mean* rates of copying and decomposition might be. Moreover certain steady states, unlike true equilibrium, are metastable and cannot stabilize themselves. There is clearly a need for a probabilistic approach to the process of self-organization in addition to the phenomenological one developed so far. This is the field of 'stochastic' theory—the extension of the theory of probability to dynamical problems (see Bartlett (1955) and Ramakrishnan (1959) for general surveys, and McQuarrie (1967) for the application of stochastic theory to chemical kinetics).

Stochastic models of fluctuation and stability

If we consider again the 'information box' (Fig. 5.1), we can ask what happens to the *rates* of formation and decomposition of an 'informational' macromolecule i when it undergoes a (positive) fluctuation, δx_i, in its number, x_i; per unit volume. That is, we ask does a positive δx_i lead to a positive $\delta(dx_i/dt)_{formation}$ and/or a positive $\delta(dx_i/dt)_{decomposition}$? If the change in the rate, whether of formation or of decomposition, has the same sign as the change in concentration (population) we shall denote this by a plus (+) sign, if the opposite sign by a negative (−) sign and if the rates are independent of the change in population by a zero (0). The possibilities are given in Table 5.1 (after Eigen 1973) in which the following are distinguished: *stable systems* (S) in which the fluctuations control themselves in that the response to a fluctuation is to compensate for it; *drifting systems* (D) in which the actual population (value of x_i) has no control on the rates of formation and decomposition which fluctuate randomly; and *unstable systems* (U) in which some fluctuations are self-amplifying. The dynamical behaviour of these three prototypes of systems can be simulated by simple selection games (cf. Eigen and Winkler 1981a). Within drifting systems are included: *indifferent* (00) systems, in which both formation and decomposition are entirely independent of δx_i; and *variable* systems which are neither certainly stable or unstable—their stability will depend on quantitative factors and they

Table 5.1

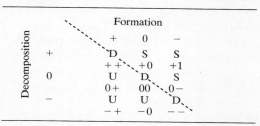

S = stable systems; D = drifting systems; U = unstable systems (see text). Amongst D, the ++ and −− are 'variable' and the 00 is 'indifferent'.

could be stable, unstable, or indifferent. Regulating mechanisms and self-organization in biology are associated primarily with the states on the diagonal indicated by a broken line.

The *stable* systems (S) correspond to the Ehrenfest urn-model (P. and T. Ehrenfest 1907)—a classical example of equilibrium fluctuations. In this model, two urns are postulated containing together a total of $2N$ spheres numbered 1 to $2N$ and arbitrarily distributed between the two urns. The 'game' is to choose a number from 1 to $2N$ at random and then to move the ball so numbered from the urn it is in to the other. Clearly the probability in each move of a ball leaving one urn to go to the other increases for an urn the more balls there are in it. So this is a self-regulating distribution and corresponds to an S-system. The probability of finding $(N+\Delta N)$ spheres in one urn and $(N-\Delta N)$ in the other is represented by a Gaussian distribution.

Drifting systems (D) can be represented by a modified version of the same game. Now the spheres are no longer numbered but there is a fixed total number distributed between the two urns, say, A and B. A coin is tossed and if 'heads' a sphere (any sphere) is transferred from A to B; if 'tails' from B to A. The direction of a fluctuation now bears no relation to the numbers in A or B, respectively, and so of the previous history of the distribution. The distribution drifts irregularly with no self-regulation and extreme distributions are quite possible; indeed all distributions (within the confines determined by the total number of spheres) are equally probable.

The third class, of *unstable* systems (U), are the only ones capable of being justifiably interpreted as parallel to the stochastic processes in selection and evolution in the system of 'reproducing' informational macromolecules that we have been discussing (Fig. 5.1). A first version (i) of an appropriate selection game (or 'life game') is devised to parallel a situation in which there are equal probabilities of duplication and removal of an informational macromolecule in a steady state. More formally, $Q_i = 1$, all A_i and D_i are the same for all i and are equal; hence $A_i Q_i = D_i$ and $A_i = D_i$, and $W_i = E_i = 0$ for all macromolecules. Phenomenologically the equation would be

$$\frac{dx_i}{dt} = (A_i - D_i)x_i = 0, \qquad (5.12)$$

and nothing should happen to the distribution in the box of 'primordial soup' we are considering. But the appropriate selection game illustrates why this conclusion is false. The game would be as follows (Fig. 5.3). Given a supply of N^2 (say, 100) numbered spheres for which each of the numbers 1 to N (1 to 10) is depicted on N (10) copies. Start with a set of N (10) spheres in a box, the spheres all being different, i.e. one each of the N (10) kinds. Now simulate a linear 'birth' and 'death' process, by alternately:

(a) taking out a sphere, and adding another of the same kind (that is with the same number, between 1 and N (1 and 10) written on it) and return both spheres to the box ('Birth');

(b) taking out a sphere and discarding it ('Death').

(a) and (b) alternate as procedures so that all the numbers (from 1 to N, 1 to 10) have the same *a priori* chance of survival. Because once a particular number has been removed by (b), it can never again be duplicated by (a), the population narrows down surprisingly rapidly to a few (or even one) kind of numbered sphere in many copies, that is, with a high redundancy (Fig. 5.3).

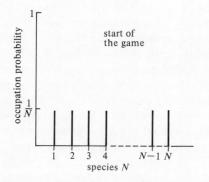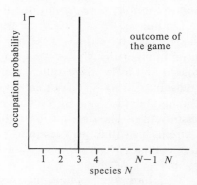

Fig. 5.3. Eigen's selection game—an example of fluctuations in the steady state (see text). Given a supply of N^2 numbered spheres in which each of the numbers 1 to N is represented on N spheres. Start with a non-degenerate set of N spheres in a box. Alternate duplication and removal of numbered spheres. Outcome: one number with N-fold degeneracy. (From Küppers 1975, Fig. 7.)

Instead of the regular alternating procedure (a) then (b), one could have *random* duplications and removals, but keeping the probabilities of each equal (e.g. by having a 'random-number selector' and duplicating, as in (a), if an even number turns up and discarding, as in (b), if odd). With this procedure a fluctuation catastrophe becomes possible in which the whole population is extinguished (after a sufficiently long, N, runs of (b)). Before such a state is reached, however, the population will, in the case of large numbers N, narrow down to a few (or even one) highly redundant numbered sphere. Notice that if we replace, in the 'game', the numbered spheres by informational macromolecules i, the system is 'closed' with respect to the addition of any new information (sequences), which can enter only via *errors* of reproduction ($Q_i \neq 1$). Furthermore, information (sequences) are lost whenever a single non-redundant information carrier (macromolecule) decomposes before it is reduplicated.

Although the chances of 'birth' and 'death' for each numbered sphere, or macromolecule i of a particular sequence and information content, were all uniformly equal (i.e. $A_iQ_i = D_i$), the distribution is metastable and fluctuations in the rate of duplication and removal are independent of each other and so are not self-regulating. Clearly the conclusion drawn from a phenomenological treatment (5.12), that nothing would change, is misleading in the absence of any stochastic analysis. This macromolecular situation with its stochastical prediction, represents the 'survival of the survivor'—the survivor is simply the one which happens to survive, and, if these were populations of living organisms, would justify the jibe of 'tautology' often directed against Darwinism. However, these selection games may be refined to allow for more realistic relations between the parameters of the population of informational macromolecules in the box of 'primordial soup'. (Only the results of Eigen (1971a) will be given briefly here.)

In a *second* selection system (ii), one could set $A_i = D_i$ for all species, but let errors occur in the reduplication so that $Q_i < 1$ and has the same value for all i. Consequently $(A_iQ_i - D_i) = (A_kQ_k - D_k)$, etc. where $i, k = 1, 2 \ldots N$; i.e., $W_i = W_k$. Q_i being less than unity allows for new sequences to be generated. Let the total rate of formation of *all* macromolecules equal their total removal rate ($\sum_i^N A_iQ_i = D_i$). Each individual species has $A_iQ_i < D_i$, and so must die out. The system can only compensate (to keep total formation and decomposition rates the same) by steadily producing new information through making errors in the copying process. The system would drift irregularly through its possible population states until it is extinguished by a fluctuation catastrophe.

Only a *third* selection system (iii) allows stable and reproducible behaviour through a finite variation of rate parameters. In this system, we have the same conditions as in the second system, viz. $Q_i < 1$ and *total* formation and decomposition rates equal, but now it is allowed that the rate parameters of formation and decomposition may differ, i.e.

$$(A_i Q_i - D_i) \neq (A_k Q_k - D_k); \qquad A_i Q_i \neq D_i$$

i.e.,
$$W_i \neq W_k; \qquad A_i \neq A_k; \qquad D_i \neq D_k.$$

This system would select the species of maximal W (supposing there is any species for which $W_k > 0$) as predicted from the phenomenological equation, already discussed. But the stochastic treatment confines this result of the deterministic theory in the sense that the species which fulfill the growth conditions do not *certainly* survive (i.e. they have a survival probability <1). For as long as the species of maximal W exists in only a few copies, it may still be in danger of decaying by fluctuation. However, the more it grows, the smaller the chance of this happening, until it finally dominates the total population. In this 'linear' growth system, there is no real 'point of no return' after which a given mutant will certainly survive; the probability of extinction decreases steadily and asymptotically with increasing redundancy, but never reaches zero. It follows from this treatment that the single molecular species is not the true target of selection but rather an organized *combination* of species with a definite probability *distribution* which emerges through selection—a combination Eigen and Schuster (1977) have called a 'quasi-species'—and which is selected against all other distributions (Küppers 1979a). (This quasi-species is closely correlated with what is called the 'wild-type' of a biological population.)

To summarize, it is clear that a stochastic treatment modifies significantly the deterministic phenomenological account of the evolution of an assembly of informational macromolecules and of the conclusions drawn from it.

5.4 Self-organizing systems of macromolecules

Hierarchies of cyclic reaction systems

In the foregoing, single species have been regarded as competing independently with each other, apart from some copying errors occasionally producing one informational macromolecule from another (term (4) in (5.1)). Now we have already seen, both as a result of thermodynamic analysis (Chapter 2) and of consideration of chemical reaction networks (Chapter 4), especially those in which diffusion is a limiting process, that autocatalytic steps in such networks play a vital role in making possible the emergence of new organization—whether this is described as that of an open 'dissipative system', in thermodynamic terms, or that of a

SELF-ORGANIZING SYSTEMS OF MACROMOLECULES

'symmetry-breaking' reaction-diffusion system (not 'symmetric' in relation to space and/or time), with reference to kinetic considerations. So the preceding treatment of selection and evolution in an informational macromolecular system must now be extended to incorporate autocatalysis if it is to provide an adequate basis for understanding the transition from non-living aggregates of matter to the minimum organization characteristic of living forms of matter (section 5.1).

'Autocatalytic' activity is attributed to a reaction *network* when the product of one reaction in the system in some way, either as catalyst or as a substrate, enhances the rate of formation of another member of the same system. It is possible (cf. Eigen and Schuster 1977) to envisage a kind of *hierarchy of cyclic reaction networks* (Fig. 5.4). The simplest level is that in which, at each step in the sequence of reactions, the products undergo further transformations and at least one of the products is itself a reactant in one, or more, of the preceding steps. The whole system, to which energy must be supplied, is then a reaction *cycle* and behaves as a whole like a catalyst—with some incoming substances acting as its substrates and other outgoing substances as its products (E in Fig. 5.4). But

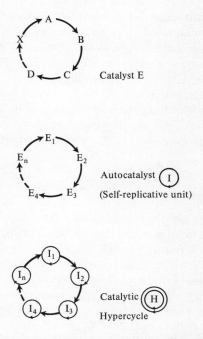

Fig. 5.4. The hierarchy of cyclic reaction networks is evident from this comparative representation. (From Eigen and Schuster 1977, Fig. 10.)

now suppose that some of the intermediate substances in the reaction cycle are themselves catalysts—and would then remain unchanged in structure during the process, even if changing in quantity. This is represented as the second level in Fig. 5.4, where all the E_i are regarded as catalysts formed by the catalytic action of its predecessor, E_{i-1}, on a substrate (S_i), not shown on the diagram, that is fed into the system. The *in vitro* template-directed replication of single-stranded RNA, although not really an enzyme, behaves like such a system (Spiegelmann 1971). *Self-replicative* units, e.g. double-stranded DNA, could themselves be linked in a higher hierarchical level of cyclic catalysis which Eigen and co-workers have called a catalytic *hyper*cycle, the next level of organization (Fig. 5.4, where 'I' enclosed by an arrowed circle denotes a *self-replicative* unit). Such systems are *hyper*cyclic with respect to the catalytic function—they connect autocatalytic or self-replicative units through a cyclic linkage. The couplings between the self-replicative I units have been denoted by single arrows in Fig. 5.4. but, realistically, if I_{i-1} units represented DNA or RNA strand(s), then the arrow to I_i might itself well represent the production of an enzyme E_{i-1} which would then catalyse the formation of I_i. It is such a system which Eigen (1971*a,b*), and Eigen and Schuster (1977, 1978*a,b*) in a more developed treatment, have proposed as the most likely kind of 'missing link' between non-living and living matter, as is discussed more fully below.

There are certain characteristics of closed loops of chemical reactions that are worth noting before proceeding any further. Thus in the 'autocatalytic' cycle, represented in the middle of Fig. 5.4, suppose there are n such catalysts $E_1, E_2 \ldots E_i \ldots E_{n-1}, E_n$, each of which catalyses, or in some way guides and enhances, the formation of its successor in the cycle, which is finally closed by the E_n influencing the formation of E_1. The formation and disappearance of E_i is governed by an equation of the form

$$\frac{dE_i}{dt} = A_i Q_i E_{i-1} - D_i E_i \tag{5.13}$$

in the absence of selection constraints and information back flow (cf. (5.1)), and the set of such equations for the interlocking cycle yields two distinct kinds of solution (Eigen 1971*a*, V.2.2): $(n-1)$ negative, possibly complex solutions representing the $(n-1)$ 'normal modes' (Eigen and de Maeyer 1963) of equilibrating relaxation processes *within* the cycle; and one positive 'normal mode' representing the *auto*catalytic growth property of the cycle as a whole. So once the cycle is internally equilibrated, it is meaningful to attribute to the cycle as a whole a 'selection equation' of the same general form as (eqn 5.1, with the 4th term omitted) when competition between different cycles is under consideration (see Eigen

1971a, V.1.1, pp. 499–501)—as it must be when assessing the selection and evolution, i.e. the 'Darwinian' behaviour, of different cycles. It can be proved that any cyclic reaction network of *self*-reproducing units, if it is going to constitute at least a link between non-living and living matter, must have functional linkages distinguished by the following qualities (Eigen and Schuster 1977, p. 564):

(a) The linkage must permit competition of each self-reproductive unit with its error copies, otherwise these units cannot maintain their information.

(b) The linkage must 'switch off' competition among those self-reproductive units which should be integrated into a new functional system and allow for their cooperation.

(c) The integrated functional system must then be able to compete favourably with any less efficient system or unit.

The potentialities of putative, feasible catalytic cycles to be self-organized systems fulfilling these requirements had therefore to be examined.

Homogeneous self-organizing cycles

The complementarity of nucleic acid chains in double-helical DNA and in the copying of RNA affords a natural possibility of self-organization in cycles containing only nucleic acids and such cycles have been examined by Eigen (1971a, Section IV; 1971b, IV.2) for their ability to meet the requirements outlined above. The reproduction process is conceived as consisting of alternation in a two-membered cycle between a template $(+i)$ guiding the formation of a complementary copy $(-i)$—complementary in the sense of nucleotide base-pairing—with the $(-i)$ strand then guiding the formation of the $(+i)$ (Fig. 5.5). It transpires that such a cycle has the advantage that it would select for the most uniform sequence of the most frequent nucleotide. But, because of its complementary instruction process, it would have to accumulate at least two different monomers and so possess the inherent capability of generating a 'code', i.e. particular sequences of monomers which could initiate specific processes, further 'molecular instructions' that can constitute functional

Fig. 5.5. (From Eigen 1971b, p. 175.)

specificities. However it also transpires that, because error-free reproduction of polynucleotides, in the absence of specific enzymes, is limited to chain lengths of 100 or less, only a very low information content can be achieved. Moreover, since different cycles have to compete with each other, such systems, without self-instructing *catalytic* help, would not be able to organize themselves to provide any type of correlated functions.

Self-organizing *enzyme* (i.e. *protein*) cycles can also be envisaged because there are enzymes which control the assembly of as many as ten aminoacids in precise sequences (e.g. the formation of the cyclic decapeptide gramicidin A (Lipmann 1971)). So proteins *can* guide the formation of other proteins without the mediation of nucleic acids. Admittedly a decapeptide is a long way from the length of polypeptide chains in active proteins, but it is conceivable that such smaller chains could be specifically linked with others under enzymic control (cf. H. A. Simons' watch assembly story of Hora and Tempus, section 1.2). Could such pure protein cycles (e.g. Fig. 5.6) have alone started the process of self-organization? Such protein cycles at first sight look more promising than the nucleic acid cycles for they have an appreciably higher precision of

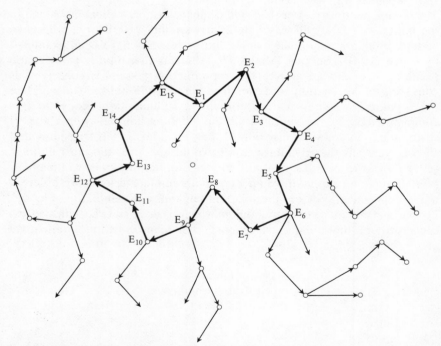

Fig. 5.6. Catalytic network of proteins, including a closed loop: $E_1 \ldots E_{15}$. (From Eigen 1971b, Fig. 3(a).)

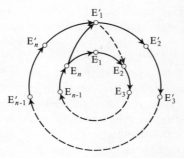

Fig. 5.7. Reproduction of mutants in a catalytic enzyme cycle. (From Eigen 1971b, Fig. 4.)

recognition of certain substrates than prevails in base recognition by nucleic acid chains; there is a higher information content involved in a multiple-step cycle as compared with a single (polynucleotide) chain of restricted length in a two-member cycle; and furthermore by branching (i.e. some enzymes having more than one catalytic function) several otherwise independent cycles could be linked together. So the net accumulation of information in such protein catalytic cycles could be considerable. However, there are counteracting disadvantages. Because specified recognition is not an inherent property of all polypeptide chains, especially in relation to a specific molecular pattern, proteins which catalyse their own reproduction via specific cycles (e.g. $E_1 \rightarrow E_n \rightarrow E_1$ in Fig. 5.7) will not necessarily reproduce their mutants (E_1') resulting from errors in copying, even when the mutants are otherwise 'advantageous' (i.e. helpful in rate competition with other cycles). Moreover such a mutant (e.g. E_1' from E_{n_1} instead of E_n in Fig. 5.7) is only advantageous if there occurs simultaneously, or if it can itself cause, specific mutations in all the sequence of intermediates leading to its own reproduction (E_2' to E_n' in Fig. 5.7). Even if such a cycle could form, much information content would be lost, so it would amount to 'revolution' instead of 'evolution'. Even the possibility of branching, which may help to enhance information content, could be a disadvantage if the branches were 'parasitic' in the sense that they conferred no selective and reproductive functions on the system.

So the lack of an inherent complementary *self*-instructive process in protein chains militates against its *evolutionary* viability. The system cannot easily utilize selective advantage in spite of, indeed because of, its load of other information not useful for selection.

From such considerations, based on more quantitative work than space has allowed to be reported here, Eigen concluded that, although both pure nucleic acid and pure protein cycles can reproduce themselves, both show extreme properties which make them unsuitable alone as the basis for self-organizing systems that could evolve.

Self-organization by encoded catalytic function: the hypercycle

It is clear from the foregoing that nucleic acids and proteins are complementary, in the general sense, in their respective advantages and disadvantages with respect to forming self-organizing, evolving cycles capable of being selected in rate competition with other cycles. On the one hand nucleic acids provide an essential prerequisite for self-organization, namely complementary instruction as an inherent property of all such chains and so code formation, but weak catalytic powers and inability therefore to couple information carriers to allow appreciable increases in information content. Proteins, on the other hand, have very great catalytic activities and functional potentialities and could build up interlocking cycles of high information content but, unlike nucleic acids, cannot easily utilize selective advantages because their recognition properties are specific and unique towards particular molecular configurations (of other proteins) and are not general and inherent in all protein chains. So, it is argued (Eigen 1971a; Eigen and Schuster, 1977, 1978a,b, 1979) a *combination* of the complementary instructional abilities of nucleic acids and the catalytic potentialities of proteins might lead to the required non-linear selection behaviour that would afford selective advantage of the 'combined' cycle, *as a whole*. The basic proposal of Eigen and Schuster is depicted in Fig. 5.8 and its legend. It consists of a number of two-member cycles of polynucleotide chains (i_+ and i_-): I_i, represented by circles (cf. Figs. 5.4 and 5.5). These have to be long enough chains to provide the information for the formation of an active polypeptide chain (or chains) denoted by E_i which catalyses the complementary instruction of the *next* polynucleotide system, I_{i+1}—and so on. Only part of the information stored in I_i and passed on to E_i needs to be relevant to the catalytic function of E_i in relation to the I_{i+1} cycle. Other parts of the information in I_i could be translated into E_i for other functions, represented by the arrows pointing away from the inner cycle. It is essential that the whole hypercycle be closed; there must be an E_n that enhances the I_1 complementary copying process. Eigen has summarized the deduced properties of this hypercycle as follows:

(1) Each cycle has autocatalytic growth properties.
(2) Independent cycles compete for selection.
(3) As a consequence of nonlinearity, selection will be very sharp, possibly resembling "all or none" behavior, if singularities are involved.
(4) With these selective properties, the system will be able: (a) to utilize very small selective advantages (which have to occur to a stochastically significant extent) and (b) to evolve very quickly (a selected system will not tolerate the nucleation of independent competitors, thus code and chirality will be universal).

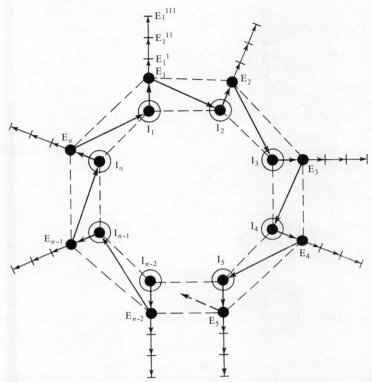

Fig. 5.8. The self-instructive catalytic hypercycle. The I_i represent information carriers, i.e. complementary single strands of RNA. Each small cycle indicates the self-instructive property of the I_i collective involving the two complementary strands. The E_i (encoded by I_i) represent catalytic functions. Each branch may include several functions (e.g. polymerization, translation, control), one of which has to provide a coupling to the information carrier I_{i+1} (e.g. enhancement of the formation of I_{i+1} by specific recognition). The trace representing all couplings must close up, i.e. there must be an E_n which enhances the formation of I_1. The hypercycle is described by a system of non-linear differential equations. (From Eigen 1971b, Fig. 5.)

(5) The cyclic coupling will provide an information capacity which is adapted to the requirements of the system. Nevertheless, the replication length of the single code unit will be small enough to ensure reproducibility.
(6) The system can evolve, i.e. improve, by utilization of selective advantages. "Genotypic" mutations, i.e. alterations in I_i, can be immediately utilized by E_{i-1} and do not have to await a correlated series of mutations in order to propagate through the cycle, as was necessary for "linear" catalytic cycles. Selective advantages can become effective via repression, derepression or promotion.

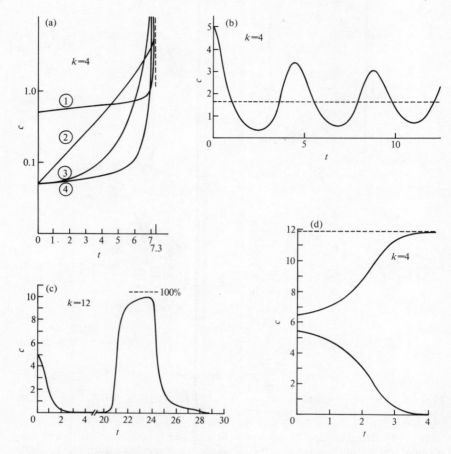

Fig. 5.9. The growth characteristics of the self-instructive catalytic hypercycle. (a) The growth of a four-membered hypercycle (four information carriers I_i encode four coupling enzymes E_i). At $t=0$, I_1 is assumed to be present in tenfold excess of I_2, I_3, and I_4. The formation rates for all four members are represented by simple second order terms: $x_i x_{i-1}$. Decomposition is neglected. The time axis is reduced and refers to a rate constant $k = 1$ (or $t = kt$). The singularity occurring at $t = 7.3$ leads to a very sharp selection, if several competing cycles are involved. (b), (c) Solutions describing the selection of k-membered hypercycles under the constraint of constant overall organization. The reaction system again is described by a simple second-order formation term—for all members identical; (cf. (a)) as well as by a first-order 'removal' term in order to maintain the condition $\sum x_k = $ constant. The solutions are shown for one member only. The 'equilibrium' value is constant for $k < 3$. For $k = 3$ the approach to selection equilibrium is represented by a damped oscillation, whereas for $k > 3$ stationary oscillations occur. This can be shown by starting from a constant distribution and introducing a small perturbation at $t = 0$. The oscillation then builds up. (Calculations by P. Schuster, unpublished results; examples: $k = 4$(b) and 12(c).) (d) Competition

(7) The system selects against parasitic branches if these have selective values smaller than that of the members of the cycle. Parasitic branches with higher selective values will not allow the cycle to nucleate if they are present from the beginning. However, if they appear after the cycle was nucleated, they will have no chance of growing, as a consequence of the nonlinear selection behavior. A cycle can reduce the number of members by constriction, if this presents any selective advantage. For coupled cycles the conditions for simultaneous existence are quite restricted.

(8) There is only one type of branches which can co-exist with the cycle, i.e. a branch whose selective value exactly matches that of the cycle. An exact matching would be possible only if branch and coding region I_i inside the cycle make use of the same promotor located in E_{i-1}. This will automatically lead to a gene and operon structure of the code system. Within the branches the system can evolve functions of general utility (e.g. polymerases, a translation system, control factors, metabolic enzymes).

(9) The system, after nucleation, has soon to escape into a compartment. Only compartmentalized systems can utilize functional branches (brought about by mutations) exclusively to their own advantage (and thus also allow evolution of the branches). By the same mechanism, the system is saved from any pollution caused by unfavourable branch mutations.

(10) A system enclosed in a compartment may "individualize" by linking its code units into a stable chain, e.g. with the help of an (evolving) ligase, and reproduce the total chain as an individual unit. In such a chain (which will be cyclic if ligases are involved) genes resulting from a given unit I_i should be localized in neighbouring positions. However, the message, for the coupling factors occurring, in I_{i-1} can be situated at a quite distant position. (Eigen 1971a, pp. 504-5.)

A more quantitative theory of this system has been worked out in considerable detail by Eigen and Schuster (1977, 1978a,b) and this forms the basis for some of the generalizations listed above. Mathematical analysis of dynamical systems using methods of differential topology (Eigen and Schuster 1978a; see also Schuster, Sigmund, and Wolff, 1979) show that only hypercyclic organizations are able to fulfill the requirements of being the best-adapted functionally-linked ensembles, of optimizing their evolution, and of having the qualities previously specified ((a), (b), and (c), p. 231). It is concluded that hypercycles that integrate self-replicative units with catalytic links, are a unique class of self-organizing chemical networks. Some of their growth characteristics, worked out by Eigen and Schuster, are summarized in Fig. 5.9. After nucleation, the system would have a selective advantage if it escaped into

---

between a three- and four-membered cycle, having the same individual rate parameters, but differing in the initial concentrations.
three-membered cycle: $\sum_k x_k^0 = 5.4$, four-membered cycle: $\sum_k x_k^0 = 6.5$. Note, that here the four-membered cycle wins though each of its species is present at a lower initial concentration than each of the species of the three-membered cycle. Plot: concentration vs. time. (From Eigen 1971b, Fig. 7.)

a compartment that prevented its members from diffusing apart and that enabled it to utilize functional branches arising from mutations. These 'compartments' could be the coacervates of Oparin (1965) or microspheres of lipids and polypeptides with mainly hydrophobic side chains (Fox 1965) and might well reach the volume of a typical bacterium (Kuhn 1972).

5.5 Reality and theory

Eigen and Schuster (1978b) have argued cogently that such an abstract hypercycle is realistically feasible given the known behaviour of polynucleotides and polypeptides, which may be presumed to be the present 'descendants' of molecular ancestors whose properties must have been closely related. At the conclusion of their series of papers, Eigen and Schuster (1978b, pp. 367 and 368) summarize their understanding, based on their physico-chemical approach, of the essential stages in the transition from the non-living to the living (see Fig. 5.10) as follows:

1. The first appearance of macromolecules is dictated by their structural stability as well as by the chemical abundances of their constituents. In the early phase, there must have been many undetermined protein-like substances and much fewer RNA-like polymers. The RNA-like polymers, however, inherit physically the property of reproducing themselves, and this is a necessary prerequisite for systematic evolution.

2. The composition of the first polynucleotides is also dictated by chemical abundance. Early nucleic acids are anything but a homogeneous class of macromolecules, including L- and D-compounds as well as various ester linkages, predominantly 2'-5' besides 3'-5'. Reproducibility of sequences depends on faithfulness of copying. GC-rich compounds can form the longest reproducible sequences. On the other hand, AU substitutions are also necessary. They cause a certain structural flexibility that favors fast reproduction. Reproducible sequences form a quasi-species distribution, which exhibits Darwinian behavior.

3. Comma-free patterns in the quasi-species distribution qualify as messengers, while strands with exposed complementary patterns (possibly being the minus strands of messengers) represent suitable adapters. The first amino acids are assigned to adapters according to their availabilities. Translation products look monotonous, since they consist mainly of glycine and alanine residues. The same must be true for the bulk of noninstructed proteins.

4. If any of the possible translation products offers catalytic support for the replication of its own messenger, then this very messenger may become dominant in the quasi-species distribution and, together with its closely related mutants, will be present in great abundance. The process may be triggered by some of the *noninstructed* environmental proteins, which in their composition reflect the relative abundance of amino acids and hence may mimic primitive *instructed* proteins in their properties.

5. The mutants of the dominant messenger—according to the criteria for hypercyclic evolution—may become integrated into the reproduction cycle,

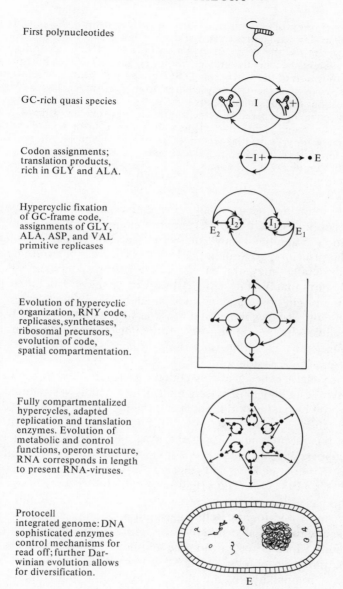

Fig. 5.10. Hypothetical scheme of evolution from single macromolecules to integrated cell structures. (From Eigen and Schuster 1978b, Fig. 63.)

whenever they offer further advantages. Thus hypercyclic organization with several codon assignments can build up. Such a hypercyclic organization is a prerequisite for the coherent evolution of a translation apparatus. More and more mutants become integrated, and the steadily increasing fidelities will allow a prolongation of the sequences. Different enzymic functions (replicases, synthetases, ribosomal factors) may emerge from joint precursors by way of gene duplication and subsequently diverge ...

6. The complex hypercyclic organization can only evolve further if it efficiently utilizes favorable phenotypic changes. In order to favor selectively the corresponding genotypes, spatial separation (either by compartmentation or by complex formation) becomes necessary and allows selection among alternative mutant combinations. Remnants of complex formation may be seen in the ribosomes.

While stressing the inevitably speculative character of such evolutionary accounts, because of the element of chance as to which mutation and fluctuation is amplified, nevertheless Eigen and Schuster also emphasize that 'the principles governing the historical process of evolution—even in the finer details—may well be susceptible to our understanding'. And it is in this spirit that Eigen (1971a) has also developed detailed proposals about the origin of the genetic code which he has related to the 'evolution experiments' of Spiegelman and those in his own laboratory. It proves to be possible—by combining the foregoing theoretical approach with studies on the replication (including *de novo* synthesis) of RNA by the replicase enzyme system of the $Q\beta$ virus that attacks *E. coli*—to discover how early RNA genes probably coded for proteins and to test ideas about early synthesis and natural selection of optimal RNA templates. These take us somewhat beyond the scope of the present work, principally on the physico-chemical principles involved, but the reader is recommended to pursue this fascinating quest in the work of Eigen and of his colleagues (see Küppers 1979b; Eigen and Winkler-Oswatitsch (1981b) on 'Transfer RNA, an early gene?'; and Eigen, Gardiner, Schuster, and Winkler-Oswatitsch (1981) on 'The origin of genetic information').

Kuhn has argued, in a series of papers (references in Kuhn 1976), that the driving force for the self-organization of matter must be seen in a specific environmental structure to be found on the Earth's surface. Such a structure which is periodic in time and heterogeneous in space could, he argues, initiate evolution and drive it towards a continuously increasing degree of complexity—by a pathway consisting of many small steps. Such a view is not necessarily in disagreement with the foregoing exposition of Eigen and his colleagues, whose main objective could be said to be that of specifying the conditions and the molecular organization for the generation, accumulation, transmission, and modification of information to be possible. What kinds of environment have allowed these criteria to be fulfilled is certainly a proper, if now historically, unanswerable question and Kuhn's approach can be regarded as an attempt to be more precise about evolutionarily stimulating environments. But it does not obviate or make any the less necessary the critical analysis of such mechanisms as the hypercycle and their quantitative assessment as potential biotic forms and precursors.

This is an appropriate point at which to clarify what is and is not purported to be explained by the kind of applications of physico-chemical concepts to the origin and evolution of biological complexity that has been, and is being, developed by Eigen and his colleagues. This is best done in his own words (Eigen 1971a, p. 515):

> *What the theory does explain* is the general principle of selection and evolution at the molecular level, based on a stability criterion of the (non-linear) thermodynamic theory of steady states. Evolution appears to be an inevitable event, given the presence of certain matter with specified autocatalytic properties and under the maintenance of the finite (free) energy flow necessary to compensate for the steady production of entropy. The theory provides a quantitative basis for the evaluation of laboratory experiments on evolution.
> *What the theory may explain* is how to construct simple molecular models representing possible precursors of "living" cells ... such models have been examined, of which only one [the protein-nucleic acid hypercycle] could be shown to fulfill all the requirements for evolution with the present state of cellular life.
> *What the theory will never explain* is the precise historical route of evolution. The "never" is a consequence of the stochastic nature of the processes involved and the tremendously large multiplicity of possible choices. This also applies to predictions of future developments beyond certain time limits. Hence: 'Whereof one cannot speak, thereof one must be silent' (L. Wittgenstein (1922) *Tractatus Logico-Philosophicus*, 7. London: Routledge & Kegan Paul).

Part III

6 The interpretation of biological complexity

6.1 Modes of biological complexity

THE intracacies of the attempts to disentangle the threads of the complexity that constitutes biological systems, whether by means of thermodynamic (Part I) or kinetic (Part II) concepts and analyses, serve to illustrate just *how* complex living systems actually are. Yet these attempts constitute only the first forays into the intricate ramifications of the biological world, an indication of which was depicted in J. G. Miller's diagrams (Figs. 1.2 and 1.3) of the relationship between his, already vastly over-simplified, nineteen sub-systems in each of the seven levels that he discerned in living systems. Such elaboration was made more explicit in the network thermodynamic formulation of the energy flow through the glycolytic pathway of skeletal muscle (section 3.7, p. 102) and reaches its apogee in the baffling complexity of the human brain with its 10^{11} nerve cells each fed by and feeding into hundreds or thousands of neurons, and making contact through $\sim 10^{14}$ synapses (Hubel 1979)—apparently the most complex piece of matter in our known universe.

There has been a tendency for philosophers of science to be concerned with the simplicity and complexity of sets of predicates or theories and for biological scientists to look upon the actual plurality of biological phenomena, theories, and models, and their overlapping domains, as being only a provisional stage before their ultimate reduction to one primary level of phenomena, to one all-encompassing theory or model. This has meant that questions concerning the relationships within this plurality have tended to be ignored 'on the supposition that all will be made clear when their relationships to the perhaps as yet unknown reducing theory are determined' (Wimsatt 1976). This situation will be examined further when we come to discuss in section 6.5 the whole question of 'reductionism' in relation to biological theories, but at this stage our concern is to stress the character of our actual descriptions of living systems and the kind of complexity manifest therein—and, in particular, to make more explicit the interrelation between different formulations of biological complexity.

Wimsatt (1976) has made a valuable distinction between what he calls 'descriptive complexity' and 'interactional complexity' and so also, corollaratively, between 'descriptive simplicity' and 'interactional simplicity'. All systems, he argues, including living ones, can be viewed from a number of different perspectives which may yield several different, non-isomorphic decompositions of the system into parts (following Kauffman 1971). Each of these different theoretical perspectives (T_i), applicable to a given living system, suggests criteria for the identification and individuation of the parts, i.e. for 'decomposition' of the system into thus-defined parts (Wimsatt calls these 'K-decompositions', presumably K for Kauffman) and they are different for each of the T_i. If all of a set of such decompositions, $K(T)_i$ of a system produce *coincident* boundaries for all parts of the system, then the latter is called *descriptively simple* relative to these $K(T)_i$, as illustrated in Fig. 6.1 (l.h.s.). If two parts from different decompositions in the set $K(T)_i$, although having a common point which is interior to at least one of them, nevertheless are not overall coincident, then there are a number of different mapping relations which can hold between their boundaries and the system is then said to be *descriptively complex* (Fig. 6.1, r.h.s.).

Particularly pertinent, is the relation between the different decompositions, among the $K(T)_i$, which apply at roughly the same spatial order of magnitude (many-to-one mappings from a micro- to a macro-level would not be surprising, or specially significant, in itself). Thus a piece of granite (Fig. 6.1) can be 'decomposed' into subregions of: constant chemical composition, $K(T)_1$; thermal conductivity, $K(T)_2$; electrical conductivity, $K(T)_3$; etc. These boundaries are at least roughly coincident so that the granite is 'descriptively simple' to these decompositions. However, 'decomposition' of a multi-cellular living organism, such as a fruit fly (Fig. 6.1, r.h.s.), into parts or sub-systems from different perspectives (e.g. by criteria based on: physical chemistry, T_1; anatomy, T_2; cell type, T_3; developmental gradients, T_4; biochemical reactions, T_5; physiological systems, T_6, etc.) results in mappings that are far from being coincident—and so this, and all other, living systems are 'descriptively complex'. (Wimsatt's diagram, Fig. 6.1, has to confine itself to a two-dimensional representation, though, of course, the complexity is three-dimensional). This kind of descriptive complexity is further compounded when one also takes into account the lack of individuation of the different theoretical perspectives in biology and their interaction with evolutionary criteria for functional analysis of organisms into sub-systems.

This account of *descriptive* complexity provides the necessary basis for formulating another kind of complexity which may be called *interactional* since it expresses the extent of the causal interactions of a system, paying attention to those interactions which cross boundaries between theoretical perspectives (Fig. 6.2). If, for a particular decomposition, the causal interactions *within* sub-systems so identified are all much stronger than those *between* the sub-systems, then we have a case (Fig. 6.2(a)) of 'near-decomposability', as Simon (1962) called it—a designation that he regarded as applying particularly to living systems (see section 1.3; such a

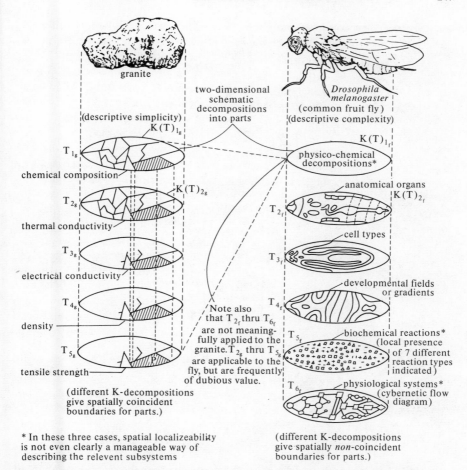

Fig. 6.1. Descriptive simplicity and complexity. (From Wimsatt 1976, Fig. 1.)

designation may be justified by comparing the energies of bonding in different entities, viz. $\sim 1.4 \times 10^6$ eV for an atomic nucleus, ~ 5 eV for a molecule, and ~ 0.5 eV for the forces holding macromolecules together and in their biologically required shapes). It is for this reason that living systems have frequently been regarded as predominantly hierarchical in character, a concept we shall discuss in the next section (6.2). Note that, unlike the case of descriptive simplicity (Fig. 6.1), we are now concerned with decomposition into sets of state variables in various perspectives, for it is causal interactions that are at issue.

If now we regard these sub-systems and the system of which they are a part as respectively identifiable only through the operation of different

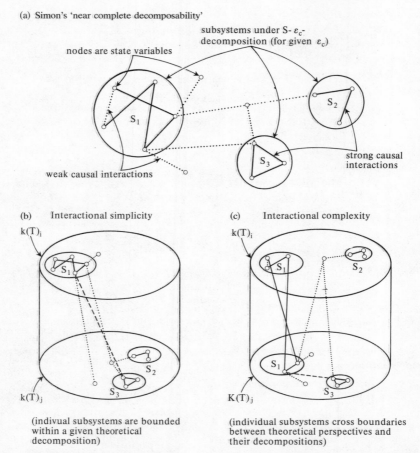

Fig. 6.2. Interactional complexity and 'near-decomposeability'. (From Wimsatt 1976, Fig. 2.)

perspectives (T_i and T_j), 'near-decomposability' can then be seen as an example of *interactional simplicity* (Fig. 6.2(b)). A system is interactionally simple if none of the sub-systems in such a decomposition cross boundaries between its different K-decompositions. Correspondingly, a system is *interactionally complex* to the extent that they do so. Clearly one can envisage a parameter ε_c = strength of *inter*-subsystemic interaction divided by strength of *intra*-subsystemic interaction. Variation in this parameter would then determine the different levels of decomposition, i, that will actually be discriminated by the different theoretical perspectives, T_i, (Fig. 6.2(c)). A system that is interactionally complex in relation to a particular value of ε_c may be interactionally simple for smaller values.

So simplicity or complexity is always relative to a particular value of ε_c. Moreover, if it is assumed that inter-subsystemic interactions are negligible ($\varepsilon_c \to 0$), then predictability of the system on this assumption under a given decomposition will diminish the larger is the actual ε_c of the system that is characteristic of that decomposition. (Wimsatt denotes the decomposition relevant to interactional complexity as $S-\varepsilon_c$ decompositions to distinguish them from the parts/whole criteria of K-decompositions.)

Such analyses of the notion of 'complexity' reinforce the intuition that systems of the complexity of the living need more than one perspective to do justice to them, for it is far from clear into what component sub-systems living organisms *should* be analysed to provide the raw material, as it were, of the most comprehensive, non-reducible theory—and this is true *a fortiori* when the component sub-systems have evolved together and are not obviously separable (Levins 1970).

6.2 Complexity and hierarchy

The concept of a hierarchy has already (section 6.1) been invoked in connection with the formulation of the idea of interactional complexity: it clearly merits closer examination in relation to living systems. Can their type of organization be properly regarded as hierarchical, as J. G. Miller (1978) implies in his massive classificatory scheme (section 1.2)?

It may come as some surprise to a biological scientist to find out how extensive and sophisticated a literature there is on the definition and elaboration of the concept of hierarchy (for which see the multi-author volumes edited by Whyte, Wilson, and Wilson (1969), by Weiss (1971) and by Pattee (1973) and especially a useful selected bibliography on *Forms of Hierarchy* by D. W. Wilson (1969)). In ordinary speech it would be taken for granted that the word 'hierarchy' would refer to some sort of system in which there was an analogy to feudal, medieval society, in which there was a delegation of authority and constraint from the top downwards. This is indeed one kind of hierarchy which H. A. Simon in his seminal paper (1962) on 'The architecture of complexity' has called a '*formal* hierarchy', that is a hierarchy in which each system is 'boss' to a set of subordinate systems. He rightly pointed out that this is too restricted an understanding of the hierarchies that we observe in the natural world and in social life. He therefore defined a hierarchical system as 'a system composed of inter-related sub-systems, each of the latter being, in turn, hierarchical in structure until we reached some lowest level elementary sub-system' (the choice of this lowest level he regarded as arbitrary). This definition of a hierarchy has been echoed in much subsequent literature and the basic conditions for a system to exist as a hierarchy have been postulated by Pattee (1969, 1970) as: (i) being autonomous, producing its own rules with no extraneous source of authority; (ii) all elements in the system obeying the normal laws of physics and chemistry; (iii) the presence of hierarchical control, that is, constraints which arise within a collection of elements and affect individual elements; (iv) production of an integrated function through collective constraints. However, more clarification of the nature of natural hierarchies is necessary if their genesis is to be elucidated and one way of doing this is to determine the parameters and features which vary as one passes from one level of natural hierarchy to another. Some of these distinguishing parameters and features proposed to classify hierarchies are as follows.

Constraints of time and space

These constraints have already been discussed (section 1.3 and Fig. 1.4) in connection with the characteristic temporal and spatial parameters of living systems. Little more needs to be added except to recall how the inter-connectedness of spatial and temporal parameters is now seen to underlie the formation of space- and time-symmetry breaking behaviour (sections 4.5 and 4.6), not least through the controlling influence of the ratio D_i/L^2 measuring the coupling between neighbouring spatial regions (section 4.5.2).

Structure–function relationships

Three levels of relative freedom of movement of structural parts in the hierarchy of living organisms have been distinguished (Gutman 1969): (i) the relative positions of the structural parts firmly fixed (e.g. a spider's web, or a bridge); (ii) the parts have a limited freedom of movement but their pathways are fixed (e.g. man-made machines); (iii) the moving parts, whose roles are well defined, have a large freedom of movement mainly within the boundary of the structure, but sometimes even crossing it, and their pathways are not fixed (e.g. multicellular living organisms). Living systems are 'organizational' structures of the third type but in their evolution and development they make increasing use of elements of types (i) and (ii). Gutman (1969) regards the activity of a part as meaningful, that is as having a function, if it contributes to the activity of the whole and constitutes a necessary link in competing a sequence of events with specific results. In this sense function has primacy over structure. There are many examples in biology in which the same end results from different beginnings and via different routes and where different structures can perform the same function. The development of parts in a whole is the result of delegation of function to specialized structures and this necessitates control.

Relationships between macromolecules, and between patterned combinations of them, have vital functional roles in living organisms. Many scientists (e.g. Elsasser (1966), Stebbins (1969), Medawar (1974), Peacocke (1976)) have noted the sequence of increasing complexity to be found in the living world: atom, molecule, macromolecule, subcellular organelle, cell, tissue organ, system of organs, multicellular functioning individual organism, population of organisms, ecosystem. This series represents a succession of levels of organization of matter in which each successive member of the series is a 'whole'. This is well illustrated by the hierarchy of complexity in muscle tissue which is constituted of parts preceding it in the series (Fig. 6.3). At each level only precise relationships between equally complex structures of different kinds allow ordered function. Stebbins (1969) has called this a principle of *relational order* that is basic to the nature of life and he defines it as follows:

> In living organisms, the ordered arrangement of the basic parts or units of any compound structure is related to similar orders in other comparable structures of the same rank in the hierarchy, permitting the structures to co-operate in performing one or more specific functions. The functions that depend upon relational order can be as diverse as muscular contraction, the transmission of nervous stimuli, the manufacture of sugar by photosynthesis, the synthesis of the protein molecules that form the building blocks of the living cell, and the transmission of heredity. Consequently, relational order is of primary importance as a unifying principle for all of life. . . . A basic feature of relational order is that it is not a series of static relationships, like those between the molecules of a chemical crystal or between the beams and rafters of a house. The ordered relationships among parts of all levels of the organismal hierarchy help organisms both to carry out certain chemical reactions and to bring about co-ordinated movements of parts at all levels. (Stebbins 1969, pp. 5 and 6.)

The constraints of time and space, mentioned above, operate jointly to produce these effects.

Control

It is possible, as in the medieval social hierarchy, for the hierarchical order to be defined as a system in which there is control from the top down so that the idea of delegation of function to specialized structures becomes transformed into the idea of control, of a 'bossing relation'. A hierarchy is then a set or system such that its components all have a single boss and, no matter how low in a hierarchy an element is, it is always under command of one particular dominant element which represents the beginning or the top (Bunge 1969). However, although taxonomic systems in biology may be regarded as hierarchies in this sense these are only formal and in fact the 'bossing', one-sided domination relationship is not what is observed in biological hierarchies. For a particular biological function tends to be distributed over much, if not all, of the systems and nothing which happens at one level can be without consequences for

252   THE INTERPRETATION OF BIOLOGICAL COMPLEXITY

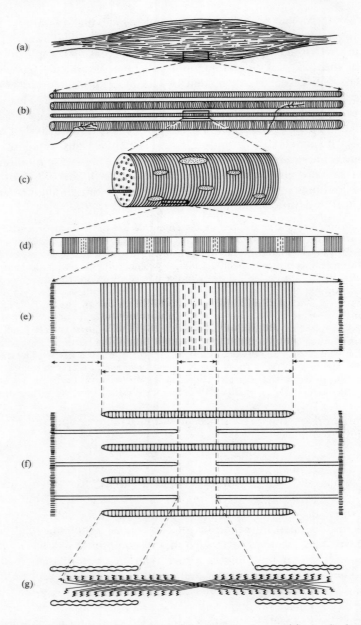

Fig. 6.3. The heirarchy of complexity in muscle tissue. (a) A single muscle, magnification ×2. (b) A group of fibres, with nerve endings attached, ×100. (c) Portion of a single fibre, ×100. (d) Portion of a single fibril, showing striations, ×10 000. (e) The indicated portion of (d), ×43 500. (f) The same, showing relative

activities at all other levels. The interaction between functional levels is reciprocal and not unidirectional. As Koestler puts it '... the functional units on every level of the hierarchy are double-faced, as it were: they act as wholes when facing downwards, as parts when facing upwards' (Koestler 1967). Hierarchical organization in living things seems to involve doubly determinate systems: systems such that an arrangement of the elements comprising the system constrains the behaviour of the elements themselves. So although we call one level 'higher' this is not meant to mean that it actually controls the 'lower' level. Indeed, it is possible to find hierarchies in the natural world to which such a concept would be entirely unapplicable, for example the hierarchical classification of physical entities in the physical and astronomical worlds which H. Shapley (1958) has made (quoted by D. W. Wilson 1969). But, as the philosopher of biology Marjorie Grene (1969) has emphasized, in all kinds of natural hierarchies there is a *ranking* of real entities, levels of organizations or levels of performance. The concept of control is not always necessary to this ranking, but this raises the question of how this ranking in natural hierarchies is to be explicated. What are the criteria of ranking within a natural hierarchy, especially a biological one? We shall only retain this term hierarchy, therefore, in the sense of a ranking of real entities without any implication that there is necessarily a 'top–down' kind of control relationship between higher and lower levels.

Specifying descriptions

A hierarchically organized system is one which is engaged simultaneously in a variety of distinguishable activities which cover a wide range and are such that different kinds of specifying descriptions are appropriate. This latter characteristic is decisive for the system to be a hierarchy: the problem is to delineate relationships between these specifying descriptions. We would like this delineation to be predictive, so that, from information at one level, information at another level could be inferred. This is the hope of the reductionist programme, which we shall discuss later, and is based on the belief that there is one 'anchor' level of greater

---

positions of the thick and thin filaments, which are, respectively, groups of myosin and actin molecules. (g) Diagram showing the arrangement of individual myosin molecules in a thick filament and portions of the rope-like structure that consists of actin molecules wound together. The 'heads' of the myosin molecules are shown as zigzag lines pointing toward the actin filaments. Magnification ×130 500. If the entire muscle shown in (a) were magnified to the same scale as the molecules shown in (g), it would be more than 5 miles long. (Based on diagrams of H. E. Huxley 1958, 1965.)

significance to which all the others can be chained (e.g. the cell in biology; or the biochemical mechanisms in genetics). This more linguistic and philosophical kind of ranking within hierarchical systems has, of course, much relevance to the philosophy of biology in general and has been particularly developed by Rosen (1969). He concludes that there is no *one* type of system description that can possibly display by itself a definite type of hierarchical structure, such as the biological, because we recognize such hierarchical structures only by the necessity for *different kinds* of description of the various levels observed. Moreover each mode of description of a system arises from a corresponding interaction of the system with ourselves, or rather with those physical instruments with which we choose to observe (Rosen 1979). This emphasizes again that the ranking of specifying descriptions characteristic of a biological hierarchy does not entail any kind of logical priority for any one of them, and indeed complexity, according to Rosen (1979), then appears as a contingent, rather than an intrinsic, property expressing interactive capabilities reflected in observation or measurement: but, one might well add, complexity' is not unusual in this respect for *all* properties in the end, thus reflect observation and measurement.

Operative constraints

There seem to be two kinds of ways in which constraints operate in natural hierarchies (Pattee 1971). In physical systems *structural* hierarchies are common in which constraints arise as the loss of degrees of freedom, of matter in solids for example, and as boundary conditions which limit the values of certain degrees of freedom, as a particle in a box. Such structural hierarchies are lifeless and living organisms seem rather to be examples of *functional* hierarchies in which processes in time require variable constraints, i.e. time-dependent boundary conditions acting on selected dgrees of freedom which are not imposed from outside, e.g. an enzyme is a time-dependent boundary condition for its substrate. Often there is a simplicity in the collective function of such a hierarchy at any particular level, quite distinct from the complexity of its details. These concepts of simplicity in the collective function and of function-producing constraints imply alternative, simplified ways of representing complex detailed motions, but this alternative record or representation must somehow be *internal* to the biological hierarchy if it is to be genuinely autonomous, as required by its basic definition (see beginning of this section, 6.2). But how can this alternative internal representation arise? It is a real aspect of the way the biological system behaves and not simply a product of an observer's viewpoint, so we have to conclude that somehow the constraints of living matter must contain their own descriptions. Indeed, it transpires that a constraint is not a definite, fixed objective structure, but an alternative simplified description of an underlying complex dynamical process. In effect, the constraint is simply some additional regularity or order, which is not explicitly found in the initial conditions. The question of the origins of such constraints is identical with the question of the origins of genetic control.

Emergence

M. Bunge (1969) has used the concept of emergence as a key to define 'level structures' devoid of any concept of domination, which he regards as an unfortunate implication of the term 'hierarchy'. In a level structure each level is characterized as an assembly of things of a definite kind (with definite sets of properties and associated laws) and such that it belongs to an 'evolutionary line', in which 'newer' means 'later'. In a level structure, qualitative novelty in process appears at each level of the hierarchy and is characteristic of it. Using this definition, based on the key concept of emergence, Bunge develops a metaphysics, epistemology, and methodology of levels in a view which he calls 'integrated pluralism', an ontology that emphasizes both the diversity and unity of the world. Of course, the possibility of genuine emergence at any one level in a level structure already implies that what emerges cannot be reduced to the level below and therefore this postulated criteria of ranking within a natural hierarchy itself depends on a resolution of the question of reductionism (see section 6.5) and is to that extent debatable.

Thus the view of biological systems as hierarchies is more multi-faceted than might be inferred from the apparent hierarchy of order, such as is illustrated in Fig. 6.3. Nevertheless all the discussion of hierarchies that are referred to above must not mislead us into thinking that, common though these relationships are in biology, all living systems are *necessarily* hierarchical. There are good grounds for thinking that this is not true of many living systems, for example, populations of animals and the human brain, which does not have one controlling master level with all other functions delegated and controlled, whether from the 'bottom up' or from the 'top down'.

6.3 Can complexity and organization be quantified?

The foregoing discussion of the hierarchical character, or otherwise, of living systems has inevitably been somewhat abstract and could understandably provoke a demand for the quantification of complexity and organization, with the hope that such quantification could lead to more precise definitions of these elusive, but nonetheless, real qualities of biological systems. If complexity and organization were to be quantified, and meaningful parameters proposed that would allow comparison between different systems, or the same system at different stages of development or evolution, then this would clearly help monitor the usefulness, or otherwise, of the physico-chemical interpretations with which we have been chiefly concerned here. Unfortunately no agreed quantification of biological complexity has yet appeared, no doubt because of the many kinds of relationship between different levels, as illustrated by the analysis of the idea of hierarchy in biology (see the preceding section). Some of the proposals for quantifying complexity are as follows.

Entropy and order

The thermodynamic quantity *entropy*, S, has been, since Boltzmann, closely associated with disorderliness (or, more loosely, 'randomness') through the Boltzmann equation which may be written as

$$S = k_B \ln W, \qquad (6.1a)$$

where k_B = gas constant (R)/Avogadro's number (N_0); and W = number of 'complexions', i.e. the number of *possible* dispositions of matter over the equiprobable, available energy states (the 'disorderliness' referred to above): or as

$$S = -k \sum_1^n p_i \ln p_i, \qquad (6.1b)$$

where $\sum_1^n p_i = 1$; and p_i = the probability of an idealized physical system being in state i of n possible equivalent but non-equiprobable states or complexions.

'Disorderliness' is, in this context, the zero value of a variable, 'orderliness', which is the extent to which 'any actual specimen of an entity approximates to the 'ideal' or pattern against which it is compared' (Denbigh 1975a). Such 'orderliness' reaches its maximum extent in the 'state of order' which is 'an ideal reference state, laid down and specified, according to certain rules or conventions, which is taken as having 100% orderliness' (Denbigh 1975a). This 'state of order' might well be that of the geometrical order of a crystal lattice at absolute zero and 'orderliness' would decrease (i.e. disorderliness would increase) according to the extent to which the atoms or molecules of the crystal were displaced from the lattice points and/or the greater were the spread of its quantized energy states relative to the ground state. For populations rather than single entities, 100 per cent order with respect to any parameter or property would be characterized by all the members of the population exhibiting the same value of that parameter, or the same property. So when, loosely, entropy is said to be a measure of 'disorder' or 'randomness' it is this kind of orderliness that is being referred to, which (to use an example of

Denbigh) is the kind of order exemplified by a perfect wallpaper pattern rather than that of an original painting. Such 'order' is therefore scarcely adequate as a measure of the complexity of biological systems (cf. Wicken 1978, 1979). Nevertheless it may be safely affirmed that the kind of *dis*order measured by entropy is incompatible with biological organization and, indeed, the state of maximum entropy, of maximum disorder in the sense defined, is equivalent to biological death. Thus we may say that a state of low entropy is a necessary but not sufficient condition for biological complexity and organization to occur.

Provided this is borne in mind, the puzzle about the compatibility of the existence of living organisms with the second law of thermodynamics can be resolved along the lines already discussed (sections 2.1 and 2.2)—namely, that the processes of living organisms, insofar as they involve a decrease in entropy, do so at the expense of an increase in the entropy of their surroundings, to which heat is transferred and which may therefore be regarded as transferring to the organisms negative entropy, or 'negentropy' (cf. Schrodinger 1944, p. 72). Furthermore the evolution criteria based on entropy that were discussed in section 2.6 for transition to new steady states, which *can* (not must) be of lower entropy, still provide sufficient, but not necessary, conditions for the formation of biological organization (cf. Section 2.6, p. 42). It must continue to be stressed that the second law is primarily about the dissipation of order and not of organization (cf. Wicken 1979).

Integrality

What we need, as K. Denbigh (1975a,b; 1981, pp. 147 ff.) has pointed out, is not so much a measure of orderliness (with 100 per cent order or disorder as its extrema), as a satisfactory measure of *organization* (cf. the discussion of P. Weiss 1971). For a system may be more highly organized and complex in this sense, e.g. a living cell, or a painting, and yet less 'ordered' in the sense expressed in the Boltzmann relation (6.1a) than, for example, a crystal or a patterned wallpaper, respectively. He has proposed the concept of *integrality* (ϕ) as a measure of organization and so of complexity. If c = the number of connections which facilitate the functions of the whole; n = the number of different *kinds* of parts; and x = other variables; then

$$\phi = cn \cdot f(x), \tag{6.2}$$

where $f(x)$ is a weighting factor taking account of the relative importance of the various connections to the actual existence of the organized system. Clearly integrality increases with increasing complexity as when, for example, an egg develops into (say) a chick—unlike 'information content' (see below) which must be remaining constant during this process. Thus integrality can increase in an enclosed system and is not conserved. It is not identical with information nor is it related to the entropy, for the degree of connectedness is not a conserved quantity and changes in it can occur at constant entropy, in a sufficiently idealized system in which switches, valves, membranes, etc. operate reversibly, or at least with only small adventitious changes in entropy.

One of the difficulties in applying this interesting suggestion comes at once in the designation of 'parts' in order to give a value to n. As we shall see below (section 6.5), when we come to discuss complexity and reduction of biological concepts, there are many different levels in a biological organism that might be chosen as the reference level for distinguishing 'parts' (the familiar sequence of molecule, organelle, cell, organ, organism, population of organisms, etc.). Although most biochemists and molecular biologists would have a predilection to choose the molecular level, there is really no 'fundamental' level of reality that has an obvious priority as a reference, especially in view of the obscurities and mysteries of particle physics (Denbigh 1981, pp. 132 and 133; for further discussion, especially with reference to the concept of 'robustness' of Wimsatt (1981), see section 6.5, below).

So the 'integrality' of an individual biological organism will always be relative to a given level of analysis but, even so, it seems clear that ϕ will rise steadily during the development of an individual organism from the zygote, will reach a plateau during maturity, and will decline to zero shortly after death.

The consideration of the total integrality of *populations* of organized systems leads to some interesting results. If ϕ_i is the integrality of a specimen of the ith class of organisms in a population, then the total integrality will not be less than $\sum \phi_i$, summed over all members of the class—and might well be greater if higher-level organization occurs within the population so increasing the number of connections, c in (6.2). If the individual members are sufficiently alike this summation could be replaced by $p_i \bar{\phi}_i$ where p_i is the population of the ith class and $\bar{\phi}_i$ is a mean integrality of each member of that class. Now the total integrality (Φ) of, say, the Earth's biosphere could be written as

$$\Phi \geqslant \sum_i p_i \bar{\phi}_i$$

where the summation is now over different classes, and a common base line for reckoning the ϕ_i is assumed to be available. The time change of Φ is then (Denbigh 1981, p. 150)

$$d\Phi/dt \geqslant \sum_i p_i \cdot d\bar{\phi}_i/dt + \sum_i \bar{\phi}_i \cdot dp_i/dt.$$

The p_i, $\bar{\phi}_i$ are both positive and, since the general level of the degree of organization has greatly increased during evolution (Simpson 1950, Chapter XV), most, if not all, of the $d\bar{\phi}_i/dt$ will be positive—and so will the dp_i/dt, since the majority of the most highly organized animals and plants have increased their populations. All of which suggests that $d\Phi/dt$ is positive, i.e. the total integrality of the Earth's biosphere as a whole has increased and is still increasing. (Denbigh, more problematically and speculatively, also tries to include, in addition, possible positive contributions to Φ of the increase in the number of artefacts in the world and of all those entities which store and transmit cultural information (World 3 of Popper (1972)). This increase in total integrality applies only to the Earth and it may well be decreasing on other planets. This stimulating suggestion does not unfortunately seem to have been followed up by other authors but, even if inadequate in itself, and perhaps using an oversimplified formulation in eqn (6.2), it does appear to be at least pointing to the significant features of organization without being too much entangled in the theoretical structures of thermodynamics or of information theory which were devised for other purposes. But to these we must now turn since they have long been associated with attempts to quantify complexity in biology.

Information

Closely related to the above use of entropy as a measure of disorder is the communication theory concept of *information* (I or H) which may be defined as

$$I = k' \frac{1}{\ln 2} \ln Z_0 = k' \log_2 Z_0, \qquad (6.3a)$$

where k' = positive constant; I = the information content (in 'bits') of one outcome or case selected out of Z_0 possible, equally probable, outcomes, or cases (if the Z_0 possible 'outcomes' or 'cases' are interpreted as the Z_0 possible complexions of a physical system (eqn (6.1a)) of which the actual state of the system is one). Information (I) so defined is related to complexity as variety.

Alternatively information may be discussed in terms of the function

$$H = -\sum_{i=1}^{i=n} P_i \log_2 P_i, \qquad (6.3b)$$

where H = information per digit in a coded message; $0 \le P_i \le 1$; $\sum_{i=1}^{i=n} P_i = 1$; and P_i = relative probability of the ith symbol generated by a source, using an alphabet of n different symbols. This H-function of Shannon implies the existence of a channel from the system to the observer and represents a missing information that we need in order to describe completely a natural system: so it measures the 'complexity', seen then as a negative quantity. The 'degree of organization' of a system has indeed been explicitly related to Shannon's H-function by defining it as the sum of H for the components of the system taken separately (and so without interaction) *minus* the H-value of the actual system including all the interactions between its components (Rothstein 1952; applied comparatively by Gatlin (1972) to DNA and proteins in various species). Denbigh (1981, p. 147) has pointed out the inadequacy of treating all interactions in a system as being on a par, as does this additivity assumption in computing H, whereas in fact some *particular* interactions or connections within an organization system are *essential* to the organization's continued existence (e.g. the spinal chord in a vertebrate). Rothstein's formulation of 'degree of organization' does not recognize this.

What this 'informational' viewpoint does indicate, however, is that a system appears complex when we do not know enough to specify it completely—even though we know enough to recognize it as a system. Atlan (1978) has expressed this 'informational' definition of complexity in the following way: 'Complexity is a negative quantity which measures our ignorance about the organization of a natural (i.e., non-man-made) system [about] which we know enough, however, to recognize [it] as an entity and to enumerate its constitutive parts.'

There is clearly a parallel between the sets of equations (6.1) and (6.3) for entropy and information respectively. Hence an increase in the disorderliness and entropy (S) of a system has often been regarded as equivalent to a decrease in information (I, H) of the system, i.e. an increase in our lack of information or knowledge of it. Conversely, a decrease in entropy has been regarded as parallel, not only to an increase in orderliness as already defined, but also to an increase in information content. The applicability of these ideas to biological systems have been widely discussed since Shannon's original formulation (e.g. Shannon and Weaver 1949, 1962; Quastler 1953; Brillouin 1956, 1962; Elsasser 1958; Theoridis and Stark 1969; Johnson 1970; Gatlin 1972; Atlan 1972; Yockey 1973; Hassenstein 1977; Wicken 1978, 1979; Pattee 1979; Atlan in press; and see the September 1966 issue of *Scientific American* on 'Information and communication') but this application has proved to be controversial (Apter and Wolpert 1965; Popper 1965, 1967; Büchel 1967; Woolhouse 1967)—even involving disputes about the actual sign (+ or −) of the relation between entropy and information (Wilson 1968). This is not surprising for the correlations between information and entropy and between information and disorderliness are only possible in situations where 'information' can have a clearly defined meaning (as it can in relation, for example, to the genetic code) and precise delineation of, for example, whether the system is being observed from within or without have sometimes been lacking (cf. Atlan 1978, 1981).

A number of authors, indeed, firmly deny that any equivalence between information and thermodynamic entropy has been proved (see Ferracin, Panischelli, Benassi, di Nallo, and Steindler 1978 and references therein; also Tonnelat 1978, Pattee 1979). The difficulty arises particularly in applying the concept of information where both new information is being generated and the system is subject to disorganizing effects from outside, i.e. to 'noise'. Can the concept of information cope with such characteristically biological problems as the development and evolution of the functional, rather than purely spatial, order of biological systems? Apter and Wolpert (1965, p. 255) concluded 'that attempts to apply information theory to development have been either faulty, meaningless or trivial'. Furthermore, one recalls, in this context, the stress of Eigen (1971a, p. 469) on the need for some independent criterion of 'valuation' of new macromolecules in an evolving system: for generation of functional orderliness depends on kinetic and stochastic considerations (amplification of fluctuations, etc.) for which the concept of 'information', unmodified, seems inadequate.

H. Atlan (1974, 1975, 1978, 1981) agrees that information content, as such, provides a very poor definition of organization but has developed the thesis that the kinetics of the *change* of information content with time, under the accumulated effects of noise from the environment, can do so. He argues (1974) that there have been two major and contradictory trends in defining organization. On the one hand, organization is seen as redundancy, based on constraint between parts, regularity or repetitive order (the 100 per cent orderliness of the crystal at the absolute zero of temperature, with its zero entropy $S = k_B \ln 1 = 0$, is regarded as optimum 'organization' from this perspective). On the other hand, under the influence of information theory, organization is regarded as variety based on non-repetitive order which is measured by an information content, i.e. a degree of unexpectedness related to variety and inhomogeneity. So, Atlan (1974) affirms, any formal quantitative definition of organization should involve a kind of optimization process so that any optimum organization would correspond to a compromise between maximum information content (and so maximum variety) and maximum redundancy. Atlan (1981, Appendix) illustrates this compromise by instancing a cultural system contained in a library. The culture exists as intermediate between (i) complete independence (no constraints or relations) between the books and (ii) mere repetition (maximum redundancy). The latter (ii) would reduce the culture to the content of only one book. The former (i) would not be compatible with what we mean by a cultural system, there being no cross-quotations, allusions, references, etc. between the books.

Shannon defined redundancy R by

$$H = H_{max}(1-R), \qquad (6.4)$$

where H is the information content of a system (or a message) with internal constraints between the parts, and H_{max} is the maximum information content computed by not taking the constraints into account, i.e. by assuming complete independence of the parts. Complexity is here seen as related to H regarded as a measure of the lack of knowledge of the internal constraints (redundancy) of the system, and R as a measure of the simplicity and order that comes from physically or deductively repetitive order (Atlan 1981). The time-variation of H under the influence of external noise from the environment is then

$$\frac{dH}{dt} = (1-R)\left(\frac{dH_{max}}{dt}\right) - H_{max}\left(\frac{dR}{dt}\right). \qquad (6.5)$$

The first term ('destructive ambiguity') is negative for the effect of noise is to destroy H_{max}, the total information transmitted to the observer, ignoring the internal constraints—the classical disorganizing effect of noise. But noise also has the effect of increasing the relative autonomy of one part of the system from another, and so of decreasing the overall redundancy of the system, i.e. dR/dt is negative and the second term in (6.5) is thus a *positive* contribution ('autonomy-producing ambiguity') to dH/dt and so increases H. dH_{max}/dt and dR/dt are different functions of time and so it is always possible for there to be an initial overall *increase* in H with time-accumulation of the effects of noise and such an increase in H is interpreted as a process of 'self-organization'. The functions dR/dt and dH_{max}/dt and eqn (6.4) then express the overall organization of the system, both structural and functional. Atlan summarizes his definition of organization as follows: (i) the *process* of organization of any system is described by the variation of information content with time, $H(t)$, and this is the combination of the decrease of H_{max} and of R with time (as in (6.5)); (ii) the kind or *state* of organization of a system is defined by the initial H_{max} when redundancy is ignored $(H_{max})_0$, the initial redundancy (R_0) and a factor of reliability (t_M), related to the overall resistance of the system to noise. $(H_{max})_0$ is a measure of the structural aspect of organization, R_0 measures both functional and structural features and t_M functional features.

So several parameters are involved in comparing the degrees of organization of different systems, or the same system at different times, and such comparisons would be best expressed vectorially in an appropriate space. Different conditions on these parameters lead to various kinds of organization. Thus a system is *self*-organizing if it is redundant enough and functions so that its redundancy decreases in time with accumulation of errors and/or noise—and yet still functions so that its information content increases. Self-organization appears, then, as a continuous *dis*organization constantly followed by a reorganization with more complexity and less redundancy. What is characteristic of living systems is the existence of different parameters allowing the initial phase of increase of complexity to be more extended and distinguishable, than in other physical systems which have a minimum redundancy and in which such an initial phase could exist in principle. The principle of organization is: that, provided its initial redundancy and functional reliability are large enough to allow for an observable period of self-organization, a system can react to random environmental stresses by an increase in complexity and variety, so that it appears to adapt itself, during this phase, to its environment.

This development by Atlan of a theory of self-organization as 'complexity from noise' is a natural extension of the 'order from noise' principle of von Foerster (1960) and is conceptually very close to the 'order through fluctuations' which lies at the centre of the thermodynamic interpretation of living organisms by the Brussels school (who are also pursuing the study of noise-induced transitions in biological systems: Lefever 1981; Horsthemke 1981). Furthermore, the idea of self-organization as a kind of optimization process of organization, involving a compromise between maximum information content (i.e. variety) and maximum redundancy, was based on an analysis of differences in the position from which observations were made (transmission of information within a system, as the functioning of the system of itself, viewed from within vs. the view of the system seen as a whole by an external observer). The observational level discriminated by the 'external' observer is then very critical for what is regarded as lacking in the information that an observer has of the system in question, so this informational theory of self-organization is closely tied in with the levels of hierarchical organization viewed by the observer—and so with the whole concept of the meaning of information in hierarchical self-organized systems. Atlan (1978) has summarized his conclusions in this regard as follows: 'What appears to the observer as an organizational noise acting in channels between different hierarchical levels, is in fact, for the system itself, the meaning of the information transmitted in these channels.'

6.4 The evolution of biological complexity

The biological complexity we actually observe now in the natural world is the product of a long evolutionary history and different levels in the

Table 6.1 The major levels of organization in organic evolution.

Level	Examples	Years ago when first appeared
8. Dominance of tool-using and conscious planning	Man	50 000*
7. Homoisothermic metabolism (warm blood)	Mammals, birds	150 000 000
6. Organized central nervous system, well developed sense organs, limbs	Arthropods, vertebrates	600 000 000 450 000 000
5. Differentiated systems of organs and tissues	Coelenterates, flatworms, higher plants	1000 000 000? 400 000 000
4. Mutlicellular organisms with some cellular differentiation	Sponges, algae, fungi	2000 000 000?
3. Division of labour between nucleus, cytoplasm, organelles	Flagellates, other protozoa (eukaryotes)	?
2. Surrounding cell membrane with selective permeability and active transport of metabolites	Bacteria, blue–green algae (prokaryotes)	3000 000 000?
1. Earliest self-reproducing organic systems	Free-living viroids (none still living)	?

* This estimate has in recent years increased by at least one order of magnitude.
From Stebbins 1969.

hierarchy of biological complexity have appeared at successive stages in the history of the Earth, as illustrated by Table 6.1 devised by G. I. Stebbins (1969). This table emphasizes the truism that today's biological structures are the result of a long process of development from earlier forms, a fact that is disregarded in the kind of calculations which try to estimate, say, the chance of a molecule of a typical protein being formed *de novo* from its constituent atoms, or even amino acids. Such calculations (*a fortiori* if complete organisms are considered) usually result in the conclusion that this probability is so low that the planet Earth has not existed long enough for such a complex assembly to appear by the random motion of atoms and molecules—whether this period be that of the total 4.6×10^9 years of the Earth's life, or the 1.1×10^9 years between the formation of the Earth (4.6×10^9 years ago) and the oldest known rocks (3.5×10^9 years ago) containing the remains of living cells (blue–green algae) found in rocks at Warrawoona, Australia. The fallacy of such calculations lies in their ignoring the actual processes whereby complex self-reproducing (initially molecular and macromolecular) systems might self-organize entirely consistently with currently known thermodynamics and chemical kinetics (as elaborated in Parts I and

II); in ignoring the role of selection of those organizations of macromolecules that have favoured reproduction rates and that, once established, irreversibly channel the evolutionary process in one particular direction (q.v. the work of Eigen, Chapter 5); and in ignoring the fundamental analyses of the architecture and evolution of complexity made by many authors, in particular H. A. Simon whose story of the two watchmakers Hora and Tempus we have already recounted in the introduction (section 1.2). What the story illustrated was the principle that the time required for the evolution of a complex form from simple elements depends critically on the numbers and distribution of potential intermediate stable forms. In the case of Hora and Tempus, if the probability was one in a hundred that they would be interrupted when adding one part to an assembly, which thereby disintegrated because of the interruption, a simple calculation (Simon 1962) shows that Tempus, who had no stable sub-assemblies for his 1000-part watches, would take 4000 times longer than Hora to complete his watches made of 10 sub-assemblies, each themselves consisting of 100 basic parts, but subdivided into sub-sub-assemblies of 10 parts each. All these less-than-total assemblies of Hora had sufficient stability to allow them to survive once completed at its own stage. More generally, if there exists a hierarchy of potential stable 'sub-assemblies' with, at each level of the hierarchy, about the same span s (=the number of components at any level in the hierarchy that combine to form a component at the next higher level; $s = 10$, in the above example) then the total number, n, of elements in the finally complete system is given by

$$n = s^{T/\phi}, \tag{6.6}$$

where T/ϕ is the number of stages (i.e. levels) in the assembly process, with T = total time of assembly, and ϕ = the time of assembly of each level, i.e. if the time of total assembly is proportional to the number of levels in the system and not to the number of component units n. (In the Hora–Tempus example, $T/\phi = 3$ and $n = 1000 = 10^3$.) So, on this assumption,

$$T = \phi \log_s n. \tag{6.7}$$

This argument shows, for example, that in such a hierarchy, the time required for a system containing 10^{25} atoms to evolve from one containing 10^{23} atoms is the same as that for a 1000-atom system to evolve from a ten-atom one. Or to put it, with Simon (1962), more broadly and only illustratively, it might well be that the time of evolution of multi-cellular organisms from single cells is of the same order of magnitude as the time required for the evolution of single-celled organisms from macromolecules. Indeed, this principle must lie behind the strong impression that evolution develops exponentially and only a logarithmic time-scale can spread its many stages in any evenly-spaced manner. The whole question of the evolution of complexity in its general aspects in relation to hierarchy theory, systems theory, information theory has been much discussed and the references quoted earlier (in the appropriate section) should be consulted, since it is a topic of major concern in all these theories (see a particularly valuable discussion in Denbigh 1975*b*). At this point, however, it is worthwhile making a few further remarks about the stability requirement of the intermediate forms.

Simon showed that complex systems will evolve from simple systems much more rapidly if there are stable intermediate forms than if there are not, and that then the resulting complex forms will be hierarchic in organization. The requirement for stability of an atomic or molecular structure reduces to the requirement that the free energy of the structure be less by virtue of its structure than that of its component atoms or molecules. Kinetically, such a new minimum of free energy is only reached by crossing an energy barrier (the activation energy) and it is the probability of this occurrence that determines the rate of the formation of the new structure. The complex structures that emerge in the evolution of the first living forms, and of subsequent forms, may be presumed to have a stability of this kind, though, since living systems are open, the stability attributed to them must not be identified directly with the net free energy of formation of their structures. The point that is essential to make here, in relation to the possibility of formation of complex forms, is that natural selection speeds up the establishment of each new stratum of stability in the succession of forms of life. Each stratification of stability is only a temporary resting place before random mutation and natural selection open up new possibilities and so new levels of (temporary) stability—rather as if the free-energy profile were a switchback with any given form stable only to the immediate environment of the stability-controlling parameters and always with a finite chance of mounting the next barrier to settle (again temporarily) in a new minimum. As we saw in section 2.7, Bronowski (1970) has pointed out that this stratified stability, which is so fundamental in living systems, gives evolution a consistent direction in time towards increased complexity. He used the metaphor of evolutionary time as a 'barbed arrow', because random change will tend in the direction of increasing complexity, this being the only degree of freedom for change with stability. Saunders and Ho (1976) have also produced a neo-Darwinian argument along similar lines—namely, that, for systems already functionally organized in relation to their environments, any random mutation will lead only to permanent increases in viability if it causes an increase in complexity, because this alone can hold out any promise of an enhanced viability, either in the prevailing environment or one yet to come. Wicken (1979, 1980) pursues the 'barbed arrow' metaphor further by arguing that it is not only that organizational constraints prevent regressive slippage but that the evolutionary arrow is propelled by the randomizing directives of the second law and the local conditions under which it operates. His approach is based on a recognition (Wicken 1978) that molecular complexity can increase in chemical reactions that involve association to more complex molecular forms, in full accordance with the second law. For any decrease in entropy that results from the decrease in the number of 'complexions' (W, eqn (6.1a)), consequent upon the loss of translational modes of molecular motion to (by and large) vibrational modes when atoms or molecules combine, is offset: (i) by an increase in the entropy of the surroundings resulting upon the heat generated by a decrease, during the chemical combination, in the potential (electronic) energy of chemical bonds; and (ii) by an increase in entropy due to the increase in configurational possibilities that occur, in spite of reduction in the number of molecules, when there is an increase in molecular heterogeneity with the formation of new chemical species (and this contribution is greater the greater the number of possible new chemical structures).

Wicken (1978, 1979, 1980) denotes (i) and (ii), respectively, as 'energy-randomization' and 'matter-randomization' to emphasize that it is the randomizing tendencies which the second law of thermodynamics formalizes, that drive forward the formation of more complex structures. The Earth's biosphere is in a steady-state of free energy flux, from the Sun to the Earth (with its biosphere) to outer space, with the rhythm of the Earth's diurnal rotation. Within the biosphere itself there is a continuous steady flow of energy through the various 'trophic levels' of the ecologist, with concomitant transfer of heat to the non-living environment and so to outer space (see Morowitz (1968 and 1974), who was a pioneer in viewing 'biological organization as a problem in thermal physics').

The potential energy stored in chemical bonds by process (i) then, entirely in accord with the second law, and in conjunction with (ii), provides the opportunity for the increase in molecular, and so organizational, complexity upon which natural selection then operates (Wicken 1979). Wicken (1980) has further argued that, for these reasons, the biosphere must evolve toward a stationary state of maximum structuring and minimum dissipation with respect to the (solar) applied free energy gradient. Wicken's general argument is quite convincing, though perhaps unnecessarily expressed (and perhaps unfortunately so in view of the caveat entered above) in terms of 'information' which, in his case, is simply the negative of entropy as thermodynamically and statistically defined, and so related to 'order' in the strict sense (and not to 'organization'). These ideas fit in neatly with the thermodynamic interpretation of the Brussels school (sections 2.7 and 2.8) that have made explicit the thermodynamic conditions for the transformation of one form of molecular organization to another regime in open systems, such as the biosphere.

6.5 Biological complexity and the reduction of biological concepts

The major part of this book has been concerned with expounding new physico-chemical concepts that have had to be developed to articulate and understand the complexities of living systems. Now the programme of molecular biology during the last few decades has typically been represented by F. H. C. Crick (1966) as: 'The ultimate aim of the modern movement in biology is in fact to explain *all* biology in terms of physics and chemistry', which, put more colloquially, is the belief that biology is 'nothing but' physics and chemistry. No one doubts that a reductionist methodology, whereby a complex living system is broken down into investigatable parts is a heuristically essential strategy in biological research. Nor is there any doubt that in a biological system there is *nothing* there other than atoms and molecules (and perhaps radiant energy). The reductionist programme, with the intentions described by Crick, has however been controversial amongst biologists many of whom have been concerned to emphasize the autonomy (that is the non-reducibility) of biological concepts *vis-à-vis* physics and chemistry; they want the theories of their science to be autonomous.

The reducibility of theories in a science applicable to higher levels in a natural hierarchy has been discussed and analysed by philosophers of science for some time. For example, the conditions of connectability of terms and concepts to the lower level and of derivability of laws at the higher levels from those of the lower, has been well understood and described by Nagel (1961). Moreover, a process at a higher level is said to be autonomous, that is not reducible, only if it is not fully determined by laws and processes occurring at the lower level. Our current understanding of natural hierarchies and, in particular, of biological organisms as instances of such, throws new light on this question (Beckner 1974; Peacocke 1976) for it now becomes clear that we have to distinguish carefully the hierarchy of natural *systems*, which we have been discussing up till now, from the hierarchy of *theories* of the sciences concerned with these systems, with which philosophers of science have been concerned in discussing the question of reductionism. Beckner has stressed the confusion which arises through failure to distinguish theories from *processes* because of the hierarchical character of natural systems, and so of the sciences appropriate to each 'level' in the hierarchy. Because of this distinction we have to be especially careful in describing and analysing the causes or connections operating between different levels. These causes or connections are not quite like the causes or connections between pairs of systems with no parts in common. It transpires that, in a hierarchy of natural systems (in a 'level-structure'), a higher level *theory* can be autonomous—that is not reducible to a lower level theory, according to the conditions of Nagel—and yet the higher level *processes* need not also be autonomous, that is, there are no processes occurring at this higher level in the system other than the processes operating at the lower level. Questions of reduction, and so of emergence, are epistemological and linguistic, they are about logical relations holding between theories, descriptions, conceptual schemes, and so on, as applied to natural hierarchies.

In view of these implications of the hierarchical nature of natural systems for the question of reduction, where does biology stand in relation to physics and chemistry—or rather are the theories of biology about biological organisms autonomous or reducible to the theories of physics and chemistry? Biologists are keenly conscious that biological organisms certainly are often hierarchies of parts making wholes at different levels and the dynamic processes are themselves interlocked dynamically in space and time in complex networks. Processes at the molecular level (e.g. enzyme-catalysed reactions) are part of the network of interlocked reactions in the metabolic web, itself distributed spatially over a structurally hierarchical framework of organelles, which are themselves interconnected by structures and by chemical messengers in a larger whole (the cell), which itself ... and so on. The biologist is bound, therefore, to stress the special concepts he has to employ to describe and understand such complexities. He finds that at each new level of biological organization new kinds of interlocking relationships emerge and these require new concepts to order them and render them coherent—as well as distinctive experimental techniques and designs of experiment. Such a biologist would therefore support the autonomy of biological concepts; that is, he is *epistemologically* anti-reductionist. It is this autonomy of biological theory in relation to physics and chemistry which numerous biologists have been concerned to emphasize. There are indeed distinctive ideas in biology which simply cannot be envisaged or translated into the conceptual terms of physics and chemistry (Medawar 1974; Jacob 1974).

In taking up this position some authors have moved over from stressing the autonomy of biological theory to basing their arguments on the fact that biological organisms evidence new complex relationships between their constituent parts and it is these relationships which make them argue for the autonomy of biological processes, in a way which does not follow from the autonomy of the theories of biology. They are then vulnerable to those philosophers for whom it is almost trivial and obvious that, as J. C. C. Smart puts it, 'new qualities emerge when simples are put together to form a complex' (Smart 1963), as when the pieces which constitute a radio set are connected in the proper sequence and then receive signals. Yet this point, regarded by such philosophers as a very modest one, is often the one which, for example, biologists want minimally to affirm when told that their subject is 'nothing but' physics and chemistry. One form of anti-reductionist argument, which makes this kind of stress on relationships, has been developed by Polanyi (1958, 1959, 1967, 1968), who says there are 'boundary conditions' which characterize even machines in relation to their components, and he then transfers the same arguments to the relation of biology to physics and chemistry. Polanyi would have regarded himself as an anti-reductionist, yet we find Schaffner (1967, 1969) making almost the same point, though he clearly sees himself as a reductionist. He defines (1969) the 'biological principle of reduction' as that, given an organism composed out of chemical constituents,

> the present behaviour of that organism is a function of the constituents as they are characterisable in isolation *plus* the topological causal inter-structure of the chemical constituents (the environment must, of course, in certain conditions be certified.)

However, this form of stress on the relationship of the parts has not always impressed the philosophers. There is a real problem here, it seems, about the logical character of biological theory and to what extent biological theories can be regarded as autonomous (i.e. not reducible in the Nagel sense) simply and only on the grounds that they have to concern themselves with the special kinds of biological interrelationships between units. This is not a question that can readily be settled for all biological theory *in toto* but is a matter for investigation in each case. Polanyi's argument has more force when he applies it not to a machine but to biological systems, for example, a living cell with its complex configuration of space and time, with its flow of constantly changing substances, both within and across the cell membrane, and with its possession of an individual life-cycle. For example, the chemical structure of DNA, its specific covalent and hydrogen bonds, are describable in terms of the categories of physics and chemistry. Within the double-helical structure there are sequences of base pairs all of which are equally permissible physico-chemically and can fit equally well into the structure (within certain minor limits which do not affect the argument). In the nucleus of any particular cell of a given organism, within the double helices of its DNA there are particular specific sequences which perform a unique set of coding functions. *This* particular base sequence in *this* DNA only has a 'meaning' when the DNA has been assembled in *that* organism and it can only have its biochemical function as the genetic 'blueprint' for the production in due order of, for example, specific proteins, when it functions in the milieu of the whole organism. Chemical processes are indeed the means whereby bases are incorporated into chains of DNA but the sequence in which the bases are assembled in the DNA is a function and property of the whole organism. As Pattee (1971) has put it

> nucleic acid molecules do not inherently possess symbolic properties. No amount of chemical or physical analysis of these molecules would reveal any message unless one presupposes the coordinated set of constraints which reads nucleic acids. The same is true for message molecules at all levels of organisation. (Pattee 1971, pp. 259–60.)

The chemical description of these message-carrying molecules and the fact that they can be described in chemical terms is an argument against process autonomy. Since no 'laws' or regularities of physics or chemistry describing the nature and stability of the chemical bonds in DNA as such can specify the actual *sequence* in any particular case, this analysis supports the kind of epistemological anti-reductionism which affirms *theory* autonomy at the biological level. To put it another way, the concept of transfer of information, at the biological level, is distinct from, and not reducible to, the concepts of physics and chemistry.

There has been a tendency to regard the level of physical and chemical description as alone referring to 'reality'. However, there are good grounds for not affirming any special priority to this level of description. Indeed it has been argued (Wimsatt 1981) that there should be a prima facie recognition of the need for a variety of independent derivation, identification, or measurement procedures ('multiple determination') for examining the existence and character of any phenomenon, object, or result with the aim of looking for what is invariant over or identical in the outcome of these procedures. What is invariant, at whatever level the procedures are directed, Wimsatt calls 'robust' and the attribution of this term (with its allusion to the detection of spatio-temporal boundaries in sense-perception of objects) is meant to convey that there is an attribution of reality to what the procedures yield at their particular level of operation. In other words 'reality' is, on this view, also an attribute of levels of description and examination of living systems other than that of the physico-chemical alone—and it is interesting to observe this move away from a throughgoing reductionism by a philosopher of biology.

Thus there does seem to be a prima facie case for arguing that some biological concepts, and so theories, might well be autonomous, and not be reducible in a strict sense; certainly biological concepts and theories do seem to be distinctive of the biological level in the natural hierarchy of systems. Because of this distinctiveness, it is not likely that all biological theories are going to prove to be reducible to physics and chemistry judging from the difficulties of reducing even chemistry to physics (Popper 1974). But this cannot be decided in advance about any particular theory or concept concerned with a particular biological phenomenon and any such will have to be examined individually with the question in mind about their autonomy, or otherwise, with respect to theories and concepts developed for lower levels in the hierarchy of complexity. As Rosen (1979, p. 183) has alternatively put it: 'It then becomes an empirical question to determine whether the observables manifest in biological interactions are distinct from those appearing in our physical description of a system.'

6.6 The thermodynamic and kinetic interpretations

Living organisms pose two groups of questions (cf. section 1.4) about their nature: (1) What is it now? How can we most accurately describe it as we see it now? and (2) How did it come to be the way it is? How did it evolve? The two groups of questions will be linked when functions are under examination for these are not only involved in the successful form of life of the organism, or part of an organism (questions (1)), but have also played a role in it becoming and evolving into what it now is (questions (2)). The thermodynamic and kinetic interpretations of living systems of Parts I and II have revealed contributions to answers to both these groups of questions.

Thus, with regard to (1), the formulations of network thermodynamics (Chapter 3) cast new light on the interchange and transmission of energy that occur in living systems and, in relation to (2), the thermodynamic formulations of the Brussels school (Chapter 2) have opened up a new understanding of the generation and evolution of complexity in chemical, and other, open systems far from equilibrium, as well as providing evolution criteria. With respect again to (1), the kinetic interpretations (Chapter 4), initially of Hinshelwood and Kacser, and later the study of oscillating networks of biological reactions, displayed how reaction processes of living systems, as they now are, interlock to manifest those features that are specially characteristic of the living—self-reproduction, self-organization, repair, utilization of matter and energy from outside, aging, defence against extraneous disturbances, etc.

Furthermore, with respect to (2), systems of reacting and diffusing molecules were shown by kinetic analysis, following the pioneering work of Turing, to be capable of time- and space-symmetry breaking in such a way as to provide plausible means by which the temporal and spatial differentiation of pattern, essential to the emergence of living matter, could have occurred. Moreover, kinetic interpretations, when linked with the recognition of the stochastic nature of sequences of reaction processes, were shown by Eigen and his colleagues (Chapter 5) to be capable of elucidating general principles of the self-organization of matter into living forms and of selection and evolution of such systems at the molecular level; and also enabled them to postulate, in the form of the hypercycle, detailed mechanisms by which self-reproduction and internal transmission of information could have occurred given the known prebiotic conditions on the Earth.

The link between the kinetic and thermodynamic interpretations has already been illustrated by the example of Eigen (1971a, Table 6; section 2.6, eqns (2.72)–(2.81)) and more generally, Nicolis (1975, pp. 44 f) has assessed in the following way the insight one gains by applying the concepts of thermodynamics to the study of living systems:

> Thermodynamics helps to eliminate kinetic models that are incompatible with the laws of macroscopic physics (e.g. coherent behaviour in, say, the form of sustained oscillations, is ruled out close to equilibrium). This role of thermodynamics is useful because much work in theoretical biology still has to be based on mathematical models since the kinetic details of many biochemical processes have not been fully elucidated.
>
> In certain problems, such as prebiotic evolution, the kinetics is so complex and poorly known that one cannot be sure about the precise form of the rate equation. It is then very helpful to be able to deduce general thermodynamic theorems (e.g. like the one about the increase of rate of dissipation per unit mass (section 2.7, see (2.83)).
>
> Fluctuations play a crucial role in all problems involving evolution, especially when the form of an initial perturbation has long-term effects on a dissipative structure. Now the probability of a certain type of fluctuation depends explicitly on thermodynamic quantities such as the excess entropy production (section 2.6). So the theory of non-linear differential equations for rates cannot by itself yield information about fluctuations unless it is supplemented with thermodynamic concepts.
>
> Irreversible thermodynamics, in particular the concept of dissipative structures provides a general framework and an appropriate language for analysing in a unified fashion, complex and extremely diverse phenomena related to the multiple aspects of biological order.

Since this 1975 assessment, the area of kinetic experiment and interpretation has become very fruitful, as we saw in Chapter 4, and the thermodynamic interpretations have remained much as they were then. As always, thermodynamics is supremely the 'science of the possible' and not prescriptive of precise mechanisms. It is the combination of the two approaches, as in the classical physical chemistry of chemical reactions, that will provide the most rounded and satisfying physico-chemical understanding of living systems and the main sections of this book provide some of the material for an assessment of progress so far.

6.7 New bottles for new wine?

The application of thermodynamic and kinetic principles to the study of the existence and evolution of something as complex as living organisms, as has been expounded in Parts I and II, exemplified repeatedly the need to stretch the characteristic concepts of thermodynamics and kinetics to situations never envisaged in classical physics and chemistry. At many points entirely new concepts had to be devised, e.g. that of dissipative structures and 'order through fluctuations' of the Brussels school and of 'selection' and the 'quality factor', as applied to an assembly of macromolecules by Eigen and his colleagues. Sometimes these involved the juxtaposition of ideas drawn from quite different sources, as when electrical-network theory and thermodynamics were combined by Katchalsky and his colleagues to create network thermodynamics with biological applications in mind. There seems to be here a prima facie case for recognizing that new non-reducible concepts are being required to allow these classical physico-chemical disciplines to cope with biological complexity. Or, to put it another way, the physics and chemsitry in terms of which living systems are being interpreted is a 'physics' and 'chemistry' so profoundly modified by the incorporation of characteristically biological concepts that it is as much a question of whether physics and chemistry have been taken into biology as whether biology has been 'reduced' to physics and chemistry. In the end, this may come only to a semantic quibble, but the developments described here certainly do not warrant a naïve reductionism. For the physico-chemical investigator of living systems is again and again baffled by the lack of conceptual resources in received physics and chemistry to deal with such complexity and with the demanding intellectual need for new ones.

In this connection it is interesting to note some of the remarks of particular investigators who, while powerfully utilizing one or other of the physico-chemical resources we have been describing, have been explicit about the nature of their task. Thus, I. Prigogine, whose work has illuminated the application of the ideas of physics to biological systems, writes as follows (1980):

> there is a microscopic formulation that extends beyond the conventional formulations of classical and quantum mechanics and *explicitly* displays the role of irreversible processes. This formulation leads to a unified picture that enables us to relate many aspects of our observations of physical systems to biological ones. The intention is not to 'reduce' physics and biology to a single scheme, but to clearly define the various levels of description and to present conditions that permit us to pass from one level to another. (pp. xiii and xiv.)

> We start with the observer, a living organism who makes the distinction between the future and the past, and we end with dissipative structures, which contain . . . , a 'historical dimension'. Therefore, we can now recognize our-

selves as a kind of evolved form of dissipative structure and justify in an 'objective' way the distinction between the future and the past that was introduced at the start. Again *there is in this view no level of description that we can consider to be the fundamental one. The description of coherent structures is not less 'fundamental' than is the behavior of the simple dynamical systems.* (p. 213, my italics.)

Similar ideas have also been expressed by F. Jacob, one of the architects of molecular biology in its crucial developments in the 1960s:

From particles to man, there is a whole series of integration, of levels, of discontinuities. But there is no breach either in the composition of the objects or in the reactions that take place in them; no change in 'essence'. So much so, that investigation of molecules and cellular organelles has now become the concern of physicists. ... This does not at all mean that biology has become an annex of physics, that it represents as it were, a junior branch concerned with complex systems. At each level of organization, novelties appear in both properties and logic. To reproduce is not within the power of any single molecule by itself. This faculty appears only with the simplest integron† deserving to be called a living organism, that is, the cell. But thereafter the rules of the game change. At the higher-level integron, the cell population, natural selection imposes new constraints and offers new possibilities. In this way, and without ceasing to obey the principles that govern inanimate systems, living systems become subject to phenomena that have no meaning at the lower level. *Biology can neither be reduced to physics, nor do without it.* Every object that biology studies is a system of systems. Being part of a higher-order system itself, it sometimes obeys rules that cannot be deduced simply by analysing it. This means that each level of organization must be considered with reference to the adjacent levels. ... At every level of integration, some new characteristics come to light. ... Very often, concepts and techniques that apply at one level do not function either above or below it. The various levels of biological organization are united by the logic proper to reproduction. They are distinguished by the means of communication, the regulatory circuits and the internal logic proper to each system. (F. Jacob 1974, pp. 406–7; italics inserted.)

Mikulecky *et al.* (1977), at the conclusion of their attempt to apply network thermodynamic modelling to coupled flows through biological structures, remark (pp. 509–10):

Living tissues are very complicated structures involving a variety of geometric forms and a multitude of interconnections. The jump from physical and chemical models to a living system has always seemed to involve concepts which are beyond simple physics and chemistry. At one time this 'extra ingredient' was explained by the "postulate" of vital spirits and vital forces. ... Eventually, the reductionist view became dominant and molecular biology began to dominate comparative anatomy, ecology, taxonomy, and the other classical subdisciplines of biology. Yet, something was still missing. Hints began to appear that

† An 'integron' is each of the units, in a hierarchy of discontinuous units formed by the integration of sub-units of the level below. An integron is formed by assembling integrons of the level below it; it takes part in the construction of the integron of the level above. (Jacob 1974, p. 302).

breaking down structure to get at pieces to study in isolation also broke down function—often those functions which were under investigation. Now we can again assign a name to that missing something—organization. Gaylord Simpson once pointed out that the reason biology was higher in the hierarchical structure of science than physics or chemistry was that it dealt with levels of organization not found among the objects of study in those disciplines. Indeed, the molecules and the physical-chemical laws are the same, but things are put together in different ways when they are alive. ... The main question of interest here is whether the network approach will be of any help in resolving this philosophical dilemma. To the extent that it provides a method for dealing with more complicated organizational patterns, it must. On the other hand, reducing a living system to a network is not far from reducing it to a collection of molecules. *The networks, as models, are more models of our theories and hypotheses about how the living system works than of the living system itself.* By creating the appropriate network, various notions we have about the workings of an organism can be quantitatively tested and a lot of hand-waving and speculation done away with. This, it seems, will be the role of network models in the next phase of understanding the nature of the living system. (Italics inserted.)

B. Hess (1980), after describing the microscopic and macroscopic spatial organization of biochemical reactions into dynamic dissipative compartments, writes as follows (pp. 89–90):

The organization of biochemical processes in time and in compartments is based on the hierarchical principle of reaction space and boundary systems. If isolated, each of the various levels ... retains its own autonomous functions and structures, although only the coupling of these functions at any level through transport and diffusion yield the physiological properties, which after all are more than the sum of its subparts. A loss of information is especially obvious in case of phenomena related to the energy dependent spatial distribution of intermediate metabolites, of ions and charges and the coupled electrical fields across or on a surface of membranes. Furthermore, clock functions and positional information, neural network functions, synchronization of processes and morphogenesis should be mentioned where research meets a complexity, which is a new challenge to science.

J. Bonner, who was one of the first to expound the unique behaviour of the slime moulds, writing in 1973 on 'Control programmes in biological development' recalled his problems with a programme for biological development in the following terms (1973, pp. 67–9):

Five years ago I played with the idea of trying to simulate a developmental process by asking myself, for each step in the development of a plant organ, what kinds of information I would need to go on to the next. Four years ago two of my colleagues and I, including Douglas Brutlag, took a further step. We wrote a program in Fortran IV, describing in detail the step by step developmental process by which a single apical cell could give rise to the cell elements which, in turn, give rise to the differentiated tissues of the stem. When one writes a program in Fortran, or in any other language, one has to think in depth and with a precision and logic that is not used in everyday thought. We found that we were unable to write, or at least to readily write, a program for causing the development of a single cell into a complicated organism unless we added, in addition to the concept of the environmental test, another hierarchical concept. This is the concept we call *developmental pathway* or *phase*. For example, when the developing cell is in the stage of making bud, it is doing all the kinds of tests of environment which are appropriate to control what a cell should do while it is inside a bud. But those same tests are not appropriate if you are a cell in a root, or flower, or someplace else. And we found, in order to write our Fortran program in rigorous detail, it was required to insert the concept that cells in the bud make tests appropriate to the bud. ...

In a somewhat more apologetic vein, justifying a more 'holistic' methodology as against a reductionist one, L. Wolpert, after explaining his ideas on 'positional information' and its role in pattern formation in biological development, writes (1978, p. 137):

It is sometimes held that no real progress has been made until a biological mechanism is placed on a firm molecular basis (until, in this case, the molecular nature of the gradients, or positional signals, is known). Such a view denies the existence of different levels of organization at which one can meaningfully investigate biological processes. Developmental biologists would like to know the molecular basis of pattern formation, but at present there is no obvious way to find it. Our problems are similar to those of any workers trying to reduce cellular behavior to molecular terms, particularly when it involves reactions within cells that lead to changes in cell state or behavior. Perhaps we should be less apologetic and remember that the study of genetics was (and is) effective at levels other than the level of DNA, and that unless we have the right phenomenology we do not know what we are trying to explain or where to look for the explanation.

J. W. Stucki (1978), at the end of his valuable and comprehensive account of stability analysis of biochemical systems, considers if such systems may usefully be classified into constitutive and functional reactions, but then continues:

It is the impression of the present author, ... that biochemical systems cannot be categorized in this simplistic fashion and that function and maintenance of the necessary energy supply to exert this function are intimately and inseparably integrated in the biochemical networks. Here the systematic application of stability analysis can open new lines of investigation. It is now quite clear that the functional order and coherence of a biochemical network cannot be understood solely on the basis of its constitutive elementary reactions. *The problem of coordination and control requires a look at a higher hierarchical level of the system.* Stated differently, the time has come to abandon the view of enzymes as isolated, albeit sophistically regulated, machines which produce some products. One has to go to a step further and try to understand *the dynamic properties of the complete integrated biochemical network* in order to understand the functions of the living cell. Biochemistry in face of this formidable problem seems to be in a severe crisis at the moment. (Stucki 1978, p. 174; first italics added; second italics, the author's.)

He goes on to express an optimistic view of the future of a biochemistry that takes account of new theoretical concepts, quoting Helmreich (1977):

> there is no cause for being pessimistic: new theoretical concepts, such as the introduction of the quality factor by Manfred Eigen and of dissipative structures by Ilya Prigogine could be the first steps towards an understanding of the principles underlying biological systems, and at last there will be a new generation of biologists, physicists and biochemists that will bring about other revolutions in cell biology. The future of biochemistry has yet to begin.

These remarks by working biological scientists have been cited at length to show that the meeting of conceptual barriers in dealing with biological complexity is not just an invention of philosophers of biology. The pressure to develop new concepts does arise from within biological research but unfortunately those involved in it do not always realize what new theoretical conceptual resources are becoming available. To bridge this gap was one motivation behind this book with its emphasis on concepts elaborated out of physico-chemical sources. This was also very much in the mind of C. H. Waddington in organizing the conferences that led to the four volumes entitled *Towards a theoretical biology* that he edited (1968, 1969, 1970, 1972), and also his last book entitled *Tools for thought* (1977) which is a valuable non-mathematical account of many new intellectual tools that biologists ought to consider using—and is largely complementary to the present physico-chemically oriented work. Amongst other things, Waddington discussed in that book catastrophe theory, which is a topological theory that finds application at the phenomenological level in biology. However its application at the molecular structural and functional is more problematic, since it does not appear to be dealing with open systems—not to mention its controversial mathematical status (the basic exposition is by Thom 1975, see also Thom 1968; the ideas have been much expounded by Zeeman 1972, 1974; see also Zahler and Sussmann 1977, for a very critical article in *Nature* and the subsequent controversies in the columns of that journal). Many of the new ideas described by Waddington, and indeed in this book, involve sudden changes in a variety of different systems and the mathematics that applies is so similar for different problems that Haken (1975, 1977*a,b*, 1978, 1980) has proposed that they are now all part of a single new subject, which he has termed 'synergetics' and which is expanding rapidly.

So the 1970s have witnessed a burgeoning of many fruitful conceptual and mathematical systems of potential importance in deepening our understanding of the special kind of complexity that occurs in the organization of living systems (for a useful account see the perspicuous short survey of these developments by Landsberg (1981) in a review of the book of I. Prigogine, 1980).

All of this may prove daunting, if not overwhelming, to the working biological scientist because such ideas have not formed part of his or her basic earlier education, or *Bildung*. Although only some of these new conceptual forays may prove fruitful, no biological scientist should ignore them. And this raises acute questions about the education of the next generation of biological scientists so that they can become critically sensitive to such developments—sceptical and testing but not dismissive and contemptuous. The growth of biological knowledge, however, is such that, in the United Kingdom at least, and I suspect it is much the same elsewhere, more and more detailed knowledge is required of students so that they can cope with the new subtleties and technology of specific biological research. I can only hope that this volume might serve to introduce biological scientists to seminal ideas, in at least physical chemistry, that are beginning to build the conceptual framework for which, I believe, biology as a science is still seeking. Let the last word be with one who has made an outstanding contribution to understanding the complexities of a particular biological system, that of glycolysis, namely Benno Hess:

> The analysis of many of [the] properties of 'large' living systems is experimentally and theoretically limited by currently available techniques. While the lower limits of reductionism in the description of biological order seem clearly defined in the concept of dissipative structures, an understanding of the function of the large complexity of biological systems is lacking. Although it is clearly recognized that biological systems are irreversibly organized (they do not explode or decay under ordinary conditions but seem to grow and evolve toward a singularity or limit cycle if energy and matter is available), the rules of their network composition, their underlying network hierarchy, and their order in time and space remain to be understood. (Hess 1978, p. 418.)

References

Alcântara, F. and Monk, M. (1974). *J. gen. Microbiol.* **85,** 321.
Aldridge, J. (1976). Short-range intercellular communication, biochemical oscillations and circadian rhythms. In *Handbook of engineering in medicine and biology.* D. G. Fleming and B. N. Feinberg (ed.) pp. 55–147 CRC Press, Cleveland.
Andronov, A. A., Vitt, A. A., and Khalkin, S. E. (1966). *Theory of oscillators.* Pergamon Press, Oxford.
——, Leontovitch, E. A., Gordon, I. I., and Maier, A. G. (1971). *Theory of bifurcations of dynamical systems on a plane.* Israel Program for Scientific Translations, Jerusalem. (Referred to by Nicolis and Portnow 1973.)
Apter, M. J. and Wolpert, L. (1965). *J. theor. Biol.* **8,** 244.
Aschoff, J. (1965). *Circadian clocks.* North-Holland Publ. Co., Amsterdam.
Ashkenazi, M. and Othmer, H. G. (1978). *J. math. Biol.* **5,** 305.
Atlan, H. (1972). *L'organisation biologique et la theorie de l'information.* Hermann, Paris.
—— (1974). *J. theor. Biol.,* **45,** 295.
—— (1975). Source and transmission of information in biological networks. In *Stability and origin of biological information.* 1st Aharon Katzir-Katchalsky Conference, Rehovot, Israel, 1973. I. R. Miller (ed.), pp. 95–118. Wiley, New York.
—— (1978). Sources of information in biological systems. In *Information and systems.* B. Dubuisson (ed.), p. 177, Pergamon, Oxford.
—— (1981). Hierarchical self-organisation in living systems In *Autopoiesis: a theory of living organization.* M. Zeleny (ed.). Basic and Applied General Systems Research Books Series, pp. 185–208. Proc. Int. Conf. on Applied Systems Research, SUNY, Binghampton, 1977. North Holland, Amsterdam.
—— (In press). Information theory. In *Cybernetics: a source book.* R. Trappl (ed.), Chapter 2. Hemisphere Publ. Co.
—— Katzir-Katchalsky, A. (1973). *Curr. mod. Biol.,* **5,** 55.
—— Panet, R., Sidoroff, S., Salomon, J., and Weisbuch, G. (1979). *J. Franklin Inst.* **308,** 297.
Auchmuty, J. F. G. and Nicolis, G. (1975). *Bull. Math. Biol.* **37,** 323.
——, —— (1976). *Bull. Math. Biol.* **38,** 325.
Ayala, F. J. (1974). Introduction. In *Studies in the philosophy of biology: reduction and related problems.* F. J. Ayala and T. Dobzhansky (ed.), pp. vii–xvi. Macmillan, London.
Babloyantz, A. and Hiernaux, J. (1974). *Proc. natn. Acad. Sci. U.S.A.* **71,** 1530.
—— —— (1975). *Bull. math. Biol.* **37,** 637.
—— Nicolis, G. (1972). *J. theor. Biol.* **34,** 185.
—— Sanglier, M. (1972). *FEBS Lett.* **23,** 364.
Bak, T. A. (1963). *Contributions to the theory of chemical kinetics.* W. A. Benjamin, New York.
Bard, J. B. L. (1977). *J. Zool.* **183,** 527.

REFERENCES

—— (1981). *J. theor. Biol.* **93,** 363.
—— Lauder, I. (1974). *J. theor. Biol.* **45,** 501.
Bartlett, M. S. (1955). *Stochastic processes.* Cambridge University Press.
Bastin, T. (1969). A general property of hierarchies. In *Towards a theoretical biology,* Vol. 2. C. H. Waddington (ed.), pp. 252–65. Edinburgh University Press.
Beckner, M. (1974). Reduction, Hierarchies and Organicism. In *Studies in the philosophy of biology.* F. J. Ayala and T. Dobzhansky, (ed.), p. 163. Macmillan, London.
Belousov, B. P. (1958). *Sborn referat. radiat. meditsin za* (Collection of Abstracts on Radiation Medicine), p. 145. Medgiz, Moscow, 1959.
Berlinski, D. (1976). *On systems analysis: an essay concerning the limitations of some mathematical methods in the social, political and biological sciences.* M.I.T. Press, Cambridge, MA.
Berridge, M. J. and Rapp, P. E. (1979). *J. exp. Biol.* **81,** 217.
Bertalanffy, L. (1950). *Br. J. Phil. Sci.* **1,** 134–65.
Blanshard, J. M. V. (1978). Unpublished work, referred to in Blanshard 1979, p. 144, and in a private communication (June 1981), from which the discussion in the text and Fig. 3.19 were extracted.
—— (1979). Physicochemical aspects of starch gelatinization. In *Polysaccharides in food.* J. M. V. Blanshard and J. R. Mitchell (ed.), Butterworths, London.
Blumenthal, R., Changeux, J.-P., and Lefever, R. (1970). *J. Membrane Biol.* **2,** 351.
Boissonade, J. (1979). *Phys. Lett.* **74A,** 285.
Boiteux, A. and Hess, B. (1971). Unpublished observations, referred to in Hess, Boiteux, and Chance (1980).
—— —— (1975). *Faraday Soc. Symp. Chem. Soc.* **9,** 202.
—— —— (1978). Visualization of dynamic spatial structures in glycolysing cell-free extracts of yeast. In *Frontiers of biological energetics.* P. L. Dutton, J. Leigh, and A. Scarpa, (ed.), Vol. 1, pp. 789–97. Academic Press, New York.
—— —— (1980). *Ber. Bunsenges. phys. Chem.* **84,** 392. (This article refers to earlier unpublished observations of spatial dissipative structures made by these authors in 1971.)
—— —— (1981). *Phil. Trans. R. Soc. B.* **293,** 5.
—— —— Plesser, Th., and Murray, J. D. (1977). Oscillatory phenomena in biological systems. In Meeting Report, *FEBS Lett.* **75,** 1.
—— —— Sel'kov, E. E. (1980). *Current topics in cellular regulation* Vol. 17. B. L. Horecker and E. R. Stadtman (ed.), p. 171. Academic Press, New York.
Bonner, J. T. (1967). *The cellular slime moulds.* Princeton University Press.
—— (1971). *Ann. Rev. Microbiol.* **25,** 75.
—— (1973). Control programs in biological development. In *Hierarchy theory: the challenge of complex systems.* H. Pattee (ed.), George Brazilier Inc., New York.
—— (1974). *On development.* Harvard University Press, Cambridge, MA.
Bray, W. C. (1921). *J. Am. Chem. Soc.* **43,** 1262.
Brillouin, L. (1956, 1962). *Science and information theory* (1st and 2nd edn). Academic Press, New York.
Britton, N. F. and Murray, J. D. (1979). *J. theor. Biol.* **77,** 317.
Bronowski, J. (1970). *Zygon* **5,** 18. (See also *Boston Studies in the philosophy of science.* R. S. Cohen and M. W. Wartofsky (ed.), Vol. XI, 1974. Reidel Publ. Co., Dordrecht and Boston.)

Brown, F. A., Hastings, J. W., and Palmer, J. D. (1970). *The biological clock.* Academic Press, New York.
Bryant, P. J., Bryant, S. V., and French, V. (1977). *Scient. Am.*, **237,** (1), 66.
Büchel, W. (1967). *Nature, Lond.* **213,** 319.
Bunge, M. (1969). The Metaphysics, Epistemology and Methodology of Levels. In Whyte *et al.* (1969), p. 17.
Bunning, E. (1960). Biological Clocks. In *Cold Spring Harb. Symp. quant. Biol.* **25**.
—— (1973). *The physiological clock* (3rd edn). Springer-Verlag, New York.
Bunow, B., Kernevez, J-P., Joly, G., and Thomas, D. (1980). *J. theor. Biol.* **84,** 629.
Bunsen Gesellschaft fur Physikalische Chemie, Discussion (1979). *Kinetik Physikalisch-Chemischer Oszillationen, Ber. Bunsenges. phys. Chem.* **84** (1980), 295–426.
Busse, H. G. (1969). *J. phys. Chem., Ithaca* **73,** 750.
Caldwell, P. C. and Hinshelwod, C. N. (1950). *J. Chem. Soc.* p. 3156.
Caplan, S. R., Naparstek, A. and Zabusky, N. J. (1973). *Nature, Lond.*, **245,** 364.
Cellular oscillators (1979). *J. exp. Biol.* **81,** 1–306.
Chance, B., Estabrook, R., and Ghosh, R. W. (1964). *Proc. natn. Acad. Sci. U.S.A.* **51,** 1244.
—— Hess, B., and Betz, A. (1964). *Biochem. biophys. Res. Commun.* **16,** 182.
—— Pye, E. K., Ghosh, A. K., and Hess, B. (1973). *Biological and biochemical oscillators.* Academic Press, New York.
Chandrasekhar, S. (1961). *Hydrodynamic and hydromagnetic stability.* Clarendon Press, Oxford.
Chen, W. K. (1976). Applied Graph Theory. In *Applied mathematics and mechanics.* H. A. Lauwerier and W. T. Koiter (ed.), Vol. 13. North Holland, Amsterdam.
Chernavskaia, N. M. and Chernavskii, D. S. (1961). *Soviet Phys. Usp.* **3,** 850.
Ciba Symposium (1975). Cell Patterning. In *Ciba Foundation Symposium No. 29.* Associated Scientific Publishers, London.
Clarke, B. (1974). *J. chem. Phys.,* **60,** 1481, 1493.
—— (1976). *J. chem. Phys.,* **64,** 4165, 4179.
—— (1980). *Adv. chem. Phys.* **43,** 1.
Cowan, J. D. (1968). Statistical Mechanics of Neural Nets. In *Neural networks.* E. R. Caianello (ed.). Springer-Verlag, Berlin.
—— (1980). Symmetry-Breaking in Embryology and in Neurobiology. In *Symmetry in sciences,* B. Gruber and B. R. Millman (ed.). Plenum Press, New York.
Crick, F. H. C. (1966). *Of molecules and man.* p. 10. University of Washington Press, Seattle and London.
—— (1970). *Nature, Lond.* **225,** 420.
Cronin, J. (1977). Some mathematics of biological oscillations. In *SIAM Rev.* **19,** 100.
Curie, P. (1908). *Oeuvres, Société Francaise de Physique* (Paris).
Dagley, S. and Hinshelwood, C. N. (1938). *J. chem. Soc.,* 1930, 1936, 1942.
Darwin, Eramus (1794–6). *Zoonomia: Or the laws of organic life,* Vol. I.
Davidson, E. H. (1968). *Gene activity in early development.* Academic Press, New York.
Davies, P. C. W. (1976). *Nature, Lond.* **263,** 377.

REFERENCES

Dean, A. C. R. and Hinshelwood, Sir Cyril (1966). *Growth, function, and regulation in bacterial cells.* Clarendon Press, Oxford.
Decker, P. (1979). *Ann. N.Y. Acad. Sci.* **316**, 236.
Degn, H. (1968). *Nature, Lond.* **217**, 1047.
—— (1972). *J. chem. Educ.* **49**, 302.
Denbigh, K. (1975a). A non-conserved function for organized systems. In *Entropy and information in science and philosophy.* L. Kubát and J. Zeman (ed.). Elsevier, Oxford and Academia, Prague.
—— (1975b). *An inventive universe.* Hutchinson, London.
—— (1981). *Three concepts of time.* Springer-Verlag, Berlin, Heidelberg, New York.
—— Hicks (Moore) Margaret and Page, F. M. (1948). *Trans. Faraday Soc.* **44**, 479.
Duysens, L. N. M. and Amesz, J. (1957). *Biochim. Biophys. Acta.* **24**, 19.
Eckman, (1981). *Rep. Progr. Phys.* **53**, 643.
Ehrenfest, P. and Ehrenfest, T. (1907). *Phys. Z.* **8**, 311.
Eigen, M. (1971a). *Naturwissenschaften* **58**, 465.
—— (1971b). *Q. Rev. Biophys.* **4**, 149.
—— (1973). In *The physicist's conception of nature.* J. Mehra (ed.), p. 594. Reidel Publ. Co., Dordrecht.
—— de Maeyer, L. (1963). In *Techniques of organic chemistry.* A. Weissberger (ed.), Vol. VIII/2. Wiley-Interscience, New York.
—— Schuster, P. (1977). *Naturwissenschaften* **64**, 541.
—— —— (1978a). *Naturwissenschaften* **65**, 7.
—— —— (1978b). *Naturwissenschaften* **65**, 341.
—— —— (1979). *The hypercycle.* Reprint as a book of Eigen and Schuster, 1977, 1978a,b. Springer-Verlag, Berlin.
—— Winkler, R. (1981a). *Laws of the game.* Knopf, New York and Allen Lane, London, 1982. (English translation of *Das spiel* by the same authors. R. Piper and Co. Verlag, Munich, 1975.)
—— Winkler-Oswatitsch, R. (1981b). *Naturwissenschaften* **68**, 217.
Eigen, M., Gardiner, W., Schuster, P., and Winkler-Oswatitsch, R. (1981). *Scient. Am.* **244**, (4), 78.
Elsasser, W. M. (1958). *The physical foundation of biology.* Pergamon Press, London and New York.
—— (1966). *Atom and organism.* Princeton University Press, Princeton, N.J.
—— (1975). *The chief abstractions of biology.* North-Holland Publ. Co., Amsterdam and Oxford; American Elsevier Publ. Co., New York.
Ermentrout, G. B. and Cowan, J. D. (1979). *Biol Cybernetics* **34**, 137.
Erneux, Th. (1981). *J. math. Biol.* **12**, 199.
—— Herschkowitz-Kaufman, M. (1975). *Biophys. Chem.* **3**, 345.
—— —— (1977). *J. chem. Phys.* **66**, 248.
—— Hiernaux, J. (1980). *J. math. Biol.* **9**, 193.
Escher, C. (1980). *Ber. Bunsenges. phys. Chem.* **84**, 387.
Faraday Society Symposium (1974). No. 9 on *The physical chemsitry of oscillatory phenomena.* The Faraday Division, Chemical Society, London.
Ferracin, A., Panischelli, E., Benassi, M., di Nallo, A., and Steindler, C. (1978). *BioSyst.* **10**, 307.
Fife, P. C. (1979). *Lecture notes in biomathematics.* Springer-Verlag, New York.
Finkelstein, A. and Mauro, A. (1963). *Biophys. J.* **3**, 215.
von Foerster, H. (1960). On self-organizing systems and their environment. In

Self-organizing systems. M. C. Yovitz and S. C. Cameron (ed.), p. 31. Pergamon Press, Oxford.
Fox, R. F. (1979). *Proc. natn. Acad. Sci. U.S.A.* **76,** 2114.
—— (1980). *Proc. natn. Acad. Sci. U.S.A.* **77,** 3763.
Fox, S. W. (1965). In *The origin of prebiological systems and of their molecular matrices.* p. 36. Academic Press, New York.
Frank, I. G. M. (1967). (ed.). *Oscillatory processes in biological and chemical systems,* Vol. I. Nauka, Moscow.
Frazier, W. A. (1976). *Trends biochem. Sci.* **1,** 130.
Garfinkel, D. and Hess, B. (1964). *J. biol. Chem.* **239,** 971.
Gatlin, L. (1972). *Information theory and the living system.* Columbia University Press, New York and London.
Gerard, R. W. (1957). Units and concepts of biology. *Science, N.Y.,* **125,** 429.
Gerisch, G. and Wick, U. (1975). *Biochem. biophys. Res. Commun.,* **65,** 364.
—— Malchow, D., and Hess, B. (1974). *Mosbacher Colloquium der Gesellschaft fur Biologische Chemie* **25,** 279.
—— Maeda, Y., Malchow, D., Roos, W., Wick, U., and Wurster, B. (1977). Cyclic AMP signals and control of cell aggregation in *Dictyostelium discoideum*. In *Development and differentiation in the cellular slime moulds.* P. Cappuccinelli and J. M. Ashworth (ed.), pp. 105–24. Elsevier, North Holland, Amsterdam.
Ghosh, A. and Chance, B. (1964). *Biophys. Res. Commun.* **16,** 174.
Gierer, A. (1974). *Sci. Amer.,* **231** (No. 6), 44.
—— (1981). *Prog. Biophys. molec. Biol.* **37,** 1; *Phil. Trans. R. Soc. Lond.* **B295,** 429.
—— Meinhardt, H. (1972). *Kybernetik* **12,** 30.
—— —— (1974). Biological pattern formation involving lateral inhibition. In *Lectures on mathematics in the life sciences* Vol. 7, p. 163. American Mathematical Society.
Glansdorff, P. and Prigogine, I. (1971). *Thermodynamic theory of structure, stability and fluctuations.* Wiley-Interscience, New York.
—— Nicolis, G. and Prigogine, I. (1974). *Proc. natn. Acad. Sci. U.S.A.* **71,** 197.
Glass, L. and Mackey, M. C. (1979). *Ann. N.Y. Acad. Sci.* **316,** 214.
Gmitro, J. I. and Scriven, L. E. (1966). A physico-chemical basis for pattern and rhythm. In *Intracellular transport.* K. B. Warren (ed.), pp. 221–55. Academic Press, New York.
Goldbeter, A. (1973). *Proc. natn. Acad. Sci. U.S.A.* **70,** 3255.
—— (1975). *Nature, Lond.* **253,** 540.
—— (1977). *Biophys. Chem.* **6,** 95.
—— Caplan, S. R. (1976). *Ann. Rev. Biophys. Bioeng.* **5,** 449.
—— Lefever, R. (1972). *Biophys. J.* **12,** 1302.
—— Nicolis, G. (1976). *Progr. theor. Biol.* **4,** 65.
—— Segel, L. A. (1977). *Proc. natn. Acad. Sci. U.S.A.* **74,** 1543.
—— Venieratos, D. (1980). *J. mol. Biol.* **138,** 137.
—— Erneux, T., and Segel, L. A. (1978). *FEBS Lett.* **89,** 237.
Goodwin, B. C. (1963). *Temporal organisation in cells.* Academic Press, New York.
—— (1965). *Adv. Enzyme Regul.* **3,** 425.
—— Cohen, M. H. (1969). *J. theor. Biol.* **25,** 49.
Greene, J. C. (1959). *The death of Adam.* Iowa State University Press, Ames, Iowa.

Grene, M. (1969). Hierarchy: one word, how many concepts?. In Whyte et al. (1969), pp. 56–8.
Griffith, J. S. (1968). J. theor. Biol. **20,** 202.
Gurel, O. and Rössler, O. E. (1979) (ed.). Bifurcation theory and applications in scientific disciplines. Ann. N.Y. Acad. Sci. 316.
Gutman, H. (1969). Structure and function in living systems. In Whyte et al. (1969), p. 229.
Hahn, H.-S., Nitzan, A., Ortoleva, P., and Ross, J. (1974). Proc. natn. Acad. Sci. U.S.A. **71,** 4067.
Haken, H. (1975). Cooperative phenomena in systems far from thermal equilibrium and in non-physical systems. Rev. mod. phys. **47,** 67.
—— (1977a). Phys. Bull. Sept. 1977, p. 412.
—— (1977b) (ed.). Synergetics. Proceedings of the International Workshop on Synergetics at Schloss Elman, May, 1977. Springer-Verlag, Berlin, Heidelberg, New York.
—— (1978): Synergetics: Nonequilibrium phase transitions and self-organization in physics, chemistry and biology Springer-Verlag, Berlin, Heidelberg, New York.
—— (1979). Ann. N.Y. Acad. Sci. **316,** 357.
—— (1980). Naturwissenschaften **67,** 121.
Hanusse, P. (1972). C.r. hebd. Séanc. Acad. Sci., Paris, **274C,** 1245.
—— (1973). C.r. hebd. Séanc. Acad. Sci., Paris, **277C,** 263.
Harrington, J. M. and Rowlinson, J. S. (1979). Proc. R. Soc. **A367,** 15.
Harris, H. and Viza, D. (1972) (ed.). Proc. 1st Int. Conf. Cell Differentiation Munskgaard Publishers, Copenhagen.
Hassenstein, B. (1977). Biologische kybernetik. Quelle and Meyer Verlag, Heidelberg.
Hastings, J. W. and Schweiger, H.-G. (1975). The molecular basis of circadian rhythms. Dahlem Konferenzen, Berlin.
Hastings, S., Tyson, J., and Webster, D. (1977). J. Different. Equat. **25,** 39.
Hawking, S. W. (1975). Communs math. Phys. **43,** 199.
Hearon, J. Z. (1953). Bull. math. Biophys. **15,** 121.
Helmreich, E. J. (1977). Trends biochem. Sci. **2,** N49.
Herschkowitz-Kaufman, M. (1970). C.r. hebd. Séanc. Acad. Sci., Paris **270C,** 1049.
—— (1973). Ph.D. Dissertation, University of Brussels.
—— (1975). Bull. math. Biol. **37,** 589.
—— Erneux, T. (1979). Ann. N.Y. Acad. Sci. **316,** 296.
—— Nicolis, G. (1972). J. chem. Phys. **56,** 1890.
Hervagault, J. F., Friboulet, A., Kernevez, J.-P., and Thomas, D. (1980). Ber. Bunsenges. phys. Chem. **84,** 358.
Hess, B. (1963). In Funktionelle u. morphologische organisation der zelle. Springer-Verlag, Berlin.
—— (1968). Biochemical Regulation. In Systems theory and biology. M. D. Mesarovic (ed.), pp. 88–114. Springer-Verlag, New York.
—— (1975). In Energy transformations in biological systems. Elsevier, Amsterdam.
—— (1978). Oscillations: a property of organized systems. In Frontiers in physicochemical biology. Academic Press, New York.
—— (1980). Organization of biochemical reactions: from microspace to macroscopic structures. In Cell compartmentation and metabolic channeling. L.

Noyer, F. Lynen, and K. Mothes (ed.), pp. 75–92. Elsevier, North Holland Biomedical Press, Amsterdam.
—— Boiteux, A. (1971). *Ann. Rev. Biochem.* **40,** 237.
—— —— (1980). *Ber. Bunsenges. phys. Chem.* **84,** 346.
—— Plesser, Th. (1979). *Ann. N.Y. Acad. Sci.* **316,** 203.
—— Boiteux, A., and Chance, E. M. (1980). Dynamic compartmentation. In *Molecular biology, biochemistry and biophysics. Vol. 32. Chemical recognition in biology.* F. Chapeville and A.-L. Haenni (ed.), p. 157. Springer-Verlag, Berlin, Heidelberg and New York. See also Hess, B., Chance, E. M., Curtis, A. R. and Boiteux, A. (1981). Complex Dynamic Structures. In *Non-linear phenomena in chemical dynamics.* S. C. Vidal and A. Pacault (ed.) p. 172. Springer-Verlag, Berlin, Heidelberg, and New York.
—— Goldbeter, A., and Lefever, R. (1978). *Adv. chem. Phys.* **38,** 363.
—— Boiteux, A., Busse, H. G., and Gerisch, G. (1975). *Adv. chem. Phys.* **29,** 137.
Hiernaux, J. and Babloyantz, A. (1976). *J. nonequil. Themodyn.* **1,** 33.
Higgins, J. (1964). *Proc. natn. Acad. Sci. U.S.A.* **51,** 989.
—— (1967). *Ind. Eng. Chem.* **59,** 19.
—— Frenkel, R., Hulme, E., Lucas, A., and Rangazas, G. (1973). In *Biological and biochemical oscillators.* B. Chance, E. K. Pye, A. M. Ghosh, and B. Hess (ed.), p. 127. Academic Pess, New York.
Hinshelwood, C. N. (1940). *The kinetics of chemical change.* Clarendon Press, Oxford.
—— (1946). *The chemical kinetics of the bacterial cell.* Clarendon Press, Oxford.
—— (1951). *The structure of physical chemistry.* p. 449. Clarendon Press, Oxford.
—— (1952). *J. chem. Soc.* p. 745.
—— (1953). *J. chem. Soc.* p. 1947.
—— Lewis, P. R. (1948). *Proc. R. Soc. Lond.* **B135,** 316.
Hofstadter, D. R. (1981). Strange attractors. *Sci. Amer.* (Nov.), p. 16.
Holmes, K. C. (1974). Selbstorganisation biologischer Strukturen. In *Verhandlhung der Gesellschaft Deutscher Naturforscher und Ärzte*, pp. 31–9.
Horowitz, B. A. (1978). Neurohumoral regulation of nonshivering thermogenesis in mammals. In *Strategies in cold: natural torpidity and thermogenesis.* L. Wang and J. Hudson (ed.), pp. 619–53. Academic Press, New York.
Horowitz, J. M. and Plant, R. E. (1978). *Am. J. Physiol.* **235,** R121.
—— Giacchino, J. L., and Horowitz, B. A. (1979). *J. Franklin Inst.* **308,** 281.
Horsthemke, W. (1981). Noise induced transitions. In *Non-linear stochastic processes in physics, chemistry and biology.* L. Arnold and R. Lefever (ed.). Springer-Verlag, Bielefeld, Z.I.F.
—— Malek-Mansour, M. (1976). *Z. Physik.* **B24,** 307.
Hubel, D. H. (1979). The Brain. In *The Brain*, p. 2. Scientific American Book, Freeman, San Francisco.
Hunding, J. (1975). *Biophys. Struct. Mech.* **1,** 47.
Hunter, S. A., Peura, R. A., Crusberg, T. C., and Harvey, R. J. (1979). *J. theor. Biol.* **78,** 499.
Huxley, H. E. (1958). *Scient. Am.* **199,** (5), 68.
—— (1965). *Scient. Am.* **213,** (6), 26.
Iooss, G. and Joseph, D. D. (1980). *Elementary stability and bifurcation theory.* Springer Verlag, New York.
Jacob, F. (1974). *The logic of living systems.* Allen Lane, London.
Johnson, H. A. (1970). Information theory in biology after 18 years. *Science, N.Y.* **168,** 1545.

Kacser, H. (1957). Some physico-chemical aspects of biological organisation. Appendix to *The strategy of the genes* by C. H. Waddington, pp. 191–249. George Allen and Unwin, London.
—— (1960). Kinetic models of development and heredity. In *Models and Analogues in Biology, Symp. Soc. Exptl. Biol.* **14,** 13. Cambridge University Press, Cambridge.
Kaimachnikov, N. P. and Sel'kov, E. E. (1977). *Biochemistry,* **42,** 490. (English translation of *Biokhimiya, U.S.S.R.*)
Karnopp, D. and Rosenberg, R. (1968). *Analysis and simulation of multiport systems.* M.I.T. Press, Cambridge, MA.
Katchalsky, A. and Spangler, R. (1968). *Q. Rev. Biophys.* **1,** 127.
Kauffman, S. A. (1971). Articulation of parts explanation in biology. In *Boston studies in the philosophy of science.* R. C. Buck and R. S. Cohen, (ed.), Vol. 8, p. 257. D. Reidel Publ. Co., Dordrecht and Boston.
—— Shymko, R. M., and Trabert, K. (1978). *Science, N.Y.* **199,** 259.
Kay, R. (1981). *Nature, Lond.* **294,** 108.
Keizer, J. (1976). *J. chem. Phys.* **65,** 4431.
—— Fox, R. F. (1974). *Proc. natn. Acad. Sci. U.S.A.* **71,** 192.
Kernevez, J.-P., Joly, F., Duban, M. C., Bunow, B., and Thomas, D. (1979). *J. math. Biol.* **7,** 41.
Koestler, A. (1967). *The ghost in the machine.* Macmillan, New York.
Konijn, T. M. (1972). *Advances in cyclic nucleotide research.* P. Greengard, G. A. Robison, and R. Paoletti (ed.), **1,** 17. Raven Press, New York.
—— van de Meene, J. G. C., Bonner, J. T., and Barkley, D. S. (1967). *Proc. natn. Acad. Sci. U.S.A.* **58,** 1152.
Kuhn, H. (1972). *Angew. Chem. Int. Ed. Engl.* **11,** 798.
Kuhn, H. (1976). *Naturwissenschaften* **63,** 68.
Küppers, B. (1975). *Progr. Biophys. molec. Biol.* **30,** 1.
—— (1979a). *Bull. math. Biol.* **41,** 803.
—— (1979b). *Naturwissenschaften* **66,** 228.
Landahl, H. D. (1969). *Bull. math. Biophys.* **31,** 775.
Landau, L. D. and Lifshitz, E. M. (1957). *Statistical physics.* Pergamon, Oxford.
Landsberg, P. T. (1981). All time is redeemable—a review of Prigogine (1980). *Times Higher Educational Supplement* 17th March, 1981.
Lefever, R. (1968). *J. chem. Phys.* **49,** 4977.
—— (1981). Noise induced transitions in biological systems. In *Non-linear stochastic processes in physics, chemistry and biology.* L. Arnold and R. Lefever (ed.). Springer-Verlag, Bielefeld, Z.I.F.
—— Nicolis, G. (1971). *J. theor. Biol.* **30,** 267.
—— Prigogine, I. (1981). In *Self-organisation.* Jean Yates (ed.) 1979 Dubrovinik Conference.
—— Nicolis, G., and Prigogine, I. (1967). *J. chem. Phys.* **47,** 1045.
Levin, S. A. (1977). Populations and Communities. In *Math. Assoc. of America, Study in Math. Biol.* Vol. II. S. A. Levin (ed.).
Levins, R. (1970). Complex Systems In *Towards a theoretical biology.* C. H. Waddington (ed.), Vol. 3, p. 73. University of Edinburgh Press.
Lewis, G. N. and Randall, M. (1923). *Thermodynamics.* McGraw Hill, New York and London.
Liesegang, R. E. (1905). *Z. phys. chem.* **52,** 185.
Lipmann, F. (1971). *Science, N.Y.* **173,** 875.

Loomis, W. F. (1975). *Dictyostelium discoideum: a developmental system.* Academic Press, New York.
Lotka, A. J. (1910). *J. phys. chem.*, **14,** 271.
—— (1920a). *Proc. natn. Acad. Sci. U.S.A.* **7,** 410.
—— (1920b). *J. Am. Chem. Soc.* **42,** 1595.
—— (1924). *Elements of physical biology* (Williams & Williams). Re-issued as *Elements of mathematical biology*, 1956. Dover Publications, New York. (Page and figure numbers in the text refer to the Dover edition.)
Mackey, M. C. and Glass, L. (1977). Oscillation and chaos in physiological control systems. *Science, N.Y.* **197,** 287.
McQuarrie, D. A. (1967). *Stochastical approach to chemical kinetics.* Methuen, London.
Martiel, J.-L. and Goldbeter, A. (1981). *Biochmie* **63,** 119.
Martinez, H. (1972). *J. theor. Biol.* **36,** 479.
May, R. M. (1974). *Stability and complexity in model ecosystems*, 2nd edn. Princeton University Press, Princeton, N.J.
—— (1979). *Ann. N.Y. Acad. Sci.* **318,** 517.
Maynard Smith, J. (1960). *Proc. R. Soc.* **152B,** 397.
—— Sondhi, K. C. (1961). *J. Embryol. exp. Morph.* **9,** 661.
Medawar, P. (1974). A geometric model of reduction and emergence. In *Studies in the philosophy of biology.* F. J. Ayala and T. Dobzhansky (ed.), pp. 57–64. Macmillan, London.
Meinhardt, H. and Gierer, A. (1974). *J. Cell. Sci.* **15,** 321.
Meixner, J. (1949). *Z. Naturforsch.* **4a,** 594.
Mees, A. I. and Rapp, P. E. (1978). *J. math. Biol.* **5,** 99.
Mesarovic, M. D. (1968). Systems theory and biology—view of a theoretician. In *Systems theory and biology.* M. D. Mesarovic (ed.), pp. 59–87. Springer-Verlag, New York.
Mikulecky, D. C. (1977). *J. theor. Biol.* **69,** 511.
—— Thomas, S. R. (1978). *J. theor. Biol.* **73,** 697.
—— —— (1979). *J. Franklin Inst.* **308,** 309.
—— Huf, E. G., and Thomas, S. R. (1979). *Biophys. J.* **25,** 87.
—— Wiegand, W. A., and Shiner, J. S. (1977). *J. theor. Biol.* **69,** 471.
Miller, J. G. (1978). *Living systems.* McGraw Hill, New York. See also: Miller, J. G. (1965). Living Systems: Basic Concepts *Behavl Sci.* **10,** p. 193; idem (1965) ibid. Living Systems: Structure and process p. 337; idem (1965) ibid. Living Systems: cross-level hypotheses p. 380.
Mimura, M. and Murray, J. D. (1978a). *Z. Naturforsch.* **33c,** 580.
—— —— (1978b). *J. theor. Biol.*, **75,** 249.
Minorsky, N. (1962). *Non-linear oscillations.* van Nostrand, Princeton, N.J.
Monod, J. and Jacob, F. (1961). *Cold Spring Harb. Symp. quant. Biol.* **26,** 389.
—— Wyman, P., and Changeux, J.-P. (1965). *J. mol. Biol.* **12,** 88.
Moore, Margaret J. (1949). *Trans. Faraday Soc.* **45,** 1098.
Morales, M. and McKay, D. (1967). *Biophys. J.* **7,** 621.
Morowitz, H. J. (1968). *Energy flow in biology.* Academic Press, New York and London.
—— (1973). Perspectives on thermodynamics and the origin of life. In *Proc. Katchalsky-Katzir Mem. Symp.* H. C. Mel (ed.). University of California Press, Berkeley, CA.
—— (1974). Energy flow and biological organization. In *Irreversible ther-*

modynamics and the origin of life. G. F. Oster, I. L. Silver, and C. A. Tobias (ed.), pp. 25–31. Gordon and Breach, New York, London, and Paris.

Murray, J. D. (1977). *Lectures on non-linear-differential-equation models in biology*. Clarendon Press, Oxford.

—— (1981). *J. theor. Biol.* **88**, 161; *Phil. Trans. R. Soc. Lond.* **B295**, 473.

Nagel, E. (1952). Wholes, Sums and Organic Unities. *Phil. Stud.* **3**, 17.

—— (1961). *The Structure of Science*, chap. 11 (Harcourt, Brace & World, New York).

Naparstek, A., Thomas, D., and Caplan, S. R. (1973). *Biochim. biophys. Acta*, **323**, 643.

Neumann, E. (1979). Network thermodynamics—an account along original lecture notes of Aharon Katchalsky. In *Molecular mechanisms of biological recognition*. M. Balaban (ed.) Elsevier North Holland, Amsterdam.

Newell, P. C. (1971). *Essays in biochemistry*. P. N. Campbell and F. Dickens (ed.), Vol. 7, p. 87. Academic Press, London.

—— (1977a). How cells communicate: the system used in slime moulds. *Endeavour NS* **1**, 63.

—— (1977b). Aggregation and cell surface receptors in cellular slime moulds. In *Microbial interactions. receptors and recognition ser. B*, Vol. 3. J. L. Reissing (ed.), p. 3. Chapman and Hall, London.

—— (1982). Attraction and adhesion in the slime mould *Dictyostelium*. In *Fungal morphogenesis*. J. E. Smith (ed.), Vol. 1. (In press) Marcel Dekker Inc., New York.

Nicolis, G. (1971). *Adv. chem. Phys.* **19**, 209.

—— (1974). Patterns of spatio-temporal organization in chemical and biochemical kinetics. In *Proc. SIAM-AMS Symp. Appl. Math., American Math. Soc.* **8**, 33.

—— (1975). *Adv. chem. Phys.* **29**, 29.

—— (1979). Irreversible thermodynamics. *Rep. Prog. Phys.* **42**, 225.

—— (1981). Non-equilibrium transitions: a mechanism of self-organisation of complex systems. In *Self-organisation*. Jean Yates (ed.), (Dubrovnik Conference, 1979). (In press.)

—— Auchmuty, J. F. G. (1974). *Proc. natn. Acad. Sci. U.S.A.* **71**, 2748.

—— Lefever, R. (1975). Membranes, Dissipative Structures and Evolution, *Adv. chem. Phys.*, **29**.

—— Malek Mansour, M. (1980). *J. stat. Phys.* **22**, 495.

—— Portnow, J. (1973). *Chem. Rev.* **73**, 365.

—— Prigogine, I. (1971). *Proc. natn. Acad. Sci. U.S.A.* **68**, 102.

—— —— (1977). *Self-organization in non-equilibrium systems*. John Wiley, New York, London.

—— —— (1979). *Proc. natn. Acad. Sci. U.S.A.* **76**, 6060.

—— Turner, J. W. (1979). *Ann. N.Y. Acad. Sci.* **316**, 251.

Noyes, R. M. (1980). *Ber. Bunsenges. phys. Chem.* **84**, 295.

—— Field, R. J. (1974). *Ann. Rev. phys. Chem.* **25**, 95.

Onsager, L. (1931). *Phys. Rev.* **37**, 405; **38**, 2265.

Oparin, A. I. (1936). *The origin of life on the earth*. Eng. Trans. 3rd rev. edn, 1957. Oliver & Boyd, Edinburgh.

—— (1961). *Life: its nature, origin and development*. Academic Press, New York.

—— (1965), In *The origin of prebiological systems and of their molecular matrices*. S. W. Fox (ed.), p. 33. Academic Press, New York.

REFERENCES

Oster, G. F. and Auslander, D. M. (1971). *J. Franklin Inst.* **292,** 1.
—— Perelson, A. S. (1973). *Israel J. Chem.* **11,** 445.
—— —— Katchalsky, A. (1971). *Nature, Lond.* **234,** 393.
—— —— —— (1973). *Q. Rev. Biophys.* **6,** 1.
Othmer, H. G. (1977). Current problems in pattern formation. In *Lectures on mathematics in the life sciences*. S. A. Levin (ed.), Vol. 9. Some mathematical questions in biology. In *Proc. 10th Symp. Math. Biol., Boston U.S.A.*, p. 57. American Mathematical Society, Providence, R.I.
—— Scriven, L. R. (1969). *Ind. & Eng. Chem.* **8,** 302.
Pattee, H. (1969). In Whyte *et al.* (1969), p. 161.
—— (1970). The problem of biological hierarchy. In *Towards a theoretical biology*, Vol. 3. C. H. Waddington (ed.), p. 117. Edinburgh University Press.
—— (1971). *Q. Rev. Biophys.* **4,** 255.
—— (1973). (ed.) *Hierarchy theory: the challenge of complex systems*. George Brasilier Inc., New York.
—— (1979). *BioSystems* **11,** 217.
Pavlidis, T. (1971). *J. theor. Biol.* **33,** 319.
—— (1973). *Biological oscillators: their mathematical analysis*. Academic Press, New York and London.
Paynter, H. (1961). *Analysis and design of engineering systems*. MIT Press, Cambridge, MA.
Peacocke, A. R. (1979). *Creation and the world of science*. 1978 Bampton Lectures. Clarendon Press, Oxford.
—— (1976). Reductionism: a review of the epistemological issues and their relevance to biology and the problem of consciousness. *Zygon* **11,** 307.
Perelson, A. S. (1975). *Biophys. J.* **15,** 667.
Peusner, L. (1970). Ph.D. Thesis: The Principles of Network Thermodynamics. Harvard University, Cambridge, MA.
Plant, R. E. and Horowitz, J. M. (1979). *J. Franklin Inst.* **308,** 269.
Polanyi, M. (1958). *Personal knowledge*. Routledge and Kegan Paul, London.
—— (1959). *The study of man*. Routledge and Kegan Paul, London.
—— (1967). *Chem. & Eng. News*, **45,** 54.
—— (1968). *Science, N.Y.* **160,** 1308.
Popper, K. R. (1965). *Nature, Lond.* **207,** 233.
—— (1967). *Nature Lond.* **213,** 320.
—— (1972). *Objective knowledge*. Oxford University Press.
—— (1974). Scientific reduction and the essential incompleteness of all science. In *Studies in the philosophy of biology*. F. J. Ayala and T. Dobzhansky (ed.), p. 259. Macmillan, London.
Prigogine, I. (1945). *Bull. Acad. r. Belg. Cl. Sci.* **31,** 600.
—— (1947). *Etude thermodynamique des phénomènes irréversibles*. Desoer, Liége.
—— (1949). *Physica* **15,** 272.
—— (1955). *Introduction to thermodynamics of irreversible processes*, 2nd edn. 3rd edn (1967), Wiley-Interscience, New York.
—— (1965). *Physica* **31,** 719.
—— (1967). Dissipative structures in chemical systems. In *Fast reactions and primary processes in chemical kinetics*, 5th Nobel Symp. S. Claesson (ed.), p. 371. Almquist and Wiksell, Stockholm.
—— (1969). Structure, dissipation and life. In *Theoretical physics and biology*, 1st Int. Conf. Theor. Phys. Biol., Versailles, 1967. M. Marois (ed.). North Holland Publ. Co., Amsterdam.

REFERENCES

—— (1980). *From being to becoming*. W. H. Freeman, San Francisco.
—— Balescu, R. (1955). *Bull. Acad. r. Belg., Cl. Sci.* **41,** 917.
—— Defay, R. (1954). *Chemical thermodynamics*. D. H. Everett (trans.). Longman, London.
—— Lefever, R. (1968). *J. chem. Phys.* **48,** 1665.
—— —— (1975). Stability and thermodynamic properties of dissipative structures in biological systems. In *Stability and origin of biological information*, 1st A. Katchalsky-Katzir Conf. Proc. I. R. Miller (ed.). Wiley, New York.
—— Nicolis, G. (1967). *J. chem. Phys.* **46,** 3542.
—— —— (1971). *Q. Rev. Biophys.* **4,** 107.
—— Lefever, R., Goldbeter, A., and Herschkowitz-Kaufman, M. (1969). *Nature, Lond.* **223,** 913.
—— Nicolis, G., and Babloyantz, A. (1972). *Phys. Today* **25,** (a) p. 23; (b) p. 38.
Quastler, H. (1953). (ed.). *Information theory in biology*. University of Illinois Press, Urbana, IL.
Queiroz, O. (1974). *Ann. Rev. Plant Physiol.* **25,** 115.
Ramakrishnan, A. (1959). *Encyclopedia of physics*. S. Flugge (ed.), **VIII/2,** 524. Springer, Berlin.
Rapp, P. E. (1975). *Math. Bio-sci.* **25,** 165.
—— (1976). *Bull. Inst. Math. Applics.* **12,** 11; *J. math. Biol.* **3,** 203.
—— (1979a). *J. exp. Biol.* **81,** 281.
—— (1979b). Bifurcation theory, control theory and metabolic regulation. In *Biological systems, modelling and control*. D. A. Linkens (ed.), pp. 10–92. Peter Peregrinus, London.
—— Berridge, M. J. (1977). *J. theor. Biol.* **66,** 497.
Ray, John. (1691). *The Wisdom of God in the works of creation*.
Riley, K. F. (1974). *Mathematical methods for the physical sciences*. Cambridge University Press.
Robertson, A. and Grutsch, J. (1974). *Life Sci.* **15,** 1031.
—— Drage, D., and Cohen, M. H. (1972). *Science, N.Y.* **175,** 333.
Rosen, R. (1969). Hierarchical organization in automata theoretic models of biological systems. In Whyte, *et al.* (1969), p. 179.
—— (1979). *Ann. N.Y. Acad. Sci.* **316,** 178.
Rothstein, J. (1952). *J. appl. Phys.* **23,** 281.
Rowlinson, J. S. and Widom, B. (1982). *Molecular theory of capillarity*. Oxford University Press.
Royal Society Discussion, London (1981), "Theories of Biological Pattern Formation" *Phil. Trans. R. Soc. Lond.* **B295** (1981) 425–617.
Sanglier, M. and Nicolis, G. (1976). *Biophys. Chem.* **4,** 113.
Sattinger, D. (1973). Topics in stability and bifurcation theory. In *Lecture notes in mathematics* Vol. 30. Springer-Verlag, Berlin.
Saunders, P. and Ho, M. (1976). *J. theor. Biol.* **63,** 375.
Schaffner, K. F. (1967). *Science, N.Y.* **157,** 646.
—— (1969). *Br. J. Phil. Sci.* **20,** 325.
Schiffmann. Y. (1980). *Prog. Biophys. molec. Biol.* **36,** 87.
Schmitz, G. (1973). *J. chim. Phys. physicochim. Biol.* **70,** 997.
Schnakenberg, J. (1977). *Thrmodynamic network analysis of biological systems*. Springer-Verlag, Berlin, Heidelberg, and New York.
Schrödinger, E. (1944). *What is life?* Cambridge University Press, Cambridge.
Schuster, P., Sigmund, K., and Wolff, R. (1979). *J. different. equat.* **32,** 357.
Seelig, F. F. (1970). *J. theor. Biol.* **27,** 197.

—— (1976a). *Z. Naturforsch.* **31a,** 731.
—— (1976b). *Ber. Bunsenges. phys. Chem.* **80,** 1126.
Segel, L. A. (1981) (ed.). *Mathematical models in molecular and cellular biology.* Cambridge University Press, Cambridge.
Sel'kov, E. E. (1967), in Frank (1967).
—— (1968). *Eur. J. Biochem.* **4,** 79.
—— (1971). (ed.). *Oscillatory processes in biological and chemical systems,* Vol. II. Nauka, Puschino-na-Oka.
—— (1972). In *Analysis and simulation of biochemical systems.* H. C. Hemker and B. Hess (ed.), pp. 145–61. North Holland Publ. Co., Amsterdam.
—— (1980). *Ber. Bunsenges. phys. Chem.* **84,** 399.
Shannon, C. E. and Weaver, W. (1949, 1962). *The mathematical theory of communication.* University of Illinois Press, Urbana, IL.
Shapley, H. (1958). *Of stars and men.* Beacon Press, Boston, MA.
Simon, H. A. (1962). The Architecture of Complexity. *Proc. Am. Phil. Soc.* **106,** 467; reprinted in *The sciences of the artificial,* H. A. Simon, Karl Taylor Compton Lectures, 1968, p. 84. M.I.T. Press, Cambridge, MA. (1970).
Simpson, G. G. (1950). *The meaning of evolution.* Oxford University Press.
Smart, J. C. C. (1963). *Philosophy and scientific realism.* Routledge and Kegan Paul, New York.
Smith, R. E. and Horowitz, B. A. (1969). *Physiol. Rev.* **49,** 330.
Spangler, R. A. and Snell, F. M. (1961). *Nature, Lond.* **191,** 457.
—— —— (1967). *J. theor. Biol.* **16,** 366, 381.
Spiegelman, S. (1971). *Q. Rev. Biophys.* **4,** 213.
Stebbins, G. L. (1969). *The basis of progressive evolution.* University of North Carolina Press, Chapel Hill, NC.
Stucki, J. W. (1978). Stability analysis of biochemical systems—a practical guide. *Progr. Biophys. molec. Biol.* **33,** 99.
Sussman, M. (1964). *Growth and development.* Prentice-Hall, New Jersey.
Sweeney, D. M. (1969). *Rhythmic phenomena in plants.* Academic Press, London.
Symposium of the Society of Experimental Biology (1973). No. XXVII: Rate Control of Biological Processes'. Cambridge University Press.
Tellegen, B. D. H. (1952). *Phillips Res. Rep.* **7,** 259.
Theoridis, G. C. and Stark, L. (1969). Information as a quantitative criterion of biosphere evolution. *Nature* **224,** 860.
Thom, R. (1968). Une théorie dynamique de la morphogénese. In *Towards a theoretical biology.* C. H. Waddington (ed.), Vol. 1, p. 152. University of Edinburgh Press.
—— (1975). *Structural stability and morphogenesis.* (Original edn: 1972) W. A. Benjamin, Reading, MA.
Thoma, J. (1971). *J. Franklin Inst.* **292,** 109.
—— (1975a). *Introduction to bond graphs and their applications.* Pergamon, Oxford.
—— (1975b). *J. Franklin Inst.* **299,** 89.
—— (1976). *Int. J. mech. Eng. Educ.* **4,** (3), 209.
Thorpe, W. H. (1974). *Animal nature and human nature.* Methuen, London.
Tonnelat, J. (1978). *Thermodynamique et biologie.* (Maloine, S. A., Editeur, Paris.)
Tornheim, K. (1979). *J. theor. Biol.* **79,** 491.
Turing, A. M. (1952). *Phil. Trans. R. Soc., Lond.* **B237,** 37.
Tyson, J. J. (1973). *J. chem. Phys.* **58,** 3919.
—— (1975). *J. math. Biol.*, **1,** 311.

—— (1976). The Belousov–Zhabotinsky reaction. In *Lecture notes in biomathematics*. S. A. Levin (ed.). Springer-Verlag, New York.
—— (1977). *J. chem. Phys.* **66**, 905.
—— Kauffman, S. (1975). *J. math. Biol.* **1**, 289.
—— Othmer, H. G. (1978). The dynamics of feedback control circuits in biochemical pathways. *Prog. theor. Biol.* **5**, 1.
van der Waals, J. D. (1894). *Zeit. phys. chem.* **13**, 657. (English trans. *J. stat. Phys.* **20** (1979), 197.)
Venieratos, D. and Goldbeter, A. (1979). *Biochimie* **61**, 1247.
Viniegra-Gonzales, G. (1973). In *Biological and biochemical oscillations*. B. Chance, E. K. Pye, A. Ghosh, and B. Hess (ed.), p. 41. Academic Press, New York.
Virchow, R. (1858). Die Mechanische Auffassung des Lebens. Lecture to the Gesellschaft Deutsche Naturforscher und Ärtze, 1858. On the mechanistic interpretation of life. Transl. L. J. Rather (pp. 102 ff.) In *Disease, life, and man*, selected essays of R. Virchow. Stanford University Press, Stanford, CA and Oxford University Press, London (1959).
—— (1862). Atome und Individuen. Vier Reden über Leben u. Kranksein, Berlin 1862. (Atoms and Individuals. L. J. Rather (transl.), pp. 120 ff. In *Disease, life and man*, selected essays of R. Virchow. Stanford University Press, Stanford, CA and Oxford University Press, London 1959).
Volterra, V. (1931). *Théorie mathématique de la lutte pour la vie*. Gauthier-Villars, Paris.
Waddington, C. H. (1968, 1969, 1970, 1972). (ed.). *Towards a theoretical biology*, Vols. 1, 2, 3, 4. University of Edinburgh Press, Edinburgh.
—— (1977). *Tools for thought*. Jonathan Cape, London.
Walgraef, D., Dewel, G., and Borckmans, P. (1980). *Phys. Rev.* **A21**, 397.
Walter, C. (1969a). *Biophys. J.* **9**, 863.
—— (1969b). *J. theor. Biol.* **23**, 39.
—— (1970). *J. theor. Biol.* **27**, 259.
Weaver, W. (1948). Science and complexity. *Am. Scient.* **36**, 536.
Weiss, P. A. (1971). The basic concept of hierarchic systems. In *Hierarchically organized systems in theory and practice*. P. A. Weiss (ed.), pp. 26 f. Hafner Publ. Co., New York.
Whyte, L. L., Wilson, A. G., and Wilson, D. (1969). (ed.). *Hierarchical structures*. Elsevier, New York.
Wicken, J. S. (1978). *J. theor. Biol.* **72**, 191.
—— (1979). *J. theor. Biol.* **77**, 349.
—— (1980). *J. theor. Biol.* **87**, 9.
Williams, K. L. and Newell, P. C. (1976). *Genetics* **82**, 287.
Wilson, A. T. and Calvin, M. (1955). *J. Am. Chem. Soc.* **77**, 5948.
Wilson, D. W. (1969). Forms of hierarchy: a selected bibliography. In Whyte *et al.* (1969), p. 287.
Wilson, H. R. and Cowan, J. D. (1973). *Kybernetik* **13**, 55.
Wilson, J. A. (1968). *Nature* **219**, 534, 535.
Wimsatt, W. C. (1976). Complexity and Organisation. In *Topics in the philosophy of biology*. M. Grene and E. Mendelsohn (ed.). Reidel Publ. Co., Dordrecht and Boston.
—— (1981). Robustness, reliability and multiple-determination in science. In *Knowing and validating in the social sciences: a tribute to Donald T. Campbell*. M. Brewer and B. Collins (ed.). Jossey-Bass, San Francisco.
Winfree, A. T. (1967). *J. theor. Biol.* **16**, 15.

—— (1972). *Science, N.Y.* **175,** 634.
—— (1973a). Time and Timelessness in Biological Clocks. In *Temporal aspects of therapeutics.* J. Urquardt and F. E. Yates (ed.). Plenum Press, New York.
—— (1973b). *Science, N.Y.* **181,** 937.
—— (1975). *Physics today,* **28** (3), 34; see also his *The geometry of biological time* (1980) (Springer, New York).
Wolpert, L. (1969). *J. theor. Biol.* **25,** 1.
—— (1971). Positional information and pattern formation. In *Current topics in developmental biology,* Vol. 6. A. A. Moscona and A. Monroy (ed.). Academic Press, New York and London.
—— (1977). *The development of pattern and form in animals.* J. J. Head (ed.). Oxford University Press.
—— (1978). Pattern formation in biological development. *Scient. Am.* **239** (Oct.), 124.
—— (1981). *Phil. Trans. R. Soc. Lond.* **B295,** 441.
Woolhouse, H. W. (1967). *Nature, Lond.* **213,** 952; **216,** 200.
Yamakazi, I., Yokota, K., and Nakajima, R. (1965). *Biochem. biophys. Res. Commun.* **21,** 582.
Yates, R. A. and Pardee, A. B. (1956). *J. biol. Chem.* **221,** 757.
Yockey, H. P. (1973). Information theory with applications to biogenesis and evolution. In *Biogenesis, evolution and homeostasis.* A. Locker (ed.). Springer-Verlag, Berlin, Heidelberg, and New York.
Zahler, R. S. and Sussman, H. J. (1977). *Nature, Lond.* **269,** 759. (See for replies to this article, which is critical of catastrophe theory: *Nature, Lond.* **270,** (1977), 381–4, 658.)
Zeeman, E. C. (1972). *Towards a theoretical biology.* C. H. Waddington (ed.), Vol. 4, p. 8. University of Edinburgh Press, Edinburgh.
—— (1974). Primary and secondary waves in developmental biology. In *Lectures on mathematics in the life sciences* **7,** 69.
Zhabotinski, A. M. (1964). *Biophysics* **9,** 306.
—— (1980). *Ber. Bunsenges. phys. Chem.* **84,** 303.
Zotin, A. I. and Zotina, R. S. (1967). *J. theor. Biol.* **17,** 57.

Appendix: Supplementary references

N.B. From the nature of the material the classification under the chapter headings cannot be precise and many articles cover more than one theme.

1. Introduction

Allen, T. F. H. and Starr, T. B. (1982). *Hierarchy.* University of Chicago Press, Chicago.
Hess, B. and Markus, M. (1985). The diversity of biochemical time patterns. *Ber. Bunsen Ges. phys. Chem.* **89,** 642.
Mercer, E. H. (1981). *The foundations of biological theory.* Wiley, New York.
Salthe, S. N. (1985). *Evolving hierarchical systems: their structure and representation.* Columbia University Press, New York.

2. The thermodynamics of dissipative structures (and chaos theory)

Abraham, R. H. and Shaw, C. D. (1983). *Dynamics: the geometry of behavior. Pt. 2. Chaotic behavior.* Aerial Press, Santa Cruz, California.
Barnsley, M. F. and Demko, S. G. (eds) (1985). *Chaotic dynamics and fractals.* Academic Press, New York.
Berry, M. V., Percival, I. C., and Weiss, N. O. (eds) (1987). *Dynamical chaos.* Papers presented at a Royal Society Discussion Meeting, Feb. 4–5, 1987, *Proc. R. Soc., Lond.* **A413,** 1ff.
Campbell, D. and Rose, H. (1983). *Order in chaos.* Proceedings of an international conference at the Center for Nonlinear Studies. Los Alamos, USA, May 24–8, 1982. *Physica* **7D** (nos. 1–3), 3ff.
Crutchfield, J. P., Farmer, J. D., Packard, N. H., and Shaw, R. S. (Dec. 1986). *Sci. Am.* **255** 38.
Davies, P. (1987). *The cosmic blueprint.* Heinemann, London.
Grebogi, C., Ott, E., and Yorke, J. A. (1987). Chaos, strange attractors and
Ford, J. (1989). What is chaos, that we should be mindful of it? In *The new physics.* Davies, P. C. W. (ed.). p. 348. Cambridge University Press, Cambridge.
Gladyshev, G. P. (1982). Classical thermodynamics, tandemism and biological evolution, *J. theor. Biol.* **94,** 225.
Gleick, J. (1988). *Chaos: making a new science.* Heinemann, London.
Goldberger, A. L. (1987). Non-linear dynamics, fractals, cardiac physiology and sudden death. In *Temporal disorder in the human oscillatory system.* Rensing, L., Heiden, U. A., and Mackey, M. (eds), Springer-Verlag, New York.
Holden, A. V. (ed.) (1986). *Chaos.* Manchester University Press, Manchester, UK, and Princeton University Press, Princeton, USA.
Kapur, J. N. and Kesavan, H. K. (1987). *The generalized maximum entropy principle.* Sanford Educational Press.
Kay, J. J. and Schneider, E. D. (1989). On the applicability of non-equilibrium thermodynamics to living systems. From the Hawkwood Institute for Evolutionary Studies, 455 Barstow Road, Prince Frederick, Maryland 20678, USA.

Lighthill, J. (1986) The recently recognised failure in Newtonian dynamics. *Proc. R. Soc., Lond.* **A407**, 35.

Mandelbrot, B. (1977, 1982). *The fractal geometry of nature.* Freeman, New York.

May, R. (1976). Simple mathematical models with very complicated dynamics. *Nature* **261**, 459.

May, R. R. and Oster, G. F. (1976). Bifurcation and dynamic complexity in simple ecological models. *The American Naturalist* **110**, 573.

May, R. M. (1986). When two and two do not make four; nonlinear phenomenon in ecology. *Proc. R. Soc., Lond.* **B228**, 241.

Moon, F. C. (1987). *Chaotic vibrations: an introduction for applied scientists and engineers.* Wiley, New York.

Olsen, L. F. and Degn, H. (1985). Chaos in biological systems. *Q. Rev. Biophys.* **18**, 165.

Pacault, A. and Vidal, C. (eds) (1979). *Synergetics*, Vol. 3, *Far from equilibrium: instabilities and structures.* Springer-Verlag, Berlin.

Peitgen, H.-O. and Saupe, D. (eds) (1988). *The science of fractal images.* Springer-Verlag, Berlin.

Peitgen, H.-O. and Richter, P. H. (1985). *The beauth of fractals.* Springer-Verlag, Berlin.

Prigogine, I. and Stengers, I. (1984). *Order out of chaos.* Heinemann, London.

Roux, J.-C., Simoyi, R. H., and Swinney, H. L. (1983). Observation of a strange attractor [in the Belousov-Zhabotinskii reaction]. *Physica* **8D**, 257.

Rosen, R. (1986). *Anticipatory systems.* Pergamon, London.

Saunder. L. M. (1986). Fractal growth processes. *Nature* **322**, 789.

Schaffer, W. M. and Koti, M. (1985). Do strange attractors govern ecological systems? *Bioscience* **35**, 342.

Schaffer, W. M. and Kot, M. (1986). Chaos in ecological systems: the coals that Newcastle forgot. *Trends in Ecology and Evolution* **1**, 58.

Schieve, W. C. and Allen, P. (eds) (1982). *Self-organisation and dissipative structures.* University of Texas Press, Texas.

Schuster, H. G. (1984). *Deterministic chaos.* Physik-Verlag GmbH, Weinheim, Germany.

Stewart, H. B. and Thompson, J. M. (1986). *Nonlinear dynamics and chaos: geometric methods for engineers and scientists.* Wiley, Chichester.

Swenson, R. (1989). Emergent attractors and the law of maximum entropy production: foundations to a theory of general evolution. *Systems Res.* in press.

Swinney, H. L. (1983). Observations of order and chaos in nonlinear systems. *Physica* **7D**, 3.

Tritton, D. (July 1986). Chaos in the swing of a pendulum. *New Sci.* **111** 37.

Ulanowicz, R. E. (1987). Increased dissipation by positive feedback, Ref. no. CBL 87-65, Chesapeake Biological Laboratory, Solomons, Maryland, USA.

Wolf, A. (1983). Simplicity and universality in the transition to chaos. *Nature* **305**, 182.

3. Network thermodynamics

Atlan, H. and Thoma, J. (1987). Solvent flow in osmosis and hydraulics: network thermodynamics and representation by bond graphs. *Am. J. Physiol.* **252**, R1182.

Fidelman, M. L. and Mikulecky, D. C. (1984). Network thermodynamic modeling of hormone regulation of sodium transport in epithelia. *Virginia J. Sci.* **35**, 135.

Fidelman, M. L. and Mikulecky, D. C. (1986). Network thermodynamic modeling of hormone regulation of active sodium transport in cultures of renal epithelium. *Am. J. Physiol.* **250**, C978.

Fidelman, M. L. and Mikulecky, D. C. (1988). Network thermodynamic analysis and simulation of isotonic solute-coupled volume flow in leaky epithelia: an example of the use of network theory to provide the qualitative aspects of a complex system and its verification by simulation. *J. theor. Biol.* **130**, 73.

Imai, Y. (1988). Modelling of biological membrane transport systems by network thermodynamics. *J. physiol. Soc. Japan* **50**, 1.

Mikulecky, D. C. (1984). Network thermodynamics; a simulation and modeling method based on the extension of thermodynamic thinking into the realm of highly organized systems. *Math. Biosci.* **72**, 157.

Mikulecky, D. C. and Peusner, L. (1985). Network thermodynamics in bioenergetics: some useful new results and their implications. *Biophys. J.* **47**, 417A.

Mikulecky, D. C. (1987). Topological contributions to the chemistry of living systems. In *Studies in physical and theoretical chemistry*, Vol. 51, *Graph theory and topology in chemistry*. King, R. B. and Rouvray, D. H. (eds), p. 115. Elsevier, Amsterdam.

Mikulecky, D. C. (1987). Geometric vs. topological information in biological networks: an approach to organization and complexity. *Fed. Proc.* **46**, 673.

Mikulecky, D. C., Sauer, F. A., and Peusner, L. (1987). The reference state: its role in network thermodynamic models of nonlinear kinetic systems. *Biophys. J.* **51**, 95A.

Mintz, E., Thomas, S. R., and Mikulecky, D. C. (1986). Exploration of apical sodium transport mechanisms in an epithelial model by network thermodynamic simulation of the effect of mucosal sodium depletion. *J. theor. Biol.* **123**, 1; 21.

Montoya, J. P., Diller, K. R., and Beaman, J. J. (1986). A network thermodynamic model for simulation and permeability analysis of coupled membrane transport. *Cryobiology* **23**, 546.

Martin, T. J. and May, J. M. (1986). Testing models of insulin binding in rat adipocytes using network thermodynamic computer simulations. *J. recept. Res.* **6**, 323.

Paterson, R. and Lutfullah, (1985). Simulation of transport processes using bond graphs 1. Gas diffusion through planar membranes and systems obeying Fick's laws. *J. membrane Sci.* **23**, 59.

Peusner, L. (1986). Hierarchies of energy conversion processes III. Why are Onsager equations reciprocal?—the Euclidean geometry of fluctuation–dissipation space. *J. theor. Biol.* **122**, 125.

Smith, G. K. (1985). A network thermodynamic model of cell volume reaction. *Biophys. J.* **47**, 316A.

Srivastava, R. C. and Ramakrishnan, P. (1985). Network thermodynamic modelling of an active transport system. *Indian J. Biochem. Biophys.* **22**, 193.

Srivastava, R. C. and Ramakrishnan, P. (1984). Network thermodynamic modelling of chemical reactions. *Indian J. Chem.* Section A. **23**, 887.

4. The kinetics of self-organization

Arcuri, P. and Murray, J. D. (1986). Pattern sensitivity to boundary and initial conditions in reaction–diffusion models. *J. math. Biol.* **24,** 141.
Babloyantz, A. (1986). *Molecules, dynamics and life.* Wiley, New York.
Clark, J. W., Winston, J. V., and Rafelski, J. (1984). Self-organization of neural networks. *Phys. Letts* **102A,** 207.
Edelmann, G. M. (1988). *Topobiology: an introduction to molecular embryology.* Basic Books, Inc., New York.
Farmer, D., Toffel, T., and Wolframs, S. (eds) (1984). Cellular automata. *Physica* **10D**.
Field, R. J. and Burger, M. (eds) (1985). *Oscillations and traveling waves in chemical systems.* Wiley-Interscience, New York.
Grosberg, A. Yu, Nechaev, S. K., and Shaknovich, E. I. (1988). Role of topological limitations in the kinetics of homopolymer collapse and in self-organization of biopolymers. *Biofizika* **33,** 247.
Haken, H. (1988). Pattern formation: thermodynamics or kinetics? In *Synergetics,* Vol. 39, *From chemical to biological organization,* Markus, M., Mueller, S. C., and Nicolis, G. (eds). Springer-Verlag, Berlin.
Harrison, L. G. (1987). What is the status of reaction–diffusion theory thirty four years after Turing? *J. theor. Biol.* **125,** 369.
Harrison, L. G. and Kolar, M. (1988). Coupling between reaction–diffusion pre-pattern and expressed morphogenesis applied to desmids and dasyclads. *J. theor. Biol.* **130,** 493.
Hess, B. (1983). Non-equilibrium dynamics of biochemical processes. *H. S. Zeit. f. Physiol.* **364,** 1.
Hess, B. and Markus, M. (1987). Order and chaos in biochemistry. *Trends Biochem.* **12,** 45 (see also **12,** 348).
Kay, J. J. (1984). Self-organization in living systems. Ph.D. thesis, University of Waterloo, Canada.
Markus, M., Mueller, S. C., and Nicolis, G. (eds) (1988). *Synergetics,* Vol. 39, *From chemical to biological organization.* Springer-Verlag, Berlin.
Markus, M. and Hess, B. (1985). Input–response relationships in the dynamics of glycolysis. *Arch. Biochem.* **18,** 261.
Markus, M., Kuschmitz, D., and Hess, B. (1985). Properties of strange attractors in yeast glycolysis. *Biophys. Chem.* **22,** 95.
Markus, M., Mueller, S. C., and Hess, B. (1985). Observation of entrainment, quasiperiodicity and chaos in glycolyzing yeast extracts under periodic glucose input. *Ber. Bunsen Ges. Phys. Chem.* **89,** 651.
Martiel, J.-L. and Goldbeter, A. (1987). A model based on receptor desensitization for cyclic AMP signaling in *Dictyostelium* cells. *Biophys. J.* **52,** 807.
Meinhardt, H. (1982). *Models of biological pattern formation.* Academic Press, London.
Murray, J. D. (1982). Parametric space for Turing instability in reaction–diffusion mechanisms: a comparison of models. *J. theor. Biol.* **98,** 143.
Murray, J. D. (1989). *Mathematical biology.* Springer-Verlag, Berlin.
Nagorka, B. N. (1989). Wavelike isomorphic prepatterns in development. *J. theor. Biol.* **137,** 127.
Nicolis, J. (1986). *The synamics of hierarchical systems.* Springer-Verlag, Berlin.
Nicolis, G. (1989). Physics of far-from-equilibrium systems and self-organization. In *The new physics,* Davies, P. C. W. (ed.), p. 316. Cambridge University Press, Cambridge, England.

Nicolis, G. and Prigogine, I. (1987). *Exploring complexity*. Piper, Munich.
Rapp, P. E., Monk, P. B., and Othmer, H. G. (1988). A model for signal-relay adaptation in *Dictyostelium discoideum*. I, II. *Math. Biosci.* **77**, 35; 79.
Ricard, J. and Cornish-Bowden, A. (eds) (1987). *Dynamics of biochemical systems*. Plenum, New York.
Schaffer, W. M. and Kot, M. (1985). Do strange attractors govern ecological systems? *Bioscience* **35**, 342.
Tyson, J. J., Alexander, K. A., Manoranjan, V. S., and Murray, J. D. (1989). Spiral waves of cyclic AMP in a model of slime mould aggregation. *Physica* **D34**, 193.
Vidal, C. and Pacault, A. (eds) (1984). *Non-equilibrium dynamics in chemical systems*. Springer-Verlag, Berlin.
Zotina, R. S. and Zotin, A. I. (1982). Kinetics of constitutive processes during development and growth of orgnisms. In *Thermodynamics and kinetics of biological process*. Lamprecht, I. and Zotin, A. I. (eds), p. 424. Walter de Gruyter, Berlin.

5. Selection and evolution of biological macromolecules

Biebricher, C. K. (1986). Darwinian evolution of self-replicating RNA. *Chem. Scripta* **26B**, 51.
Biebricher, C. K., Eigen, M., and Luce, R. (1986). Template-free RNA-synthesis by Q-beta replicase. *Nature* **321**, 89.
Corliss, J. (1989). The dynamics of creation: the emergence of living systems in Archean submarine hot springs. *Proc. R. Swedish Acad. Sci.* in press.
Eigen, M. (1984). The origin and evolution of life at the molecular level. *Adv. Chem. Phys.* **55**, 119.
Eigen, M. (1985). Macromolecular evolution: dynamic ordering of sequence space. *Ber. Bunsen Ges. Phys. Chem.* **89**, 658.
Eigen, M. (1986). The physics of molecular evolution. *Chem. Scripta* **26B**, 13.
Eigen, M. (1987). New concepts for dealing with the evolution of nucleic acids. *Cold Spring Harbor Symp.* **52**, 307.
Eigen, M., Gardner, W., Schuster, P., and Winkler, R. (1981). The origin of genetic information. *Sci. Am.* **244(4)**, 88.
Eigen, M. and Schuster, P. (1982). Stages of emerging life: 5. Principles of early organization. *J. Mol. Evol.* **19**, 47.

6. The interpretation of biological complexity

Allen, P. M. (1985). Ecology, thermodynamics and self-organization: towards a new understanding of complexity. In *Ecosystem theory for biological oceanography*. Ulanowicz, R. E. and Platt, T. (eds), *Canadian Bull. Fish Aquatic Sci.* (Dept. of Fisheries and Oceans, Ottawa) **213**, p. 3.
Brooks, D. and Wiley, E. (1986, 1988). *Evolution as entropy: towards a unified theory of biology*. University of Chicago Press, Chicago.
Campbell, J. H. (1985). An organizational interpretation of evolution. In *Evolution at a crossroads: the new biology and the new philosophy of science*. Depew, D. J. and Weber, B. H. (eds). MIT Press, Cambridge, Mass.
Ciba Foundation Symposium 144 (1989). *Cellular basis of morphogenesis*, Wiley, Chichester.

Haken, H. (ed.). *Synergetics*, Vols. 1–40. Springer-Verlag, Berlin.
Jantsch, E. (1980). *The self-organizing universe*. Pergamon, Oxford.
Johnson, L. (1981). The thermodynamic origin of ecosystems. *Can. J. Fish Aquatic Sci.* **38**, 571.
Lewin, R. (1984). Why is development so illogical? *Science* **224**, 1327.
Lovtrup, S. (1983). Victims of ambition: comments on the Wiley and Brooks approach to evolution. *Syst. Zool.* **32**, 90.
Markus, M. Mueller, S. C., Plesser, T., and Hess, B. (1987). On the recognition of order and disorder. *Biol. Cybern.* **57**, 187.
Rosen, R. (ed.) (1985). *Theoretical biology and complexity*. Academic Press, New York.
Rosen, R. (1985). Organisms as causal systems which are not mechanisms: an essay into the nature of complexity. In *Theoretical biology and complexity*. Rosen, R. (ed.). Academic Press, New York.
Saunders, P. T. and Ho, M. W. (1981). On the increase of complexity in evolution, II. *J. theor. Biol.* **90**, 515.
Stuart, C. I. J. M. (1985). Bio-informational equivalence. *J. theor. Biol.* **113**, 611.
Schneider, E. D. (1988). Thermodynamics, ecological succession and natural selection: a common thread. In *Entropy, information and evolution: new perspectives on physical and biological evolution*. Weber, B. H. *et al.* (eds), p. 107. Bradford Book, MIT Press, Cambridge, Mass.
Ulanowicz, R. E. (1986). *Growth and development: ecosystem phenomenology*. Springer-Verlag, New York.
Ulanowicz, R. E. (1989). *A phenomenology of evolving networks*, in press.
Ulanowicz, R. E. and Hennor, B. M. (1987). Life and the production of entropy. *Proc. R. Soc., Lond.* **B232**, 181.
Weber, B. H., Depew, D. J., and Smith, J. D. (eds) (1988). *Entropy, information and evolution: new perspectives on physical and biological evolution*. Bradford Book, MIT Press, Cambridge, Mass.
Weber, B. H., Depew, D. J., Dyke, C., Salthe, S. N., Schneider, E. D., Ulanowicz, R. E., and Wicken, J. S. (1989). Evolution in thermodynamic perspective: an ecological approach. *Biology and Philosophy*, **4** in press.
Wickens, J. S. (1987). Entropy and information: suggestions for a common language. *Phil. Sci.*, **54**, 176.
Wickens, J. S. (1983). Entropy, information and nonequilibrium evolution. *Syst. Zool.* **32**, 438.
Wicken, J. S. (1984). The cosmic breath: reflections on the thermodynamics of creation. *Zygon* **19**, 487.
Wicken, J. S. (1984). On the increase of complexity in evolution. In *Beyond neo-Darwinism*. Saunders, P. T. and Ho, M. W. (eds.), p. 89. Academic Press, London.
Wicken, J. S. (1984). Autocatalytic cycling and self-organization in the ecology of evolution. *Nature and System*, **6**, 119.
Wicken, J. S. (1987). *Evolution, information and thermodynamics: extending the Darwinian program*. Oxford University Press, New York.
Wiley, E. O. and Brooks, D. R. (1982). Victims of history—a nonequilibrium approach to evolution. *Syst. Zool.* **31**, 1.
Wiley, E. O. (1983). Nonequilibrium thermodynamics and evolution: a response to Lovtrup. *Syst. Zool.* **33**, 209.

Zotin, A. I. (1972). *Thermodynamic aspects of developmental biology.* S. Karger, Basel.
Zotin, A. I. (1985). Thermodynamics and the growth of organisms in ecosystems. In *Ecosystem theory for biological oceanography.* Ulanowicz, R. E. and Platt, T. (eds). *Can. Bull. Fish and Aquatic Sci.* (Dep. of Fisheries and Oceans, Ottawa), **213,** 27.

Name index (excluding supplementary references)

Aldridge, J. 137, 152, 189
Amesz, J. 138, 191
Andronov, A. A. 144, 182
Apter, M. J. 260, 261
Aschoff, J. 181
Ashkenazi, M. 182
Atlan, H. 90, 92, 93, 100, 101, 260, 261, 262, 263
Auchmuty, J. F. G. 139, 172
Auslander, D. M. 73, 82, 83
Ayala, F. J. 13

Babloyantz, A. 57, 59, 66, 67, 68, 169, 184, 186
Bak, T. A. 153, 157
Balescu, R. 44, 127
Barkley, D. S. 209
Bard, J. B. L. 130, 133, 134, 135, 186
Bartlett, M. S. 224
Bastin, T. 10
Beckner, M. 268
Belousov, B. P. 40, 93, 137, 197
Benassi, M. 261
Berlinski, D. 12
Berridge, M. J. 152, 189, 190, 191, 193, 195, 209, 210, 211
Bertalanffy, L. 4
Betz, A. 138, 191
Blanchard, J. M. V. 104, 105, 106
Blumenthal, R. 68
Boissonade, J. 148
Boiteux, A. 138, 151, 152, 191, 192, 193, 197, 198, 199, 202, 213
Boltzmann, L. 1, 18, 19, 20, 24, 33, 42, 50, 56, 145, 256, 257
Bonner, J. 182, 203, 204, 209, 276
Borckmans, P. 146
Bray, W. C. 137
Brillouin, L. 260
Britton, N. F. 179
Bronowski, J. 61, 266
Brown, F. A. 137
Brutlag, D. 276
Bryant, P. J. 183
Bryant, S. V. 183
Büchel, W. 260
Bunge, M. 251, 255
Bunning, E. 181, 189
Bunow, B. 178, 186

Busse, H. G. 152, 197

Caldwell, P. C. 116
Calvin, M. 138
Caplan, S. R. 152, 170, 171, 191
Chance, B. 138, 151, 191, 192, 193, 197
Chandrasekhar, S. 199
Changeux, J. P. 63, 68
Chen, W. K. 91
Chernavskaia, N. M. 172
Chernavskii, D. S. 172
Clarke, B. 157, 158
Cohen, R. S. 182, 183, 184
Cowan, J. D. 129, 183, 187
Crick, F. H. C. 184, 268
Cronin, J. 138, 152
Crushberg, T. C. 102
Curie, P. 34

Dagley, S. 114
Darwin, C. 19, 112, 113, 214, 216
Darwin, E. 113
Davidson, E. H. 182
Davies, P. C. W. 18
Dean, A. C. R. 114, 118, 119
Decker, P. 138, 158, 159, 160, 161, 172
Defay, R. 28
Degn, H. 137, 164, 165
Denbigh, K. 129, 142, 153, 256, 257, 258, 259, 260, 266
De Vault, D. 193
Dewel, G. 146
de Donder, Th. 19
Drage, D. 182
Duban, M. C. 179
Duhem, P. 19, 28, 32
Duysens, L. N. M. 138, 191

Eckmann, J. P. 138
Ehrenfest, P. 225
Ehrenfest, T. 225
Eigen, M. 20, 52, 54, 57, 127, 149, 216, 217, 218, 220, 221, 222, 223, 224, 227, 228, 229, 230, 231, 232, 233, 234, 235, 237, 238, 239, 240, 241, 261, 265, 272, 273, 274
Einstein, A. 50, 51, 71, 145, 146, 147
Elsasser, W. M. 251, 260
Ermentrout, G. B. 187
Erneux, T. 130, 138, 145, 177, 178, 186, 209

NAME INDEX

Escher, C. 138
Estabrook, R. 138, 191
Euler, L. 106, 107

Ferracin, A. 261
Field, R. J. 41, 137
Fife, P. C. 138, 171
Finkelstein, A. 107
Foerster, H. von 263
Fox, R. F. 54, 55
Frank, I. G. M. 137, 151, 165
Frazier, W. A. 205, 210
French, V. 183
Frenkel, R. 163
Friboulet, A. 179

Gardiner, W. 240
Garfinkel, D. 191
Gatlin, L. 260
Gebben, V. D. 94
Gerard, R. W. 3, 8
Gerisch, G. 152, 205, 209, 210, 211
Ghosh, A. M. 138, 151, 191, 192, 193
Giacchino, J. L. 96
Gibbs, J. W. 1, 24, 28, 29, 30, 32, 79
Gierer, A. 59, 183, 184, 185
Glansdorff, P. 19, 20, 28, 29, 30, 31, 32, 42, 43, 46, 52, 54, 55, 68, 91, 92, 127, 129, 161, 181, 199
Glass, L. 188, 189
Gmitro, J. I. 130, 144, 171
Goldbeter, A. 63, 64, 143, 144, 152, 166, 167, 168, 169, 170, 176, 177, 179, 180, 181, 191, 199, 201, 209, 210, 211
Goodwin, B. C. 6, 129, 162, 163, 181, 182, 183, 184
Gordon, I. I. 144
Greene, J. C. 113
Grene, M. 253
Griffith, J. S. 163
Groot, S. R. de 19
Grutsch, J. 205
Gurel, O. 138
Gutman, H. 250

Hahn, H. S. 171
Haken, H. 138, 146, 278
Hanusse, P. 138, 144
Harrington, J. M. 25
Harris, H. 182
Harvey, R. J. 102
Hessenstein, B. 260
Hastings, J. W. 137, 189, 212
Hawking, S. W. 18
Hearon, J. Z. 153
Helmreich, E. J. 278
Herschkowitz-Kaufman, M., 41, 138, 144, 172, 176, 177, 178
Hervagault, J. F., 179
Hess, B., 11, 54, 64, 138, 151, 152, 179, 191, 192, 193, 195, 196, 197, 198, 199, 201, 202, 205, 213, 276
Hicks, M. 129
Hiernaux, J. 59, 145, 184, 186
Higgins, J. 137, 155, 156, 157, 161, 163, 164, 166, 193
Hinshelwood, C. N. 3, 111, 114, 115, 116, 117, 118, 119, 120, 123, 272
Ho, M. 266
Hofstadter, D. R. 138
Holmes, K. C. 13
Horowitz, B. A. 96, 97, 98, 99
Horowitz, J. M. 96, 99
Horsthemke, W. 147, 263
Hubel, D. H. 245
Huf, E. G. 107
Hulme, E. 163
Hunding, J. 164
Hunter, S. A. 102, 103, 104
Huxley, H. E. 253

Iooss, G. 138

Jacob, F. 66, 67, 68, 186, 269, 275
Johnson, H. A. 260
Joly, F. 179, 186
Joseph, D. D. 138

Kacser, H., 119, 120, 122, 123, 124, 125, 272
Kaimachnikov, N. P. 166
Karnopp, D. 80
Katchalsky (Katzir-), A. 19, 34, 73, 90, 92, 106, 107, 108, 171, 274
Kauffman, S. A. 182, 186, 246
Kay, R. 205, 210
Keizer, J. 54, 55
Kernevez, J. P. 179, 186
Khalkin, S. E. 144, 182
Koestler, A. 253
Konijn, T. M. 204, 209
Kuhn, 238, 240, 241
Küppers, B. 217, 218, 226, 228, 240

Landahl, H. D. 164
Landau, L. D. 146
Landsberg, P. T. 278
Lauder, I. 130, 133, 134, 135
Lee, I. Y. 193
Lefever, R. 57, 58, 64, 68, 69, 129, 143, 144, 151, 152, 166, 167, 168, 169, 172, 174, 179, 263
Leontovitch, E. A. 144
Levin, S. A. 171

NAME INDEX

Levins, R. 249
Lewis, G. N. 18, 115, 116
Liesegang, R. E. 125
Lifshitz, E. M. 146
Lipmann, F. 232
Loomis, W. F. 205
Lotka, A. J. 126, 127, 128, 129, 142, 150, 153
Lucas, A. 163
Lyell, C. 214

McKay, D. 162, 163
Mackey, M. C. 188, 189
McQuarrie, D. A. 224
Maeda, Y. 210
Maeyer, L. de 230
Maier, A. G. 144
Malchow, D. 205, 210
Malek-Mansour, M. 147, 148
Martiel, J. L. 170, 209
Martinez, H. 69
Mauro, A. 107
Maxwell, J. C. 24, 30, 73
May, R. M. 189
Maynard Smith, J. 130
Medawar, P. 13, 251, 269
Meene, J. G. C. van de 209
Mees, A. I. 164
Meinhardt, H. 59, 184, 185
Meixner, J. 19, 153
Mela, L. 193
Mesarovic, M. D. 5, 11
Mikulecky, D. C. 107, 108, 275
Miller, J. G. 7, 9, 10, 245, 249
Mimura, M. 179, 187
Minorski, N. 31, 138, 142
Monod, J. 63, 66, 67, 68, 167, 186
Moore, M. J. 129, 153
Morales, M. 162, 163
Morowitz, H. J. 10, 88, 95, 96, 267
Murray, J. D. 138, 151, 171, 179, 186, 187

Nagel, E. 8, 13, 268, 269, 270
Nakajima, R. 138
Nallo, A. di 261
Naparstek, A. 171
Neumann, E. 106, 107, 108
Newell, P. C. 203, 204, 205, 206, 207, 208, 209, 210, 212
Nicolis, G. 19, 20, 21, 31, 36, 37, 39, 43, 44, 45, 47, 50, 51, 55, 56, 57, 62, 65, 66, 67, 68, 71, 94, 129, 130, 136, 137, 138, 139, 140, 142, 144, 145, 146, 147, 148, 149, 150, 151, 152, 158, 166, 167, 169, 171, 172, 174, 175, 176, 177, 181, 273
Noyes, R. M. 41, 137

Onsager, L. 19, 34, 36, 37, 38, 39, 43, 44
Oparin, A. I. 3, 214, 238
Oster, G. F. 73, 78, 79, 82, 83, 84, 85, 86, 87, 88, 90, 92, 94, 107
Othmer, H. G. 130, 138, 144, 152, 171, 182, 183

Page, F. M. 129
Palmer, J. D. 137
Panet, R. 100
Panischelli, E. 261
Pardee, A. B. 162
Pasteur, L. 214
Pattee, H. 12, 249, 254, 260, 261, 271
Pauli, W. E. 13
Pavlidis, T. 182
Paynter, H. 80, 83
Peacocke, A. R. 12, 251, 268
Perelson, A. S. 73, 88, 107, 108
Peura, R. A. 102
Peusner, L. 107, 108
Planck, M. 17
Plant, R. E. 75, 95, 96, 99
Plesser, T. 151, 197
Polanyi, M. 270
Popper, K. R. 259, 260, 271
Prigogine, I. 19, 20, 21, 24, 28, 29, 30, 31, 32, 33, 34, 36, 37, 39, 41, 43, 44, 45, 46, 47, 50, 51, 52, 54, 55, 56, 57, 58, 65, 68, 69, 71, 91, 92, 127, 129, 130, 136, 138, 140, 144, 146, 147, 151, 158, 161, 171, 172, 174, 175, 176, 177, 181, 199, 274, 278
Portnow, J. 137, 144, 149, 150, 152, 171
Pye, E. K. 151, 193

Quastler, H. 260
Queiroz, O. 137

Ramakrishnan, A. 224
Randall, M. 18
Rangazas, G. 163
Rapp, P. E. 138, 139, 152, 164, 188, 189, 190, 191, 193, 195, 209, 210, 211
Ray, J. 112
Riley, K. F. 27
Robertson, A. 182, 205
Roos, W. 210
Rosen, R. 254, 272
Rosenberg, R. 80
Rossler, O. W. 138
Rothstein, J. 260
Rowlinson, J. S. 25
Rutherford, E. 1

Salomon, J. 100
Sanglier, M. 66, 186
Sattinger, D. 139, 186

NAME INDEX

Saunders, P. 266
Schaffner, K. F. 270
Schiffman, Y. 138
Schmitz, G. 142
Schnakenberg, J. 73
Schrödinger, E. 22, 257
Schuster, P. 149, 228, 229, 230, 231, 234, 236, 237, 238, 239, 240
Schweiger, H. G. 189
Scriven, L. E. 130, 144, 171
Seelig, F. F. 153, 165, 179
Segel, L. A. 138, 209, 211
Sel'kov, E. E. 137, 151, 152, 165, 166, 172, 190, 213
Shannon, C. E. 259, 260, 262
Shapley, H. 253
Shiner, J. S. 107
Shymko, R. M. 186
Sidoroff, S. 100
Sigmund, K. 237
Simon, H. A. 6, 10, 232, 246, 248, 249, 265, 266
Simpson, G. G. 259, 276
Smart, J. C. C. 269
Smith, R. E. 96, 130
Snell, F. M. 153, 154, 155, 161
Sondhi, K. C. 130
Spangler, R. A. 34, 153, 154, 155, 161, 171
Spiegelmann, S. 230, 240
Stark, L. 260
Stebbins, G. L. 251, 264
Steindler, C. 261
Stucki, J. W. 44, 52, 138, 157, 158, 171, 277
Sussman, M. 278
Sweeney, D. M. 182

Tellegen, B. D. H. 90, 91, 92, 93
Theoridis, G. C. 260
Thom, R. 278
Thoma, J. 80, 94, 95
Thomas, D. 107, 108, 171, 179, 186
Thorpe, W. H. 13
Tonnelat, J. 261
Tornheim, K. 193
Trabert, K. 186
Turing, A. M. 69, 130, 131, 132, 133, 134, 135, 136, 137, 171, 182, 183, 186, 272
Turner, J. W. 148
Tyson, J. J. 137, 138, 152, 164, 172, 182, 189, 197, 212

Venieratos, D. 170

Viniegra-Gonzales, G. 163
Virchow, R. 3, 113, 114
Vitt, A. A. 144, 182
Viza, D. 182
Volterra, V. 127, 128, 129, 142

Waals, J. D, van der 25
Waddington, C. H. 278
Walgraef, D. 146, 148
Wallace, A. R. 214
Walter, C. 163, 170
Weaver, W. 1, 2, 260
Webster, D. 212
Weisbuch, G. 100
Weiss, P. A. 249, 257
Whyte, L. L. 249
Wick, U. 209, 210
Wicken, J. S. 257, 260, 266, 267
Widom, B. 25
Wiegand, W. A. 107
Williams, K. L. 203
Williamson, G. 193
Wilson, A. G. 249
Wilson, D. W. 249, 253
Wilson, H. R. 138, 187
Wilson, J. A. 260
Wimsatt, W. C. 13, 245, 246, 247, 248, 249, 258, 271
Winfree, A. T. 182, 189, 197
Winkler (-Oswatitsch), R. 127, 217, 224, 240
Wittgenstein, L. 241
Wolff, R. 237
Wolpert, L. 182, 183, 184, 260, 261, 277
Woolhouse, H. W. 260
Writing, P. 63
Wurster, B. 210
Wyman, P. 63

Yamazaki, I. 138
Yates, R. A. 162
Yockey, H. P. 260
Yokota, K. 138

Zabusky, N. J. 171
Zahler, R. S. 278
Zeeman, E. C. 278
Zhabotinsky, A. M. 40, 41, 93, 137, 173, 176, 197, 217
Zotin, A. I. 59
Zotina, R. S. 59

Subject index

Page numbers in **bold** type indicate the principal reference when several are given.

Acrasiales 70
across-variables 74ff.
activation by product 166, 172
affinity, of a chemical process 27, 86, 219
allosteric enzyme action 166–70, 179–81
asymptotic stability 31
ATP reactions 100–4, 166
autocatalysis 50, 52–4, 60, 66, 126–7, 153, **158–61,** 219, 229; second-order-in-products (ASOP) 158–61
autonomy, of processes 269ff.; of theories 269ff.
autosynthesis 114ff.

Belousov–Zhabotinsky reaction **40–1,** 93, 137, 217
Bénard phenomenon 41–2
bifurcation 112, **138–42,** 146
biological clocks 137
biology, molecular 2; central problem 11–14; epistemology 12, **268–72**
Boltzmann relation **18–20,** 42, 145, **256**
bond graphs **75–81;** for chemical reactions with heat production and flow 96; and conventional biochemical representations 88; of reactions 87–8
brain, human 245
branch, thermodynamic 49, 178
bridge problem (of Euler) 106–7
brown fat cells, heat generation in, bond graph representation 96–100
'Brusselator' 160, **172–8,** 216

capacitances 73ff.
catabolite repression 66
catalysis, with cross-coupling 153–7
catastrophe theory 278
causality, representation in bond graphs 82–4
cellular oscillators, an atlas 188–91
cellular systems, as dissipative systems 68–70
chaotic behaviour 145
chick embryo, heat and entropy production in 59
circadian rhythms 136, 152, 189, 190

coat colour distribution 186–7
complexity, biological 2, 3, 13, **245–9;** descriptive 246–7; disorganized 1; evolution 263–8; formation, driven by Second Law 267; and hierarchy 249–54; interactional 248–9; organized **1–3,** 255–63; quantification 255–63
constitutive relations 75ff.
control 5, 6; in hierarchies, 10, 249, **251–3,** 277
covalent enzyme modification, and oscillations 170–1
critical length 180–1
cycles, heterogeneous and self-organizing 234–8; homogeneous and self-organizing 231–4; hyper- 234–8; nucleic acid 231–2; protein 232–3

differentiation 182ff.
diffusion 14, 26, 35, 54, 64; coupling 85
disorder (disorderliness) 18, 40, 256
dissipation, increase in during evolution 57–9; and energy storage 75–81
dissipative structures (systems), and bifurcations 137–48; in biology 62–70; in the 'Brusselator' system 174ff.; and evolution 56–9; and evolutionary feedback 56–9; and 'order through fluctuations' 40–2; thermodynamics of 17–72
drifting system (stochastic model) 225
Drosophila wing discs 186

effort source 75ff.
Einstein formula 50–1
electrical circuits 73–4
emergence 255, 269
energy bonds 79–81
energy-randomization 267
energy storage and dissipation 75–81
entropy, in classical thermodynamics 17–19; density 25; excess, production 29–32, **45–50,** 54; excess, second order 50; and information 260–1; and order 256–7; in open systems 20–3; production **23–8,** 44, 48–50, 59–61, 64; production, under linear conditions 36–40; and statistical

SUBJECT INDEX

thermodynamics 17–19; and the steady state 21–3
epigenetic oscillations 162–3
equilibrium, and classical thermodynamics 17–19; local **23–8,** 43, 147; stability of 28–32
evolution, chemical 217; criteria of **42–4,** 70; and dissipative structures 56–9; and levels of organization 263–4; from macromolecules to cells 238–41; of macromolecules 214–41; and stability 59–62

feedback 5, 54, 150; enzyme reactions with 62–8; evolutionary and dissipation 56–9; negative 164, 178; positive 66, 164
feed-forward 5, 54
flow source 75ff.
flows (fluxes) and forces, linear relation 32–5; non-linear relation 41–2
fluctuations (perturbations) 20, 28, 29, 34, 62, 71, 273; imposed 171; initiating 217; and instability **50–2,** 57–9; in non-linear systems 40, **48–50;** role of 145–8; about the steady state 39

Gibbs formula 24, 30
glucose metabolism (in skeletal muscle), bond graph representation 102–4
glycolysis, dynamic compartmentation in 197–202; oscillations in 138, 152, 166, 190, **191–5**
goal-seeking 5

heat generation in biological systems, network thermodynamic interpretation of 94–100
hierarchies, biological 6, 8–11, 12, 73, 75, 217, **249–54,** 279; of catalysis 123; and complexity 249–54; conditions for existence of 249–50; and constraints in time and space 250; control in 10, 249, **251–3,** 277; and evolution 265–6; of evolutionary span 10; formal 249; information in 263; and operative constraints 254; of spatial size 10; and specifying descriptions 253–4; and structure–function relationships 250; sub-assemblies 61, **265;** symmetry-breaking in 187; theories about and processes in 268–9; time of response of 10, 11, 250
Hill coefficient 169
holism 12, 112
Hora and Tempus 6–7, 265
Hydra 136, 184
hypercycle 234–8

ideal elements 75–9
ideal junctions 80–1

incremental quantities 75–6
inductances 73ff.
information, change of content of (and organization) 261–3; and complexity (and organization) 259–63; definitions of 259; and entropy 260–1, 267; flux 153; theory 216; transmission by macromolecules 216
inhibition, by end-product **161–4,** 172; lateral 184; by substrate **164–6,** 172, **179**
instability, chemical 52–4; controlling parameter values 69; creative force of 213; and self-organization 61
integrality 257–9
integrated pluralism 255
ionic transport 100–4
irreversible processes, linear thermodynamics of 32–5; non-linear thermodynamics of 40–56

kinetic approach, pioneers of 115–36
kinetics, of biological self-organization 14–21; of complex reaction systems 24; and living systems 111–14; non-linear 24; and thermodynamics 272–3
Kirchhoff's current law 74; and Tellegen's theorem 90–1
Kirchhoff's voltage law 74; and Tellegen's theorem 90–1

lac operon-galactosidase-permease system 66
Liesegang rings 125
limit cycle 142, 178
linear graph theory 107
living matter 114–24
Lotka–Volterra model 127–9
Lyapunov function 31, 39, 47, 54, 55

macromolecules 10, 56; biological, selection and evolution of 214–41
macrospace 196–7, 202
matter-randomization 267
Maxwell–Boltzmann distribution 24
membrane, network thermodynamic account of 77; depolarization 68, 69; diffusion through (bond graph representation) 81–2; oscillations 171
memory, of fluctuations 178
messenger RNA 67, 162
microspace 195–6, 202
mitochondrial respiration, oscillations in 188, 193
mitosis 182
modularization 6
molecular biology, *see* biology, molecular
Monod–Jacob model for induction and repression 66–8

morphogenesis 69–70; chemical basis of 130–7
morphogens 59, 69, 130, 183, 186

near-decomposability 246–8
negentropy 22, 40
network, evolution 92–4; properties 90–4; theorem 119; theory 73, 75, 79
network thermodynamics **73–108,** 272; application and assessment of 94–107; and heat generation in biological systems 94–100
neuron nets 187
noise, effects of external (on stability) 147; order and complexity from 263

Onsager reciprocity relation 34, 36
open chain 153
open systems **20–3,** 120ff., 216
order, and entropy 256–7; through fluctuations 40–2; relational 251
organization, biological and thermal physics 267
origin of life 214–17
oscillations, in the absence of diffusion 141–4; chemical 189; of clinical interest 189; coupled 181–2; kinetic models of 151–87; source of 157, 189–90; and symmetry-breaking 125–36
'oscillatory rule of thumb' (Higgins) 157

patchiness (predator–prey systems) 171, 187
pattern formation 182–7
peroxidase-catalysed oxidation, oscillations in 138
perturbations, *see* fluctuations
phase transitions 146
phenomenological coefficient 33, 34, 36
phenomenological (deterministic) treatment of evolution of macromolecules 218–24
phosphofructokinase **62–6,** 142, 166, 190, **193**
photosynthesis, oscillations in 138, 172, 188, 193
physicochemical ideas, contribution to biology 11–14
Poisson distribution 146–8
ports 76–9
positional information **183,** 277
pre-biotic stage (of evolution) 56, **214–17**
pre-patterns 183
'primordial soup' 218
product inhibition 165
productivity of macromolecule formation, excess 220; mean 220, 222
propagating waves 144, 145, 175
protein synthesis 5, 66, 100–4

randomness 18, 256
reaction–diffusion processes 14, 26, 35, 59, **144–5,** 171, 186, 272; spatial and temporal symmetry-breaking in **171–87, 195–212;** two-dimensional 138
reaction(s), coupling between 85; open systems of 129–30; systems 86; as transducers 84–9
reduction(ism) 12, 13, 114; of biological concepts **268–72,** 274–8; epistemological 269ff.; of processes 269ff.; of theories 269ff.
redundancy 262–3
regulatory gene 66–8
repressor 66
resistances 73ff.
'robustness' and reality 271

scalar phenomena 34
selection, criteria for in macromolecular evolution 222; Darwinian, of macromolecules 222ff.; equilibrium 221–3; games 225–8; pressure 60; of macromolecules, under constraint 220–1
selective value 223
self-copying, of macromolecules 219
self-organization 61, 69, 216; biological kinetics of 111–213; and change of information content 262; as complexity from noise 262–3; conditions for 148–9; and the environment 240; of homogeneous reaction cycles 231–4; of heterogeneous reaction cycles 234–8; macromolecular model of 218ff., 272; and oscillations 149–51; spatial examples of 195–212; spatial, models of 171–87; temporal, examples of 189–94; temporal, models of 152–71
simplicity, descriptive 246–7; interactional 248–9; problems of 1
slime mould (*D. discoideum*) 70, 170, **203–12;** life-cycle 204; process of aggregation 205
stability, asymptotic 31; of chemical reaction networks 157; and evolution 59–62; of networks 91–2; stratified **61,** 266
stability criteria, analysis 171; Glansdorff–Prigogine 47, 50, 54–6, 63; for non-equilibrium states 44–50
stable systems (stochastic model) 225
standing waves 145, 178
starch gelatinization, bond graph representation 104–6
stationary waves 134, 144
steady (stationary) states, close-to-equilibrium 35; multiple 144, 145; non-

equilibrium, linear range 35–40; unstable 144
stochastic approach to evolution of macromolecules 224–8
stochastic models of fluctuations and stability 224–8
stochastic process 72, 146
stochastic theory 112, 217
'Struggle' bead game 127
switch mechanism (kinetic) 123
symmetry-breaking, in neuron nets 187; and 'patchiness' (predator–prey systems) 187; spatial 144–5, **171–87, 195–212,** 272; temporal 142, 145, 150–1, **152–71, 189–94,** 272
synergetics 278
systems, general theory **3–8,** 12; generalized 8; living 3, **5–8;** thermodynamics and living 17–20

Tellegen's theorem 90–2

thermodynamics, and the Boltzmann relation 18, 20; classical 17, 18; and electrical circuits 73–4; irreversible 19; irreversible and biological interpretation 70–2; and kinetics 272–3; linear, of irreversible processes 32–5; and living systems 17–20; network 73–108; non-linear of irreversible processes 40–56; second law of 18, 19
through-variable 74ff.
topological graph representation 106–7
transducer 84–90
tri-molecular reactions 178
trophic levels 267

Uniformitarian principle, 214
unstable systems (stochastic model) 225

vector phenomena 34

Zhabotinsky reaction 40, 41, 93, 137, 217

DATE DUE	
APR. 3 0 1999	
APR 2 8 2000	
JUN 1 4 2002	
APR 2 5 2003	

DEMCO, INC. 38-2931